Construction Quality

Construction Quality

Do It Right or Pay the Price

Glen Copeland, PE
Copeland Geotechnical & Associates, Inc.

George C. Frank, CFC, FABFEI
Frank Consulting, Ltd.

Joseph A. Gervasio, PE
Gervasio & Associates, Inc.

Edward Holdsworth
SEMTEC Laboratories, Inc.

Wellington R. Meier Jr., Ph.D., PE,

David C. Tierney, Attorney
Sacks Tierney P.A.

Craig Walling, Architect

Jesse R. Wyatt, PE, SE

Prentice Hall

Boston   Columbus   Indianapolis   New York   San Francisco   Upper Saddle River   Amsterdam   Cape Town
Dubai   London   Madrid   Milan   Munich   Paris   Montreal   Toronto   Delhi   Mexico City
Sao Paulo   Sydney   Hong Kong   Seoul   Singapore   Taipei   Tokyo

Editorial Director: Vernon R. Anthony
Acquisitions Editor: David Ploskonka
Editorial Assistant: Nancy Kesterson
Director of Marketing: David Gesell
Executive Marketing Manager: Derril Trakalo
Senior Marketing Coordinator: Alicia Wozniak
Senior Marketing Assistant: Les Roberts
Project Manager: Holly Shufeldt
Art Director: Jayne Conte
Cover Designer: Karen Salzbach
Cover Image: Dreamstime
Full-Service Project Management and Composition: Integra Software Services, Pvt. Ltd.
Printer/Binder: Edward Brothers
Cover Printer: Lehigh Phoenix Color
Text Font: Minion

Credits and acknowledgments borrowed from other sources and reproduced, with permission, in this textbook appear on the appropriate page within the text. Unless otherwise stated, all artwork has been provided by the author.

Copyright © 2012 Pearson Education, Inc., publishing as Prentice Hall, One Lake Street, Upper Saddle River, New Jersey, 07458. All rights reserved. Manufactured in the United States of America. This publication is protected by Copyright, and permission should be obtained from the publisher prior to any prohibited reproduction, storage in a retrieval system, or transmission in any form or by any means, electronic, mechanical, photocopying, recording, or likewise. To obtain permission(s) to use material from this work, please submit a written request to Pearson Education, Inc., Permissions Department, One Lake Street, Upper Saddle River, New Jersey, 07458.

Many of the designations by manufacturers and seller to distinguish their products are claimed as trademarks. Where those designations appear in this book, and the publisher was aware of a trademark claim, the designations have been printed in initial caps or all caps.

Library of Congress Cataloging-in-Publication Data
  Construction quality : do it right or pay the price / Glen Copeland ... [et al.].–1st ed.
    p. cm.
  Includes bibliographical references and index.
  ISBN-13: 978-0-13-217151-9 (alk. paper)
  ISBN-10: 0-13-217151-1 (alk. paper)
  1. Building–Quality control.   2. Building–Superintendence.   3. Liability for building accidents.
  I. Copeland, Glen.
  TH438.2.C55 2012
  624.068'4–dc22
                                                                                                        2011005965

10 9 8 7 6 5 4 3 2 1

www.pearsonhighered.com

ISBN 10: 0-13-217151-1
ISBN 13: 978-0-13-217151-9

Contents

Preface ix
About the Authors xi
Foreword xv

SECTION ONE Professional Expectations (We Are What We Project) 1

Jump-Starting Section 1 1

Professional Ethics 2

Professional Organizations – Expectations and Responsibilities 3
- Regulatory (Mandatory) Organizations 3
- Professional (Voluntary) Societies 3

Ethics & Architecture 3

Ethics & Engineering 4

Ethics & Forensic Investigation 5

Ethics & Contractors 6

Legal Actions—A Fact of Life 7
- Contract Liability 7
- Tort 7
- Negligence 7
- Gross Negligence 7
- Intentional Wrongs 7
- Strict Liability for Reasons of Public Policy 8

SECTION TWO Contractually Speaking (What Did I Agree To?) 9

Jump-Starting Section 2 9

Contract Terms & Conditions 9
- Examples of Implied Conditions 10
- Precedent, Subsequent, and Concurrent Conditions 10
- Time and Performance Conditions 10

Design Contracts 10

Construction Contracts 11
- Fixed Price or Stipulated Sum 11
- Cost Plus a Fixed Fee with Guaranteed Maximum 11
- Cost Plus an Incentive Fee with a Guaranteed Maximum Price 11
- Critical Path Scheduling 11
- General Provisions 11
- The Owner 12
- The Architect/Engineer 12
- The Contractor 12

Subcontract Agreements 13

What to Look For—From the Owner's Perspective 13
- The 20 Key Provisions in the Construction Contract 13

A Matrix to Better Understand AIA Contracts 14

Key Changes to the 2007 AIA Documents 18

Insurance and Indemnification 22
- Surety bonds 22
- Indemnity 23
- Insurance 24

Alternate Dispute Resolution 26

Mediation 26

Arbitration 27

Special Master 27

SECTION THREE Government & Public Issues (Working with Rules, Regulations, & Uninformed People) 28

Jump-Starting Section 3 28
- Public Perception 28

Federal Law & Its Influence 29

State Law & Its Influence 31
- State Board of Technical Registration 31
- Contractor State License Board—Registrar of Contractors 32
- Minimum Workmanship Standards 33

County Ordinances & Their Influence 33

Local Government/Municipal Law & Its Influence 36

Who Catches Defects Before, During, & After? 40

Warranty Issues versus Normal Maintenance 40

Statute of Limitations—What's Included? 41

Home Owners & Home Owner Associations 42

The Home Inspector, Public Hero or Trouble Maker? 42

Indemnification 43

The Public Adjuster versus the Insurance Adjuster 43

Poor, Biased, & Misleading Reporting to the News Media 43

SECTION FOUR Myths and Facts (Issues That Lead to Misunderstandings & Confrontation) 45

Jump-Starting Section 4 45
- Starting Out With Simple Facts 45

Soil Myths 46

Concrete Myths 47

Framing Myths 49

Masonry Myths 50

Stucco Myths 51

Roofing Myths   52
Roads & Paving Myths   53
General Myths   54

SECTION FIVE   Architectural Issues & Construction Defects (Is It Always a Contractor Responsibility)   55

Jump-Starting Section 5   55

Early Problems   56
Risk Management Is More Than Contract Language   56

Professional Communication & Coordination   57
Client Responsibility   57
Architect Responsibility   57

Environmental Durability & Maintenance   58
The Importance of Maintenance Manuals   59
Natural Factors   59
Physical Factors   59
Human Factors   60

Lack of Knowledge on Specified Products   60

Specifications   61
Specifications General   61
Specifications "Boilerplate"   61
Standard Specifications   61
Master Specifications   62
Insufficient Specification Detail   62

Cover Sheets   62

General Notes   62

Code References   63

Details   63
Waterproofing Details   64
Flashing Details   64
Windows & Flashing Details   64

Do's in the Practice of Architecture   65

Don'ts in the Practice of Architecture   67

SECTION SIX   Engineering & Construction Defects (Engineers Don't Make Mistakes)   69

Jump-Starting Section 6   69

Company Philosophy   69

Insufficient Initial Project Information   70
Responding to the Request for Proposal (RFP)   70
Preparing the Scope of Services   70
Establishing the Fee   71

Lack of Adequate Design Information   72
Schedule Requirements   72
Existing Site Conditions   72

Lack of Communication & Coordination with Others   73
Design Concepts   73
Client's Expectations   74

Design   74
Competence   74
Assignment   74
Quality Control   75

Engineering Issues That Promote Construction Defects   75
Failure to Properly Determine Loads & Forces   75
Overreliance on Computers   76
Failure to Understand the Limitations of Computers   76
Failure to Transfer the Results of the Design Calculations to the Drawings   77
Lack of Judgment   78
Failure to Coordinate Documents   78
Failure to Complete the Last 5 Percent of the Construction Documents   78
Specifications Must Be Consistent with the Drawings   78

The Construction Phase   79
The Engineer During Construction   79
Post-Construction & the Engineer   79

Do's in the Practice of Engineering   80

Don'ts in the Practice of Engineering   80

SECTION SEVEN   Construction Issues (Defects & Litigation—Why Me!)   81

Jump-Starting Section 7   81

Uneducated, Unaware, or Incapable Field Supervision   81
Insufficient Knowledge of the Work   82
Inexperienced Supervision   83
Little If Any Knowledge of Quality Standards   83
Lacks Understanding of Value Engineering   84
Inability to Coordinate Work Activities   84
Lack of Scheduling Experience   85
Unrealistic & Improperly Prepared & Monitored Schedules   85
Failure to Complete or Properly Update Schedules   85

The Fallout from Poor Field Supervision   86
Cannot or Will Not Follow Plans & Specifications   86
Allows Others to Control or Approve the Work   87
Cheats on Inspection & Testing Requirements   87
Covers Up Mistakes   88
Plays Architect & Engineer Rather Than Seeking Professional Assistance   89

Other Factors That Affect Job Harmony   89
  Heavy-Handed Tactics with Subcontractors & Suppliers   89
  Failure to Make Use of Pre-Construction Meetings   90
  Failure to Include Design Professionals in Weekly Progress Meetings   90
  Failure to Understand the Value of Inspection & Testing Services   90

Lack of or Insufficient Documentation   91
  Incomplete or Lack of Daily Reports   92
  No Weekly Progress Meetings   92
  Requests for Information (RFIs)   92
  Documentation & Construction Claims   92

Quality Assurance/Quality Control   93
  Quality Control Testing   94
  Quality Assurance Testing   94

Federal, State, & Municipal Law   95

Lack of Skilled Tradespeople   96

Key Points to Remember   96
  Important Do's   96
  Important Don'ts   97

SECTION EIGHT   Care, Custody, & Control (Who's Watching the Store)   98

Jump-Starting Section 8   98

Gone in 40 Minutes!   98

Dormitory Fire Delays Occupancy!   99

Fire Levels Site, Fire Engine, & Surrounding Homes!   102

Concrete Isn't Waterproof?   103

Waterproofing & Drainage! Who Needs It?   104

A Bow String Truss That Lost Its Bow!   105

Who Needs an Engineer for Shoring?   105

Falling Like Dominos!   106

Superman Syndrome!   106

Who Needs a Body Harness?   107

Don't Fence Me In!   108

Where Did All the Supervision Go?   109

It's Only a Sewer Line Trench! What's the Beef?   112

Compaction, Who Needs Compaction?   112

Care, Custody, But No Control!   113

Compaction Around Underground Storage Tanks Too!   114

It's Only a Small Leak!   117

Out of Sight, Out of Mind!   118

Building in a Bathtub Does Not Make One Clean!   119

A Slide Plane by Design!   122

You Mean There Is a Difference in Contractors?   124

Select the Right Contractor or You May Pay the Price!   126

SECTION NINE   Specialized & Investigative Services (Finding the Needle in the Haystack)   129

Jump-Starting Section 9   129

Forensic Investigation in the Construction Industry   130

The Preliminary Report   130

Pre-Investigation Considerations   131
  Client & Team Player Information   131
  What Is the Investigation About   132
  Project Fees   132

Project & Work Task Scheduling   133

Project-Related Documents   133
  Original Project Documents   133
  Legal Case Documents   133

Work Flow & Document Processing   134
  Incoming Information   134
  Incoming Materials   134

Forensic Analysis, Intrusive Testing, & Failure Analysis   134
  Chain of Evidence & Protection Thereof   135

Special Considerations in Forensic Testing   135

Evaluation of Field-Testing Procedures   136
  Soils   136
  Concrete   137
  Masonry   138
  Stucco   138
  Exterior Insulation Finish System (EIFS)   139
  Windows & Doors   139
  Structural Investigations   141
  Asphalt Pavement Mixtures   144
  Examples of Other Recognized Construction Field Tests   146
  Testing Caveats   148

Laboratory Testing   148
  Microscopic Examination   148
  Typical Examples of Microanalysis   150
  Computer-Generated Analysis   153
  Other Laboratory Tests   153

Forensic Photography   155

SECTION TEN The Expert Witness (What to Look For—What to Expect) 158

Jump-Starting Section 10   158

Introduction to the Expert's Role   159

What to Look for in Selecting an Expert   161
- *Curriculum Vitae   161*
- *Specialization   161*
- *Case History   161*
- *References   161*
- *Demeanor   161*
- *Where to Look   161*

The Non-Testifying "Consulting" Expert   161

The "Hurried" Expert   162

The "Careless" Expert   162

Expanding Risks for the Testifying Expert   165

Preventing Expert Opinion & Testimony   166

Shooting Down the Expert   167

Conclusion   168

Exhibits   168

Glossary of Terms   170

Index   185

Preface

The contributors to this book have a combined construction industry experience of over 400 years. Working as experts in the fields of architecture, engineering, construction, and law, they are daily confronted with problems affecting the construction industry. These problems are more commonly referred to in the legal profession as contributory negligence, fraud, product defects, design defects, construction defects, and defective workmanship. When discovered, these problems often lead to costly litigation, business failures, increased insurance rates, loss and/or suspension of professional licenses, jail time, and loss of professional reputation, as well as the effects on one's family, clients, and/or the general public. Every year billions of dollars are paid out in judgments that are handed down as a result of these problems.

Historically, judgments have ranged from fair to totally outrageous. Whether or not the judgment was just, the rising cost of litigation cannot be ignored. Increased liability insurance rates traceable to litigation make it difficult for many in the construction industry to maintain a quality level of service at competitive rates. To support these added costs, one is faced with the need to either pass them down to the consumer or cut back on services to remain competitive. In response to competitive pressures, the reduction in the quality of services leads to problems on the job and subsequent hidden costs.

Here are three fundamental points to remember:

1. In the American judicial system, anyone can be sued by anybody at any time for anything.
2. The best defense against being sued is to do your work well.
3. Good record keeping and photo documentation are essential for your defense.

Our intent is that the information contained in this book will serve as a constant reminder and mandate to every construction professional to look seriously at his or her operation concerning what is needed to bring quality and performance into everyday activities. In other words: "Do It Right or Pay the Price."

We have endeavored to bring to the reader's attention specific areas of concern, which we believe are undermining the credibility of our respective professions, and our industry in total. Throughout this book, our concern is illustrated with actual cases that reflect the results of "Not Doing It Right the First Time."

There are numerous "how to" books, architectural and engineering handbooks, technical journals, and so on covering everything from setting up your own business to conducting finite element analysis on structures. However, little is said about the consequences if one "Fails to Do It Right." Because of the importance of "Doing It Right," we dedicate this book to all architects, engineers, contractors, attorneys, owners, building officials, developers, insurance carriers, manufacturers, suppliers, sureties, and testing agencies throughout the construction industry, as well as those teaching young members of society who are preparing to enter the profession.

While the authors have attempted to address many of the everyday litigated and complex issues affecting the construction industry, this book is only a beginning. The approaches to the numerous issues presented herein are based on the combined experience of the authors, as are the "do and don't" recommendations. One must keep in mind when confronted with situations similar to those covered in this book that not all situations are the same. Conditions may look similar, but often can and do vary greatly depending upon local conditions, local laws, geographic location, climatic conditions, and numerous other controlling factors encountered at the time an issue arises. The scientific method is an excellent system to use in addressing a problem. It's the authors' intent that the principles discussed in this book be applied in general, as they will not cover every specific case.

ACKNOWLEDGMENTS

We wish to take a moment to acknowledge those who have motivated, aggravated, stirred, caused, encouraged, inspired, assisted, and put up with us in the development of this book. We hereby extend our thanks and appreciation.

- To all of the situations involving every type of problem imaginable, and some unimaginable, that *motivated* the contributors to write this book.
- To all the irrational, uncooperative, irritating people who totally *aggravated* us when we tried to arrive at an amicable resolution to the problem being considered.
- To the emotions of those who lived through disastrous effects on life and property as a result of not taking the time to Do It Right and Paid the Price that *stirred* our desire to write this book.
- To the failure of our respective professions to recognize the need for improvement and change that *caused* us to band together in developing the various sections of this book.
- To our colleagues, who have experienced many of the problems affecting our industry and *encouraged* us to do this book and have eagerly awaited its publication.
- To the bantering back and forth of opinions, facts, and ideas gained through experience by all of the contributors that *inspired* each of us to dig a little deeper in the preparation of each section.
- To the staff at First Watch who put up with us during our breakfast meetings; to Jan and others at Sacks Tierney, P.C., who ensured that lunch was there for

our Saturday review and editing sessions; to Brad Pfeifle, who diligently typed our edits, while we argued our point and chased around the text projected on the screen with lasers; to Steve Trail, P.E., and John Denney, P.E., who added their engineering expertise and experience; and to all of the others who **assisted** us, from professionals, workers, and clients whom we consulted and questioned and helped provide additional authenticity to this book.

To our wives and families who **put up with us** not being present two Saturdays each month and many evenings over the past two years as we worked on the various sections of this book.

To our peer reviewers, we wish to express our very special thanks for their **honesty** in helping authenticate the efforts and needs we believe in, and wish to share with you the reader.

THANK YOU

About the Authors

The following biographical sketches of the authors who have devoted their energies to make this book possible truly believe in its title, *Do It Right or Pay the Price*. We have worked as a team over the past two plus years, each contributing our knowledge, experience, concern, and convictions in hopes that the information provided will have a positive effect on our readers: to improve their professional relationships and understanding with clients and colleagues, by continually seeking quality within one's respective profession, and to generate pride in oneself and one's business, spawning greater recognition from one's peers, and, ultimately, fulfilling the highest expectations of one's clients. We firmly believe that *doing it right* pays back in personal satisfaction as well as in higher levels of professional success, with satisfied clients inspiring greater success in one's business.

Glen Copeland, P.E., President, Copeland Geotechnical & Associates, Inc., specializes in forensic investigations and expert testimony related to soils' issues affecting completed construction. Glen attended Arizona State University and received BSCE in 1967 and MSCE in 1970. Prior to 1970, he practiced as an engineer-in-training at the civil engineering firm of Holmquist and King, Consulting Engineers. While at Holmquist, Glen had a diverse training experience: starting as a rodman and chainman for a survey party, draftsman and design engineer, and project manager. He worked on all types of civil engineering projects consisting of industrial, commercial, and residential developments. Roads, culverts, bridges, utility infrastructure, grading and drainage design, improvement of district plans and specifications, and cadastral and design surveying were all part of his daily routine and experience. In 1970, he moved into the geotechnical engineering area and over the next 35 years has participated in or written approximately 5000 geotechnical engineering reports in the states of Arizona, New Mexico, Utah, and California. Since the founding of Copeland Geotechnical Consultants in 1993, Glen has also participated in or managed about 600 soils-related construction failures and/or defects reports, including potential slope failures, grading and drainage challenges or failures, expansive and collapsible soils issues, below grade soil moisture and/or groundwater conditions impacting structures, pavement failures, pipeline backfill failures, excessive lateral earth pressures against retaining walls, and standard of care reviews. He specializes in on-site soil exploration and sampling, laboratory testing of recovered soil samples, analysis of field and laboratory test data, use of specialized computer programs to depict original and "as-built" facility or soil conditions, and preparation of engineering reports and rendering of expert opinions. Glen is an associate member of the Structural Engineers Association of Arizona and a Fellow and Life Member in the American Society of Civil Engineers. He served as a state director for the ASCE in the 1960s and was one of the original founding and participating members of the ASCE Younger Members Forum in Phoenix, Arizona.

George C. "Joe" Frank CFC, FABFEI, President, Frank Consulting, Ltd., specializes in forensic investigations and expert testimony related to construction. Joe attended California State Polytechnic College where he majored in agriculture. He was the qualifying party for licenses in general building and heavy construction, plumbing, boilers steamfitting, and process piping in the state of Arizona for over 40 years, and he currently remains licensed in that state as a commercial contractor. His projects resume, since entering the construction field in 1957, includes supervision in both design and construction on over $36 billion worth of completed construction in 12 states and four foreign countries. Projects included crude oil pipelines and pumping stations, bridges, roads, manufacturing and processing plants, hospitals, hotels, prisons, high- and mid-rise office buildings, apartments, homes, and entire new cities. Since entering the field of construction consulting, in 1981, Joe has conducted over 900 investigations in eight states and the territory of Guam, covering construction failures and defects, including earthquakes, landslides, grading and drainage failures, expansive and collapsible soils issues, structural steel and concrete problems, failures and defects in light gauge and wood framing, water intrusion issues associated with stucco and exterior insulation finish systems, and fire damage assessment. Specialization includes on-site testing, reconstructive laboratory testing and failure analysis, and using finite graphics to support the findings and conclusions rendered. In addition to his forensic endeavors, he serves as a neutral on the Panel of Commercial Arbitrators for the American Arbitration Association and is a member of its local Construction Advisory Committee. He was awarded Fellow status with the American Board of Forensic Examiners and is an advisor on the American Board of Forensic Engineering & Technology.

Publications attributable to the talents of George C. "Joe" Frank, CFC, include the following:

- **Forensic Investigation & Remediation of Expansive Soils:** *American Board of Forensic Examiners Seminar, 1999.*
- **The Construction Consultant in Discovery:** *Arizona State Bar Construction Law Seminar, 1995.*
- **The Construction Expert in Arbitration:** *American Arbitration Association Seminar, 2000.*
- **The Expert's Role in Construction Defect Litigation:** *Arizona State Bar Construction Law Seminar, 1994.*
- **Construction Accident Investigation and Analysis:** *Associated General Contractors of America, Seminar, 1990.*
- **Higher Construction Profits at the International Level:** *Operations Manual for Morrison Knudsen International Division, 1977.*

- **A Project Management Organization for Transportation Agencies:** *Operations Manual for the Arizona Department of Transportation 1981.*
- **Project Scheduling for Transportation Agencies:** *Operations Manual for the Arizona Department of Transportation, 1981.*

Joseph A. Gervasio, P.E., President of Gervasio & Assoc., Inc., a 40-person consulting firm offering services in civil and structural design, forensic engineering, and architectural investigations. His firm designs approximately 200 projects per year and conducts approximately 300 forensic investigations per year. Joseph obtained his civil engineering degree from the University of Arizona in 1957. While attending school, he was a surveyor with the Southern Pacific Railroad. Upon graduating from collage, he was commissioned as a first lieutenant in the U.S. Air Force where he served as a sanitation and industrial hygiene engineer. He returned to Phoenix and joined the structural engineering firm of Magadini & Assoc., starting out as a structural inspector on a 10-story cast-in-place concrete flat slab structure and then joined the design staff to design institutional, educational, industrial, and commercial structures over the next six years. It was during this time that Joseph received his registration as a civil and structural engineer in Arizona and California. In 1968, he joined the architectural firm of Haver, Nunn & Jenson and started its engineering department. His designs included the civil engineering of the South Campus of Northern Arizona University and structural engineering for a series of furniture stores from Minneapolis to San Francisco. In 1970, he joined Sullivan & Mason, an MPES consulting firm, where he started the Civil Engineering Department and shortly thereafter became partner in charge of structural and civil engineering. During this time, he supervised the design on a 32-story moment-resisting concrete framed U.S. Bank building in Phoenix, Arizona. In 1976, he started Gervasio & Assoc., first as a civil and structural consulting design firm. In 1989, Gervasio & Assoc. purchased a leading four-person forensic firm, which has continued to grow by not only emphasizing but providing quality services throughout the industry.

Ed Holdsworth, General Manager and Chief Executive Officer of SEMTEC Laboratories, Inc., graduated from Arizona State University with a degree in chemistry in 1966. From 1966 to 1968, he served in the U.S. Army, where he operated an electron microprobe analyzer at the Nuclear Defense Laboratory in Edgewood Arsenal, Maryland. From 1968 through 1976, he was in charge of the Electron Microbe Laboratory at the Arizona State University Department of Chemistry. In 1976, Ed founded SEMTEC Laboratories, Inc., a micro-analytical laboratory providing services to the industry with emphases on scanning electron microscopy and micro-x-ray techniques. For the past number of years, he has been involved in numerous forensic investigations related to the construction industry where he provides general failure analysis services to other forensic consultants, including legal and insurance interests. In his spare time, he donates weekends as a volunteer ski patroller and is a certified instructor for the National Ski Patrol in outdoor emergency care.

Wellington R. Meier, Jr., Ph.D., P.E., specializes in construction materials, primarily in the transportation field. This has involved forensic investigations and expert testimony related to design, construction, and performance issues. His employment has included 15 years with the Nebraska Department of Roads; 7 years with the Bureau of Indian Affairs, Navajo Area Roads Office; and 20 years with Western Technologies Inc. in Phoenix, Arizona. Wellington received his B.S. degree in civil engineering and M.S. in engineering mechanics from the University of Nebraska and a Ph.D. with a major in civil engineering from the University of Arizona. He obtained registration as a professional engineer in Nebraska, New Mexico, Arizona, and California. He has held membership and presented papers for the Transportation Research Board and the Association of Asphalt Pavement Technologists.

David C. Tierney is a litigator practicing primarily in commercial construction law. He has been a partner in the Phoenix firm of Sacks Tierney P.A. since 1974. Mr. Tierney received his B.A. degree from Brandeis University and his J.D. degree from Harvard Law School in 1965. He was clerk to Justice Reardon, Massachusetts Supreme Court; a Peace Corps volunteer (still fluent in Spanish); and former chairman of the Governor's Task Force on Computers in Education. David was the chairman of the Board of Trustees for Risk Management (self-insurance for Maricopa County); a founder of the Horace Rumpole Inn of Court; a past chairman of the State Bar Construction Law Section; one of two editors of the *Practice Manual* for the State Bar Construction Law Section; the chairman of the Construction Advisory Council to the American Arbitration Association in Phoenix; the chairman of the Restorative Justice Resources Council, Inc.; the chairman of the Irish American Cultural Institute Chapter in Phoenix; and the treasurer of the Arizona Coalition for Tomorrow (a Head Start–related entity). In October 1998, he received the "Hon Kachina" Award for volunteer service to the community. In September 2003, the National Association of Probation Executives presented the William Faches Award to him. David is a former adjunct professor at ASU Law School teaching construction law. He was selected for American Bar Association Fellows 2006. Also in 2006, he received the Supreme Court of Arizona Award for Improving the Legal System through Public Service. David has participated in several Lorman seminar presentations, including the March 15, 2007, "Managing Construction Projects"; as well as the November 2003 and November 2004 "Expert Witness Procedures in Arizona" presentations. He has chaired and presented at all-day continuing legal education programs for the Arizona State Bar and for the American Arbitration Association on such topics as ethics, alternative dispute resolution (ADR), the Revised Uniform Arbitration Act, construction bidding process, and contractual requirements on governmental and private jobs.

Publications attributable to the talents of David Tierney include the following:

- "Arizona Corrections, Accomplishments and Unfinished Business," 8/78 *Arizona Bar Journal* (14, No. 2), p. 28, by David C. Tierney and Tom Irvine.
- "Prison Crisis, A Need for Rehabilitative Programs," 5/1/79 *Weekly Gazette*, Convention Issue, by David C. Tierney and Tom Irvine.
- "Alternatives to Incarceration," 1/22/80 *Scottsdale Progress Saturday Magazine*, David C. Tierney.
- "Our Prisons, Upside Down & Backwards Priorities," 7/15/84 *Tempe Daily News*, David C. Tierney.
- "Controlling the Means of Construction; New Exposure for Architects on Construction Jobs," January 1985 issue of *Southwest Contractor Magazine*.
- "'Unless and Until' Clauses, Slow Pay for Subcontractors." Annual Issue of *Arizona Contractors Association Almanac*, 1989, and a related article in the March 1990 State Bar Construction Law Section Newsletter.
- One of five contributors to and editors of the Maricopa County Bar Association's "Litigation Manual," sold by the Bar to law firms in Maricopa County, Arizona. Judge Lankford was the chief editor (1992–1993).
- Article, "Changes Coming in Business Dispute Resolution, Arbitration—Mediation" for the *Business Journal* in May 2000.
- One of two editors of the Arizona State Bar Construction Law Section "Practice Manual," a 300-page book released June 9, 2003, for sale by the Arizona State Bar.
- "Limitations on Closing Arguments," 11/03 *Arizona Attorney*, p. 22, David C. Tierney.

Craig Walling, Architect, has since 1985 limited his practice and specialized in roofing consultation, which also includes waterproofing. A 1957 graduate of Arizona State University with a bachelor of science degree, he worked for other architectural firms until 1969. Throughout most of his apprenticeship, he was the firm's technical person producing project specifications and checking contractor's work in the field. Craig obtained his registration as an architect in 1966; in 1969, he joined the Tempe firm of T.S. Montgomery and upon Mr. Montgomery's passing continued that conventional architectural practice with two different partners. In 1985, he went solo with his practice and began specializing in roofing consultation. Active in two professional organizations over the years, the American Institute of Architects and the Construction Specifications Institute, he served in various offices in both organizations, including president, Arizona Society of AIA; two terms as president, Phoenix Chapter CSI. He received the Arizona Architects Medal in 1982. Craig's public service includes 18 years as a member of the Arizona State Board of Technical Registration Architectural Enforcement Advisory Committee that functioned as a screening panel for the State Board. The scope of his roofing consultation practice today includes investigating and making recommendations, providing plans and specifications for re-roofing, consulting with owners and other architects in producing good roofing and waterproofing, investigating roofing claims for insurance companies, and providing unbiased findings.

Jesse R. Wyatt, P.E, S.E. Faci, President of Jesse R. Wyatt, P.E., Ltd., specializes in failure investigations and trouble shooting on construction and serves as an expert witness in civil litigation involving construction-related issues. He received his bachelor of science degree in civil engineering from the University of Texas in 1958 and is registered as a civil and structural engineer in Arizona and as a civil engineer in Texas, New Mexico, Colorado, Utah, Nevada, and California. Jesse was named a Fellow in the American Concrete Institute and received the Peter Courtois Memorial Award from the Tilt-up Concrete Association for Outstanding Contributions to the Advancement of Tilt-up Concrete Construction. His experience includes working for the Texas Highway Department as a design engineer in El Paso and Austin. He also worked as a design engineer for Parkhill Smith & Cooper, Consulting Engineers, in El Paso, Texas, and W. C. Cotton, Consulting Engineer, in Austin, Texas. He spent 18 years with the Portland Cement Association, advancing to the regional structural engineer's position. Duties with the Portland Cement Association including serving on General Design, Fire and Life Safety, and Seismology committees of the International Conference of Building Officials. He was the liaison to the Structural Engineers Association of California Seismology Committee. Duties also included educational programs on Portland cement and concrete. He has lectured extensively in seminars and as a guest lecturer at universities on a wide range of topics, including structural design, tilt-up design and construction, building codes, seismic design, fire resistive design, concrete fundamentals, architectural concrete, and concrete construction practice. Publications include lecture notes on a wide range of topics and journal papers on specifications and tilt-up wall design and construction. He authored the wall chapter of the 1973 and 1978 editions of "Notes on ACI 318" published by the Portland Cement Association.

Jesse established Jesse R. Wyatt, P.E., Ltd in 1983. In addition to trouble shooting on construction and forensic investigations, he has continued to lecture on topics primarily related to cement and concrete. Notable projects include investigation and repairs for the London Bridge in Lake Havasu City, Arizona, and the cooling towers for the Palo Verde Nuclear Plant west of Phoenix, Arizona. Other projects include buildings, parking structures, prestressed and post-tensioned concrete, stucco, masonry, retaining walls, bridges, power plants, concrete pipe, pavements, water and wastewater treatment plants, tanks, and reservoirs.

Foreword

Broaching the topic of construction failures and their consequences, *Do It Right or Pay the Price*'s authors build upon their years of experience to provide a distinctive perspective on professional ethics and professional responsibility within the industry. We all know that the construction process involves numerous variables affecting the successful design and erection of a structure. Apart from the traditional expectations of functionality, builders and owners alike must approach the process with a keen eye for the pitfalls and tempting shortcuts that all too often characterize the construction and engineering industries. While these challenges are not new, they have evolved with the field.

With the strong focus on professional responsibility and the ensuing fallout if that responsibility is neglected, *Do It Right or Pay the Price* reminds everyone—from contractors to technicians to engineers and academicians—that the success or failure of a construction project rests upon the preparedness and ethics of approach. Speaking from their respective positions as long-time practitioners in the field, the authors of this book advocate for humility on the part of major players—an approach that they contend will ultimately minimize costly errors, miscalculations, or unexpected failures.

While the authors' strong emphasis on personal experience may undermine some discussions of the value of new technology in the field, this book accomplishes its task of revealing the bigger picture issues at stake when considering the true cost of work in the construction industry. The authors readily recognize that even the most thorough preparedness will not prevent every possible failure; however, their approach is intended to curtail negligence and prepare builders for the potential legal actions that may arise.

By highlighting these sorts of considerations and by describing the legal accountability for such work, *Do It Right or Pay the Price* presents a legally minded approach to ethics and responsibility in the construction and engineering industries that is written for a broad audience of participants in the construction industry. Although each reader may well find specific points in the book to disagree with, almost all readers in the construction industry will derive value from the following pages.

Sandra Houston Ph.D., P.E.
Arizona State University

SECTION ONE

PROFESSIONAL EXPECTATIONS

or

We Are What We Project

1.0 JUMP-STARTING SECTION 1

A "professional" is a person who is engaged in one of the learned professions, generally referred to as an occupation requiring special education and skill, characterized by conformance to the highest ethical standards. If this be true, the reader should approach this section armed with the following points to better understand the importance of the issues presented:

1. One's professional reputation and ethics are largely forged within the first few years of professional employment.
2. When you join a professional organization, choose well.
3. Codes of conduct are normally found in the bylaws of the professional organization and need to be considered as minimal levels of performance.
4. In your chosen profession, your ethical standards are the quality standards you set for yourself and should be set higher than the minimum normally expected.
5. A prime example of such a professional is the forensic investigator who examines physical evidence in preparation for factually testifying to his or her findings and how they apply to the case in question. As an officer of the court, the forensic investigator is held to a higher level of ethical conduct, well beyond the duty owed to any employer.
6. Legal actions are a fact of life in the construction industry. Consider legal actions seriously no matter how frivolous they may seem at first. Like those who come to you for advice because they lack experience in your field, do the same: Retain a qualified attorney for legal advice. That puts the best person in the right position and clears your attention for what you do best.
7. Involve the attorney early; don't wait until shortly before a legal deadline to retain one. You can't design and construct a good structure in a day; don't expect your attorney to build a good case in a day either.
8. The field of law, as are the fields of engineering and construction, is extremely specialized. Find a competent firm in the area of construction law that fits your needs and keep that number handy.
9. A basic understanding of relevant legal definitions and actions should be attained by those working within this industry. Examples include the definitions of tort, criminal, negligence, gross negligence, and contract liability.

There is a refreshing change in public works projects that is worth mentioning here. Bidding procedures now give value to quality and responsibility rather than simply to low cost. The ability to properly manage, schedule, coordinate, and provide true-value engineering to a project now increase a contractor's ability to be selected for a public works project.

Ethical behavior is an integral part of being a professional. It's just that simple. Defining "ethical behavior," however, is not easy. You cannot simply list 12 steps to ethical behavior. Ethical behavior was defined within most of our families as we grew up, most especially by our parents. Ethical behavior is also influenced by our religious associations, by laws as prescribed by various legislative bodies, and by the examples of people for whom we have great respect. Ethical behavior is, as noted, the sum of many factors; it is strong enough to withstand daily assault while supple enough to include a wide swath of backgrounds and experiences. Those of us in the professions need to lead by example in this very important area. Greed, speed, and the need to please do not constitute ethical behavior but, unfortunately, these three gremlins run rampant throughout the construction industry.

Keep in mind that your professional ethics are grounding philosophies and not a list of legal "got yahs."

Just as some professionals design only to the code minimum in terms of the quality of their work, some architectural, engineering, and construction entities seem to believe that it is "ethical" to merely comply with the licensing statutes and regulations. However, "legal" is not a synonym for "ethical."

As professionals within an industry that plays a major role in the well-being of the general public, we are charged with certain moral responsibilities to utilize our respective knowledge in a manner that benefits society. If we look back in time, we can see several instances when ethical disasters have forced revisions to rules of ethics. Here are a few:

- Reaction to Enron's collapse forced the Sarbanes-Oxley legislation on corporate ethics.
- Reaction to Arthur Anderson's problems forced revisions to the Certified Public Accountant (CPA) ethical standards.
- The excesses in appraised values in Phoenix, Arizona, in the mid-1970s relating to the Mountain Preserve led to the formation of the State Board of Appraisers.

- The complete lack of ethics and the overriding greed found throughout the financial collapse starting in 2007 showed many in high positions to be acting more like cheap Chicago gangsters of the 1930s than responsible business titans. "Doing It Right" was never in their minds. Unfortunately, few of them have "Paid the Price" for their activities.

In each of these cases, it had been generally *assumed* that people holding so much responsibility would actually perform properly. That they fell so far short of the mark was shocking to many people. Public outrage encouraged legislation that at least attempts to force a level of basic ethical behavior. "Good Luck!" Real ethics come from within, not too-little-too-late legislation accompanied by public breast thumping. The fact that such legislation was passed is, in itself, an indicator of the pervasiveness of poor to missing ethical standards as practiced by some.

1.1 PROFESSIONAL ETHICS

As early as the 1940s, the Engineers Council for Professional Development stated the following relating to professional ethics: ***"Honesty, justice and courtesy form a moral philosophy which, associated with mutual interest among men, constitutes the foundation of ethics. The engineer should recognize such a standard, not in passive observance, but as a set of dynamic principles guiding his conduct and way of life."*** Although these statements were made relative to the engineering profession, the principles apply equally to other professions.

Individuals must not only demonstrate experience, knowledge, and competence and obtain registration or licensing, but must also continue with education and development to maintain competence in the continually changing world of construction.

It is essential for the betterment of one's self, and the construction industry as a whole, to participate and to teach our fellow professionals that knowledge, understanding, and truthfulness will bring about better quality and performance. The contractor who knowingly submits a bid at cost on the premise that the contract documents were so incomplete that "a killing" can be made on change orders not only exposes him- or herself as a fraud, but cheats other bidders and the owner as well.

The architect who fails to provide proper details and directions in the contract documents, attempting to transfer the responsibility to the contractor, is cheating the client as well as the contractor.

The engineer who fails to consider all of the contributing factors and relies solely on the data from a canned computer program is cheating the client and all who are relying on the accuracy of his or her work.

We've already mentioned that the first few years of one's professional life have a great effect on the still-developing ethics of that person. That is because the young professional sees how it's done for the first time in "real life." Seeing a complete lack of ethics may lead an observer down the same path. Those early formative years create the foundation of one's professional reputation. This developing reputation will determine how this individual will likely be known throughout his or her career, notwithstanding the possibility that minor corrections may be made as time goes by. It is vital that these young people start out right, and it is our responsibility as mentors to demonstrate through our actions the right way. If we don't, who will?

Corporate ethics have been satirically defined by Robert Jackall in *Moral Mazes* as "What is right in the corporation is what the guy above you wants from you." That statement is considered by many to be the only way to survive in an organization. But that statement cannot, and should not, be used as a moral or ethical justification for actions that are wrong.

How many times have we heard someone attempt to justify a self-serving unethical act with the expression, "It's the American way, isn't it?" No, it is not the American way, and we should stand up against it.

The actions of those in our business who make it a practice of conducting themselves as noted above leave an indelible stain on the profession. It is the responsibility of all within a profession to become involved in extinguishing such practices by reaching out to those who have fallen below the established ethical standards. This can be accomplished by implementing programs within organizations and/or through legislative action that would bring about sanctions or penalties for those who refuse to abide by the rules.

For those organizations that have failed to adopt a "code of ethics" and for those who believe that there is really no need for such a code, think again. Litigation in the construction industry is at an all-time high. Professional liability and general liability insurance costs are at an all-time high, if insurance can be obtained at all, or the insurance is available only with excessive deductible amounts.

Legislation continues to be enacted that places more controls on how we do business. Continual changes in technology, lack of qualified personnel, higher material costs, and accelerated schedules requiring quick decisions have led to a breakdown in the quality of services.

When poor workmanship results in costly redoes, disguised as something else, the owner often has to foot the bill or at the minimum enter into negotiations over the costs or seek legal restitution. Unfortunately, the good construction people who had no part in the incident see their insurance premiums take another jump. Keep in mind that we all pay for poor work and unjustified cost overruns when our fellow professionals just can't "Do It Right."

All of these factors have had and will continue to have a major and costly effect on our industry, as long as we allow ourselves to be "taken down" by the actions of those who just don't concern themselves with professional ethics. Unfortunately, too many in our industry have pointed to these problems as an excuse to cut back their services, quality, performance, and duty. Instead of taking it upon themselves to become a part of improving our industry by taking an active role in making things better, they tend to sit back passively and allow these negatives to proceed unabated.

1.2 PROFESSIONAL ORGANIZATIONS—EXPECTATIONS AND RESPONSIBILITIES

Design professionals and contractors must deal with two primary types of organizations; one is regulatory (mandatory), the other professional (voluntary). The regulatory (mandatory) body issues registration or licensing on the basis of knowledge, experience, and capability. The other consists of organizations whose members share information and training within their specific fields and provide information to encourage proper conduct.

Regulatory (Mandatory) Organizations

These bodies governing design professionals generally exist at the state level and are mandated by statutes. Their overall purposes are to safeguard life, health, and property and to protect public welfare. The state boards that implement statutes have various titles and are organized in a variety of ways. They are generally titled as "Boards of Registration" or "Boards of Licensure." Boards for some states may include a wide range of professional disciplines, while others may have several boards, each of which controls registration of a single or a small number of professional disciplines. The functions of these boards are discussed further in Section 3, Government & Public Issues.

Similar bodies exist for contractors. To be licensed in a particular area of construction, contractors must demonstrate experience, capability, and knowledge within that field of construction. License requirements exist not only at the general contractor tier but also for subcontractors performing various tasks under the general contractor. Construction performed by licensed contractors must comply with industry standards and project requirements. If there are questions raised regarding a contractor's performance, those issues may be referred to the licensing board for resolution. Similarly, design professionals whose performance is lacking may be referred to their licensing board. Statutory requirements for registration and licensing of contractors and design professionals are also further discussed in Section 3.

Professional (Voluntary) Societies

Many voluntary societies are available to professionals practicing in the design and construction fields. Paramount among these is the American Institute of Architects (AIA) and the National Society of Professional Engineers (NSPE). The Associated General Contractors of America (AGC) provides a similar function for contractors.

The two most prominent societies that address forensic investigation related to the architectural, engineering, and construction fields are the National Academy of Forensic Engineers (NAFE) and the American College of Forensic Examiners International (ACFEI). It is the mission of all these organizations to promote skill and integrity among members. This is done to improve the accuracy and factual presentation of findings to the trier of fact or other interested parties and to thereby protect the public interest.

Among the objectives of these organizations are continuing education for members; improving the architectural, engineering, and construction methodology by encouraging advancement and application of new technologies; and promoting the highest standards of professional conduct. All these organizations function at the national level with memberships served by state and local branches, sections, or chapters.

A number of societies are dedicated to advancement and development of professional practices within specific disciplines. The Engineering Founder Societies include the American Society of Civil Engineers (ASCE); the American Society of Mechanical Engineers (ASME); the Institute of Electrical and Electronic Engineers (IEEE); the American Institute of Mining, Metallurgical, and Petroleum Engineers (AIME); and the American Institute of Chemical Engineers (AIChE). These societies are organized on the national, state, and local levels, as required. They often incorporate committees or units that function in specific areas within their disciplines, to further serve their professional development goals.

It should be the duty of all professionals to continually work on personal improvement and with the regulatory and professional organizations to affect the standards that govern and educate their respective professions. This is accomplished by first understanding the standards and goals of the organizations and then becoming an active voice in changes that will maintain, improve, and embrace these standards and goals, and by speaking out in favor of enforcement.

1.3 ETHICS & ARCHITECTURE

The AIA has long "soft peddled" the creation of serious ethical rules, due in part to some economic challenges it faced years ago, and which it still faces. In general, the AIA's *Code of Ethics and Professional Conduct* includes five basic principles of conduct: General Obligations, Obligations to the Public, Obligations to the Client, Obligations to the Profession, and Obligations to Colleagues. Every member architect must consider these principles before taking on an assignment. However, the burning question is: Do these five principles effectively address the issues facing the architect today? It is thought by many in the profession that architecture and ethics are synonymous because at times they merge.

It is said that ethics in architecture consist of physically defining the person, the society, the religion if appropriate, and the business within the edifice being constructed, whether it be a single home, a church, or an entire city. There is truth in characterizing architecture and ethics as being somewhat synonymous, as both are complex disciplines that embody some historical and cultural theories that can vary in time and place. However, do ethics really matter from a professional standpoint? As a professional, should one have a duty to clients to remove the blinders and look at a broader picture of just what the responsibilities are? And should we take into consideration our own communal and moral values? Architecture, on the other hand, is viewed by many as shaping our physical habitat in an ambience suited to the human needs of the client. When an architect takes on an assignment, one wants to believe that the broader picture and not only the obvious needs of the client have been considered.

Clearly, many of the ethical issues affecting our lives today have been reduced to laws and standards, including regulations ensuring that consideration is given the handicapped, protecting the environment, creating a safe work place, and so on. Professional ethics do not end after ensuring that the final design meets established laws and standards.

Creativity does have its place, as do monuments, but should not overshadow what is right and just for the client. Designing a school with exceptional grandeur and no money left over for proper classrooms does not speak well of doing what's right for the client.

The following three points were taken from the publication *Ethics and the Practice of Architecture* and, according to the author, should be used as a means to objectively evaluate the purpose of the architect in the overall scheme of building in today's world. These points are quite real and have been the "guiding light" for many an architect.

- The essence of architecture is not in its *usefulness*—the purely practical solutions it offers to the human need for shelter—but in the way it meets the much more profound spiritual need to *shape our habitat*. In our culture, architecture transcends the mere physical substance of buildings by endowing constructed forms with aesthetic, emotional and symbolic meanings which elevate them to symbols of civilization.

- A work of architecture is an image, a symbolic expression of the limitations, tensions, hopes and expectations of a community. I also believe that architecture is an ethical discipline before it is an aesthetic one. . . . This moral dimension is legitimized when architecture is presented . . . as something concrete and practical which each individual citizen . . . can relate to in a practical way.

- When we build we have not just a responsibility to ourselves and our clients, but to those who came before and those who will come after. . . . architecture transcends local issues. Questions of space, light and material, what makes a great building, are separate from client and site. Yet they are realized in a specific way, according to a genius loci.

Many think that these points reflect clear statements of architecture's most basic and clearly understood purposes.

The question becomes how far does one take this approach in arriving at what is ethically right or wrong. Is it ethically right for an architect to specify something because it aesthetically fits into the overall ambience and symbolic meaning of the structure but creates a major problem with the overall performance of the structure? Is it ethically correct for an architect to provide incomplete details, because they do not apply to the aesthetic value of the structure, and rely on the builder to determine how the details are to be provided? Is it ethically right when defects occur as the direct result of the architect's unwillingness to accept responsibility for the above noted actions (**utilizing the excuse that the builder has the responsibility to determine the means and methods of construction, as long as the finished work meets the overall design intent**)? In the field of professional ethics, such actions are morally wrong. Further discussion of architectural issues that promote construction defects can be found in Section 5, Architectural Issues & Construction Defects.

1.4 ETHICS & ENGINEERING

Engineering is a major part of the construction industry and has often been referred to as one of the most responsible and learned professions. Engineers, therefore, are expected to reflect the highest standards of honesty, impartiality, fairness, and integrity in their professional behavior. As such, professional engineers are charged with the following *Fundamental Canons*:

- Hold paramount the safety, health, and welfare of the public.
- Perform services only in areas of one's competence.
- Issue public statements only in an objective and truthful manner.
- Act for each employer or client as faithful agents or trustees.
- Avoid deceptive acts.
- Conduct oneself honorably, responsibly, ethically, and lawfully so as to enhance the honor, reputation, and usefulness of the profession.

In order to maintain the integrity of the profession, the National Society of Professional Engineers (NSPE) has maintained a code of ethics that is monitored by its Board of Ethical Conduct. This board reviews cases in which the society's members have been charged with a violation of the code of ethics or responds to questions from its members when confronted with issues that may put them in violation of the code.

Most professional societies have adopted functionally equivalent codes of ethical conduct and have established boards to review the ethical conduct of their constituents. The questions are: Do all their members live by them, and what is our professional responsibility to bring violations to the forefront, and to whom do we report these violations so they can be effectively acted upon? The next question to be addressed is: Will there be justice if such violations are reported or will the violator receive a "slap on the wrist" upon promising not to do it again? Unfortunately, in most cases, the only penance the violator receives is a slap on the wrist. It is essential to the credibility of these professional societies to adopt, and enforce, policies that censure and properly punish offenders. Continued leniency to irresponsible acts of a member may lead to disastrous consequences that discredit the society and all of its members.

How often do we hear in our daily lives that *"It's not my responsibility,"* particularly when problems arise. If not ours, then whose responsibility is it? Obviously, one may look at responsibility in various forms. Responsibility, in its truest form, has a series of meanings:

- Moral and forward-looking responsibility is defined as a best-effort duty to others to achieve a quality result. Despite one's top effort, the expected or hoped for result may not be achieved, but knowing you did your best makes you a better professional.
- Obligatory responsibility can address issues such as appointing someone to take notes at a meeting and to distribute them to all concerned.
- Official responsibility at times is often associated with job descriptions or generating prescriptions of personal conduct that at times may require the person to act unethically. Failure to comply with official responsibilities may affect a person's status within the company framework under which such responsibilities were formulated.
- Professional responsibility, on the other hand, is much like moral responsibility in that it arises from the special knowledge one possesses that bears directly on the well-being of others.

As an engineer, it is essential to take on responsibilities both morally and professionally. If the organization or the environment in which you work expects less or proposes that one's ethics be compromised for the benefit of others, such actions will destroy not only your ethical standards but also your integrity and reputation. Being a "hired gun" for an attorney to make a case, compromising your responsibility to thoroughly review calculations in order to expedite a project out the door, failing to take sufficient density tests on a mass fill and certifying that it meets Code or design intent, or sealing a set of plans that has not been thoroughly reviewed will soon result in one's demise as a respected member of the engineering profession. Further discussion on engineering issues that promote construction defects can be found in Section 6, Engineering Issues & Construction Defects.

1.5 ETHICS & FORENSIC INVESTIGATION

Forensic investigation, in essence, encompasses the application of the natural and physical sciences to the resolution of conflicts within a legal setting. A forensic investigator, also sometimes referred to as *expert witness,* examines physical evidence and then testifies about the results. The investigator is, by Federal and State Evidence Rule 701, considered an officer of the court, although paid by one party. It is the responsibility of one serving as an expert in the subject field to provide professional testimony to the "trier of fact,"[1] so the trier of fact may better understand and evaluate the evidence. Forensic investigation in construction has become a full-time profession for many in the recent past. A more thorough description of the responsibilities and duties of a forensic investigation and expert witness is presented in Section 9, Specialized and Investigative Services, and Section 10, The Expert Witness.

Responsibility placed on the shoulders of the forensic investigator requires credentials above reproach and elevated professional ethics. There can be no misrepresentation, either to the client or the trier of fact, that the forensic investigation and subsequent opinions will be anything but the truth, fully supported by the evidence uncovered, professionally evaluated and scientifically presented. The forensic investigator can only be an advocate for the truth, whose sole purpose is to discover and publish the facts, be they to the advantage or disadvantage of the client.

The American College of Forensic Examiners International is the largest professional society in the field of forensic examination and has adopted the following code:

"As a Forensic Examiner, my role is to serve justice by using my expertise to further the doctrine of fairness. Therefore, I pledge:

- To maintain the highest standards of professional practice.
- To remain totally objective and to use my ability so that justice is served by an accurate determination of the facts involved.
- To thoroughly examine and analyze the evidence in a case, to conduct examinations based on established scientific principles, and to render opinions, which have a demonstrably reasonable basis for my conclusion.
- Not to intentionally withhold or omit any findings or opinions discovered during a forensic examination that would cause the facts of a case to be misrepresented or distorted.
- Never to misrepresent my credentials, education, training, experience or membership status.
- To refrain from any conduct that would be adverse to the best interests and purpose of the American College of Forensic Examiners International (ACFEI).
- To be forever vigilant of the importance of my role and to conduct myself only in the most professional manner at all times."

This code of conduct does not override fundamental canons, code of ethics, or other such pledges one has to one's respective profession, as those principles only increase the value and credibility of the forensic investigator.

---

[1] Often, the trier of fact is collectively a jury, but it could also be a judge, a hearing officer, or other impartial decider of an issue.

1.6 ETHICS & CONTRACTORS

Professional ethics continue to be a topic of discussion throughout the construction industry. There has been, however, a dramatic increase in interest in applied ethics as they relate to construction. This new interest has been brought about, in part, by the liability insurance crisis; the increasing rate of litigation; skyrocketing awards given plaintiffs by the courts; and, sadly, the actual growth in construction. One may ask: Isn't growth good for construction? Normally, the answer is "yes." However, when the rate of growth is so high that the experienced supervisors and qualified workers, including design professionals, are pushed too hard, a loss of quality may result. Marginal workers at all levels enter the field, along with fly-by-night companies existing only for the quick profit that then close up before the defects in design and workmanship become obvious.

In high-growth times, it becomes virtually impossible for a responsible company to compete against a rogue company that cuts corners on materials, worker safety, taxes, and wages. Only an owner who is responsible and willing to pay more will obtain a quality final product.

Society is demanding improved standards of professional competence and performance from contractors. It is time for contractors to become aware of their social responsibilities and prepare themselves to reflect critically on their moral obligations to the public. The public must also be prepared to accept its own responsibilities where those of the contractors end. However, the public cannot and should not assume the blame for what has happened to the industry. The day of the handshake and "I'll take responsibility for that" has been replaced by "You can't prove it's my responsibility" or "Forget it we don't have the time" philosophy.

Some contractors have recognized the importance of professional ethics and have mandated within their own organization a code of ethics they expect their employees to honor. This is admirable on their part and a worthy act to follow; however, if the construction industry is to succeed and reduce litigation and skyrocketing insurance premiums, professional ethics in construction must become an everyday practice. To accomplish this, organizations such as the Associated General Contractors of America (AGC) and the American Subcontractor Association (ASA) need to take an active role in promoting professional ethics. The AGC is probably the most recognized contractor organization and for years has professed its dedication to improving the construction industry through education, skills training, promoting the use of the latest technology, and advocating building the best-quality projects.

The AGC is committed to its three tenets of industry advancement and opportunity: *Skill, Integrity, and Responsibility.* The ASA is one of the leading subcontractor associations and promotes integrity and leadership in ethical and equitable business practices, quality construction, and a safe and healthy work environment. While these attributes are all well and good, it is essential for the industry as a whole to adopt a professional code of ethics that is both meaningful and enforceable. The construction industry cannot afford to utilize weasel-worded agreements that push liability onto others. The industry must take up the challenge and demand that all members become accountable for their actions. Too often the industry tends to rationalize unethical behavior by saying, "That's the way it is today." Making up excuses and closing one's eyes to what others are doing does not promote professionalism:

- "Don't you realize that it's a major and costly effort to shovel concrete? By adding water we save on labor, and who's to know?"
- "Running this framing all the way to the roof makes no sense and only adds time and money to the job."
- "Having density tests every 8 to 12 inches in elevation is ridiculous and only slows down the earthwork and delays our starting on the structure."
- "We didn't hire that subcontractor; they were hired directly by the owner, so it's the owner's responsibility not ours, to watch over that work."
- "We'll just pad our estimate for this change order as we know it will be cut anyway."
- "The plans are so bad, we'll cut out the profit on the bid so as to get the job and we will make it up on change orders."
- "That electrical panel that came off this job will do on that other project and save us a bunch."
- "Just push that trench backfill over the sewer line—no one will know the difference after the slab on-grade has been placed."
- "Just throw a piece of loose plywood over that duct opening in the roof—the air conditioning guy will be in tomorrow."
- "We don't need a lift—we can bull it into place."

Being aware of problems in what's going on around you but not reporting them to proper authority makes one just as guilty as the person who created the problems. Employers making a point to each of their employees to take on a moral and forward-looking responsibility toward their work will go a long way in developing the professional ethics needed in our industry. There is no question that a great deal of arrogance exists throughout the industry, which is why it is so important that the construction industry learn to regulate itself, by raising the bar on behavior of its members, through both legislation and organizational action.

With the demise of the philosophy that "Low Bid Is King" and the arrival of "Design-Build" and "Construction Manager at Risk" on the public scene, we are seeing evidence of a reduction in the importance of "low" dollar over quality on public works. This change in attitude that gives value for quality of performance rather than simply rewarding the cheapest performance is both smart and ultimately rewarding to the public. As such, it should be applauded by all in our industry.

Unfortunately, this refreshing change can easily be derailed by abuse. The first hint of influence and scandal could imperil this valuable achievement and drop the public works back into that "Low Bid Is King" mentality. It is, therefore, essential for those folks appointed to or working on selection panels to fully understand the need for exercising professional ethics in seeing that the panels function properly and in the best interests of the public.

1.7 LEGAL ACTIONS—A FACT OF LIFE

It doesn't take long in today's world to find oneself in litigation with a client or being pulled into litigation as a third party at fault. It matters not if you performed your work in a forthright and ethical manner. The chances of ending up in litigation have increased tenfold in recent times to the point where many professionals are "folding their tents" because they cannot afford the high cost of liability insurance, if they can even obtain it. While embracing the responsibilities demanded under professional ethics will reduce the chances of becoming involved in litigation, it does not eliminate them. It is therefore essential for today's professional to at least have an understanding of the types of legal actions that may arise.

Legal actions against the architect, engineer, or contractor by the owner take many forms, depending on the terms of the agreement between parties and the issues involved. These include the type of complaint (a courthouse or arbitration filing or an administrative law action before the Registrar of Contractors), the type and location of the project, and the project financing (governmental or private). The following are several relevant factors.

Contract Liability

Contract liability refers to actions that are confined to the parties who have executed an agreement, which generally affords the one claiming to be harmed some means of protection of rights under that agreement.

Tort

The word *tort* is a "Law-French" term with which people in the field should be familiar. Basically, torts are "social wrongs" and may be committed either intentionally or unintentionally. A tort is distinguished from a criminal act in that the tort is a private injury to person or property. A criminal act is an act that violates statutes designed to protect the public. Tort liability will normally include the following:

- Negligence
- Gross negligence
- Intentional wrongs
- Strict liability for reasons of public policy

Tort actions may be filed by a plaintiff against a number of parties. If an individual is sued, that person as defendant has the right to bring in others who may be responsible for all or part of the plaintiff's damage claims. While the plaintiff may bring a series of actions (claims for relief) regarding a single defined event or harm, the plaintiff will only be able to collect damages once. The following may provide a better understanding of the process as it usually relates to the construction field.

Negligence

Negligence can best be defined as a defendant's breach of the standard of care that our society requires be exercised toward others, such as causing harm through any act or omission that brings injury or others, even though the person committing the act had no intent to cause such harm or omission. In other words, negligence regarding architects, engineers, and construction professionals would consist of not exercising the standard of care expected of one in the performance of a duty (which a professional owed toward another), which conduct caused harm to a person or property that should have been foreseen and therefore should have been prevented. Based on the issues involved, there are various forms of negligence, including contributory negligence, imputed negligence, and gross negligence. In most cases, negligence is predicated on facts; however, one can also be found negligent as a matter of law ("negligent per se") by not following procedures required by statutes, codes or regulations.

The burden of proof in any negligence case rests with the plaintiff who must show the defendant owed a duty of care (under an agreement or due to a relationship) and breached that duty of care, which resulted in a loss involving injury to a person or property, be the property realty or personal, or harm to a person's reputation, feelings, or personal well-being. In developing such proof, the plaintiffs will most likely address the standard of care defined in the respective professional's code of ethics, in conjunction with the duties as spelled out in the agreement or required by the relationship between the parties. One should expect, if being sued for negligence, that there will most assuredly be testimony by others in the profession (expert witnesses) defining the standard of care and establishing how the defendant adhered to or departed from these standards. It must also be pointed out that such peer testimony, when presented in front of a jury, can be disastrous for a professional if a proper defense has not been prepared.

Gross Negligence

Gross negligence is what it sounds like, a very substantial departure from the standard of care. It can never be precisely defined but the distinction is meant to reflect that the deviation from the standard of care was so great as to involve some near-intentional harm to others, a risk of harm that was so great that society will punish it more severely than mere negligence.

Intentional Wrongs

Again, intentional wrongs are what they sound like—deliberate placing of others "in harm's way," thus violating their rights in a fashion so reprehensible that the society will impose punitive damages (over and above compensatory damages).

Strict Liability for Reasons of Public Policy

There are situations in which some activity is so inherently dangerous for a person to have engaged in it that society will require damages to be paid for the harm that resulted from the activity. In days of old, strict liability was imposed when a property owner created a millpond by damming up a stream and then the water escaped (even if by no fault of the millpond owner). He was strictly liable (with no defense). These days, the modern equivalent is the auto manufacturer who sets loose among the public an automobile with a defectively manufactured part that harms someone. Society will require the maker of the defective consumer product to pay all damages resulting therefrom and will call it "strict liability" in tort.

To avoid such actions, those of us who have chosen this rewarding profession of construction must dedicate ourselves to the advancement of professional ethics throughout our industry by demanding them first from ourselves, then our employees, and finally from our constituents.

SECTION TWO

CONTRACTUALLY SPEAKING

or

What Did I Agree To?

2.0 JUMP-STARTING SECTION 2

Understanding the importance of and need for clear contracts within the construction industry is key to the successful relationship among all parties involved. Important points that the reader should absorb in reviewing this section include the following:

1. The contract must clearly delegate responsibility to each party.
2. For example, an owner–architect contract must state the expectations of both parties as to the budget for and functionality of the work.
3. For example, an architect–engineer contract must state the intersection of the design responsibilities of each professional and which professional has the obligation to review submittals and observe the work.
4. For example, an owner–general contractor contract must state the obligations of the owner to provide the general contractor with site information and access.
5. For example, a subcontract must be clear in parceling out obligations to subcontractors so that there are no gaps in the scopes of work.

This section has to do with how one interprets a contract or how one might write a contract, in order to create a good structure by which to manage the project. However, it must be said that the industry is cursed with a failure to read contracts before signing and, most of all, a failure to observe the terms of the contract.

The best contract in the world can be neutralized and made inconsequential if the contractor or the design professional makes or ignores deviations from the contract's terms. A good example of this problem for the industry is the change order clause found in almost every construction contract. By the third week of almost every job, the clause has been ignored and waived by the conduct of the participants in the project. The clause may call for a written, signed change order before any work is performed. Very often, the design professional has verbally ordered work, the contractor has performed that work, the design professional and the owner have paid for the change order work performed on oral direction, and the change order clause is waived and no longer in effect.

Using the change order clause as an example, if the design professional has not been vigilant, a contractor can argue in six different ways that a change order performed on oral direction (but unknown or unapproved by the owner) must be paid because of the existence of one or more of the following:

- Implied authority
- Custom
- Apparent authority
- Ratification
- Waiver
- Estoppel

The watchword for those on a construction job is to learn the terms of the contract and perform according to those terms, so that terms are not modified, waived, or the subject of an estoppel (conduct showing that a clause has been, in effect, modified). If it becomes necessary in some extreme circumstance to act in a manner inconsistent with the terms of the contract, then the affected participant should send a letter as soon as possible expressly *reinstating* the terms of the contract so that no waiver will occur.

How often has one really reviewed or truly thought about an agreement, written or verbal, before committing oneself to it? How many times have we asked ourselves: Why did I ever agree to take this job, or how do I get out of this mess? We have all, at one time or another, been faced with some type of contractual issue that takes us back to the question: What did I agree to? In this section, we will explain some of the pitfalls within various types of agreements found throughout the construction industry today. Two of the most important things to remember and understand in reviewing any agreement are how the agreement is structured and how it will be interpreted. All agreements are structured around a series of conditions that may be either expressed or implied in fact or implied in law and affect the rights and duties of the parties involved.

2.1 CONTRACT TERMS & CONDITIONS

Terms and conditions vary based on the type of agreement, ranging from a complicated design-build contract to simple purchase orders. Terms will normally address scope, quality, time, and money, while the conditions operate to suspend, rescind, or modify certain obligations under given circumstances. The following provides a quick review of the basic conditions one might find in a written agreement:

Examples of Implied Conditions

- The continued legality of any contract is implied.
- The owner impliedly warrants that there is access to the construction site.
- The availability of goods to be sold in any contract is implied.
- The life of both parties is implied in an employment contract.

Precedent, Subsequent, and Concurrent Conditions

- A **condition precedent** occurs when payment or other compensation is predicated on a defined event's taking place. If the event doesn't take place, then there is no obligation to pay or perform. Such condition precedents are often found in surety bond agreements, that is, obligation of the surety occurs only if the principal fails in a defined event. A "pay when paid" clause or a "pay if paid" clause is a good example of a condition precedent.
- A **condition subsequent** is one that relieves the promisor of a liability to pay for an issue that has already occurred. This type of condition is found in various forms in subcontract agreements in which the subcontractor must make any claim for added cost within 14 days after the event that created the added cost or lose his or her right to make the claim.
- A **concurrent condition** is one in which both parties have agreed to do certain things concurrently; for example, a hotel owner has agreed to shut down one bank of rooms at a time to allow the contractor to perform and complete work, after which the next bank of room will be released to the contractor.

Time and Performance Conditions

- Time and performance conditions are usually a requirement under the contract. If no stipulated time is provided under the contract, it will be perceived that the effort will be completed in a "reasonable time" period. On the other hand, as is usually the case, if a time for performance is stated, then time is deemed "of the essence" and the job must move forward with dispatch. Irrespective of how the time is perceived, the requirements for performance will be based on all applicable industry standards, which will incorporate governmental rules, regulations, codes, and trade practices.

2.2 DESIGN CONTRACTS

A **design contract** is an agreement between an owner and a design professional. This design agreement provides for and outlines the responsibilities of the two parties in order for the designer to prepare the contract documents on a project for the owner's use. It includes the responsibilities and benefits for both parties. The design agreement may be prepared by either the design professional or the owner. In either case, the design agreement's provisions are of utmost importance to both parties and to the project itself.

Some owners or their representatives have considerable experience and are highly skilled in the development of design agreements. This may be helpful to the design professional; however, it is imperative that the contract provisions ensure that the design work is well defined and fully covered and that all exclusions to the work are included. It is further essential that the interests of both parties are fairly and adequately protected.

Regardless of which party prepares the contract agreement, certain points must be clearly presented and agreed upon. The following paragraphs cover these basic points and include conditions that the design professional should consider when entering into a design agreement.

A definition of the project and its purpose must be included. The project's name and its location (address and legal description) and the business names and addresses of the parties to the agreement must be given. It is very important to use the correct name of entities that may be involved and to check the names with the Corporation Commission or County Recorder. Having a contract with the wrong entity is a really bad idea.

The owner's contribution to the design effort should be defined. The owner's responsibility should include providing good title and access to the project. In addition, the owner should provide an American Land Title Association (ALTA) survey with a topographic survey and a geotechnical examination of the site. When these are provided, the designer must review these documents for adequacy and reserve the right to require further information.

The design professional should verify that the contract agreement includes the complete scope of services to be performed. This should include the ability to negotiate service contracts with other design professionals, such as civil, mechanical, electrical, structural, and geotechnical firms. The design agreement should be clear as to whether the work is completed when a final design has been provided or that services also include contract administration during construction. The design professional must ensure that the agreement provides reasonable limits of liability, with no responsibility for other party's actions.

The agreement must specify the fees for the work, how the fees are determined, and how and when they are to be paid. The agreement should include provisions for payment for additional work or changed conditions. It further should provide for payment of additional work for unforeseen circumstances, for example, to respond to subpoenas should the project become involved in litigation.

If contract administration and surveillance of work during construction are included in the contract, the design professional should maintain control of inspections and testing for acceptance. All design agreements, including those with consultants, should include provisions for utilizing design services in the event of revisions or changed conditions during construction.

2.3 CONSTRUCTION CONTRACTS

Construction contracts can vary greatly from a verbal understanding (not advisable), to a single-page agreement, to one that is well over 100 pages. Various types and forms of construction contracts are used today. Normally the job requirements (and perhaps the competence of the design team) will dictate the type of contract that will be used:

Fixed Price or Stipulated Sum

Fixed price or stipulated sum contract agreements are normally used when the scope of work is known and the contract documents have been completed. In as much as the contract documents have been completed, it is not unusual to have a stipulated completion date incorporated into the contract documents, along with liquidated or actual delay damages for failing to complete the work on time.

Cost Plus a Fixed Fee with a Guaranteed Maximum

Cost plus a fixed fee with a guaranteed maximum contract agreements are normally used when the scope of work is known, but the design effort has not yet been fully completed and the remaining effort entails cooperation between the contractor and the design professional to work together in a team arrangement. Under most instances, the general contractor becomes the team leader in this undertaking and will develop a guaranteed maximum price once a fixed scope of work has been determined.

Cost Plus an Incentive Fee with a Guaranteed Maximum Price

Cost pus an incentive fee with a guaranteed maximum price contract agreements are normally used under a "fast track" schedule with time being of the essence and the scope of work is fairly well defined. The general contractor is the "team leader" under this type of agreement and will be responsible for completing the project by a stipulated date and will receive an incentive fee for every day the project is brought in ahead of that date. These contracts require very close coordination among all parties from the owner down to the smallest subcontractor and supplier.

Critical Path Scheduling

Critical path scheduling is a methodology that develops a sequence of activities that must be completed on schedule (time) for the entire project to be completed on schedule (time). This is the longest duration path through a defined workplan. If an activity on the critical path is delayed by one day, then the entire project will be delayed by one day (unless another activity on the critical path can be accelerated by one day.

Examples of these various types of contracts can be obtained through the Internet from the AIA, AGC, and other organizations.

General construction contracts have been developed with years of experience. In many respects, the general conditions of a general construction agreement have been standardized by various organizations and governmental agencies, which include the following:

- Probably the most widely used set of general conditions is "General Conditions of the Contract for Construction," current AIA Document A201, The American Institute of Architects, Washington D.C. (This document often is modified for use by various tribes, state, and municipal agencies to comply with respective administrative codes and/or regulations.) A revised set of the AIA General Conditions (2007 Version) is now being accepted by most governmental agencies. The AIA has some 50 contract forms and related documents.

- The Associated General Contractor of America (AGC) continues to be very active in developing contractual forms and currently has over 75 different types of forms available to its membership or to anyone who wants to purchase them. These forms include everything from the AGC 200 Standard Form of Agreement and General Conditions Between Owner and Contractor—*Where the Contract Price Is a Lump Sum* (which is compatible with the AIA 101 & A201), to the AGC 260 Performance Bond, to such specialty agreements as AGC 655 Standard Form of Agreement Between Contractor and Subcontractor—*Where the Contractor and the Subcontractor Share the Risk of Owner Payment.* The AGC is a good source of documents relating to construction management and design-build contracts.

- "Standard Conditions of the Construction Contract," EJCDC No.1910-8 (Current Edition). Engineers Joint Contract Documents Committee. Published by the National Society of Professional Engineers, Alexandria, Virginia.

The articles contained in these documents (many of which have been jointly developed among the various professional associations) define what is generally recognized as acceptable terminology by the construction industry and cover the following issues.

General Provisions

General provisions will normally define what makes up the agreement between parties and include

- *The Contract Documents,* including the Agreement between parties, Conditions of the Contract, General & Supplemental Conditions, Drawings, Specifications, Addenda to the Agreement, and so on.

- *The Contract,* represents the entire agreement between the parties and supersedes any prior representations, written or oral.
- *The Work,* constitutes the construction effort and services as defined under the contract documents.
- *The Project,* consists of the total construction effort necessary to complete the work in whole or in part, which may include work performed by others.
- *The Drawings,* constitute the graphic and pictorial portion of the contract documents, showing the elements, materials, locations, and dimensions of the work.
- *The Specifications,* are the written portion of the contract documents, which define the various segments of the work, including the materials, standards, type, manufacturer, equipment, workmanship, and performance requirements.
- *The Project Manual,* will normally be an assembly of instructions to bidders, sample forms, Conditions of the Contract, Addenda, Special Requirements (such as local codes and ordinances, geotechnical requirements, Native American requirements, etc.) along with the project specifications.
- *Execution, Correlation, and Intent,* informs the contractor that once the contract is executed, the contract documents are binding, confirming that the contractor has visited and inspected the site and fully understands the meaning and intent of the work as defined in the contract documents.

The Owner

The owner or the owner's representative is defined as the person to whom the contractor is bound by the contract and is responsible for providing the funding to cover the work as defined in the contract documents. Under most agreements, the owner is responsible for the following:

- Having legal title to the property upon which the project is being constructed.
- Providing ALTA surveys, plats of the property, and documents showing the location of all known utilities, easements, or other characteristics of the property.
- Providing access to the property.
- Providing all of the contract documents.
- Providing all funding in a timely manner to support the work as defined by the contract documents.
- Providing final approval for any changes in the work or the contract documents.
- Providing special inspection services as may be required to confirm that the work is being performed in accordance with the contract documents.
- Providing builder's risk insurance and owner's general liability insurance.

In general, the owner will have the right to stop the work, to carry out the work with his or her own forces, or contract the work to others, in the event of a breach by the contractor.

It is essential for the contractor to fully understand the role of the owner as it applies to the contract documents. Particular attention should be directed to the type of insurance provided by the owner as many agreements are modified to transfer all insurance risk to the contractor. Likewise, the indemnity clause must be carefully scrutinized.

The Architect/Engineer

The architect/engineer is usually the party who prepared all or a portion of the contract documents with particular attention given to the drawings, specifications, special inspections and changes to the contract documents. The architect/engineer is called upon to interpret the contract documents in an unbiased manner. This duty of the design professional to act in an unbiased manner is a fundamental and crucial feature of essentially all contract documents.

The Contractor

The contractor has the responsibility to construct the work in accordance with the contract documents. This is a simple statement, but one with significant meaning, particularly when the contract is a standard or modified AIA Document A201. To better define these responsibilities, every contractor should take note of the following issues:

- Upon executing the Agreement, the contractor warrants that he or she has read and fully understands the Agreement, which may include:
  - An obligation to construct the work for an agreed upon fee.
  - Being totally familiar with the site conditions upon which to construct the project, for example, surface conditions, and all structures and obstructions both natural and artificial and all surface water on the site and surrounding area.
  - Being totally familiar with the general area and location in which the contract is to be constructed, including without limitation the availability, quality and conformance of local labor, materials, and equipment to the work as defined in the contract documents.
  - Having thoroughly reviewed all of the contract documents and notified the owner of any errors or omissions and by not providing timely notice of such errors and omissions may be required to assume a proportionate share of the cost to correct such errors and omissions. The requirement of reasonableness does not increase the standard of ordinary care required of the contractor.
  - The contractor shall be responsible for all acts and omission of the contractor's employees, subcontractors, their agents and employees, and other persons performing work under a contract with the contractor. In other words, the contractor will be held accountable for all

work covered under the contract agreement whether or not such work was performed by contractor's forces.

Although verbal agreements and single-page contracts may have little substance and fail to define the duty and responsibility of the contractor, most states have statutes and case law that define the duties and responsibilities the contractor has to the client. In general, any construction contract, verbal or otherwise, requires the contractor to perform properly or be subject to damages and perhaps civil penalties, depending on the state in which the work is being performed. These minimum performance standards[1] will normally require the contractor to

- Carry a current contractor's license in the field of work being undertaken.
- Perform the work in accordance with industry standards.
- Not use false or misleading documents for the purpose of inducing a person to enter into a contract or to pay money for work to be performed.
- Fail or neglect to properly apply funds received.

2.4 SUBCONTRACT AGREEMENTS

Subcontract agreements, service contracts, and purchase orders are prepared by the contractor to secure goods and services in support of the construction effort. Each of these documents will normally contain the terms and conditions under which the supplier of these goods and services is bound to the contractor. In nearly all instances, these agreements are developed by the contractor and written in such a manner that attempts to totally protect the contractor and transfer all liability for the goods and services defined in the agreement to the subcontractor or the supplier.

By contrast, the latest edition of the AIA Document A401, Standard Form of Agreement between Contractor and Subcontractor, has been approved by many state and municipal authorities and may be used by the contractor when performing work under General Conditions of the Contract for Construction, AIA Document A201. The AIA Document A401 is generally more even handed than a contractor-prepared agreement.

In general, the AIA A201 document binds the subcontractor to the contract documents, and the subcontractor assumes toward the general contractor all obligations and responsibilities that the general contractor assumes by the contract documents toward the owner.[2] This is ordinarily called a "flow down" clause. In essence, these obligations and responsibilities would include the following:

- Preserve and protect the rights of the Owner as defined under the Contract Documents with respect to the Work as defined in the Subcontract Agreement so that the performance of such subcontract work does not prejudice the owner's rights.[2]
- Allow the Subcontractor, unless provided otherwise in the Subcontract Agreement, the benefit of all rights, remedies and redress against the Contractor that the Contractor, by the Contract Documents, has against the Owner.[3]

It is essential, when entering into any agreement, that a thorough review of the rights and remedies be undertaken, as it is not uncommon to find that by signing the agreement one may have forfeited all rights and remedies. In such a case, one may be left at the mercy of the contractor as to how one is protected when changes in the work, delays in the work by others, and/or other such impacts affect the performance of the subcontractor. An example of such a clause to look for in the agreement before entering into it is

- Subcontractor agrees to conform to the General Contractor's schedule as it is announced to the Subcontractor from time to time.

The flow down clause must be looked at carefully particularly in addressing unexpected site conditions. A number of court rulings have been handed down in favor of the owner.

2.5 WHAT TO LOOK FOR— FROM THE OWNER'S PERSPECTIVE

The 20 Key Provisions in the Construction Contract

Any owner reviewing a construction contract should look at the contract for these key points:

- *Scope of Work.* What does the scope really include? What does it exclude, for example, all landscaping, all off-site improvements like curbing, gutters, site drainage, and pavement?
- *Time for Performance.* Is it a "killer" short schedule or a "leisurely" job?
- *Schedule Requirements.* Is the contractor required to meet whatever changes to the schedule the owner proposes without additional cost? Are there substantial (costly to perform) updates of the schedule required every week or every month? Does the contractor have experience with the software that is required by the contract? Does the owner's staff have experience with that software?
- *Time Extensions.* Will time extensions be given on an equitable basis if there are changes, or is there a "must complete by a date certain" timeline? See A.R.S. §

---

[1] Civil penalties taken in part from Arizona Revised Statutes & Rules.

[2] Taken in part from AIA Document A201, paragraph 5.3.1.
[3] Taken in part from AIA Document A201, paragraph 5.3.1.

34-221(F), providing that an equitable adjustment must occur in public contracts when an unreasonable unexpected delay occurs due to owner.[4]

- *Site Condition Provisions.* What are the concealed site conditions provisions? The *Spearin* doctrine requires that the plans provided by the owner to the contractor are warranted by the owner to be correct and adequate to build the job. In this event, the owner will be unable to sue the contractor for the results. Contracts usually provide that if the conditions on or under the job site are different from what the plans show (or what a person would reasonably expect to find on site), then the owner, not the contractor, will be liable for the resulting expense.
- *Delay Damages.* Is there a limitation on damages that can be recovered if a party causes delay, or, perhaps, a waiver of all "consequential damages"?
- *Liquidated Damages.* Are there liquidated damages awarded if a schedule completion date (or milestone) is missed? Are they so minor as to be ineffective, fair, or oppressive?
- *Cost.* If it is a "cost plus" contract, how is "cost" defined and who determines it? What overhead is included?
- *Discounts.* Who gets to keep discounts for any early or accelerated payment to vendors? Are there any prior approval requirements that impede selection of subcontractors or incurring expenses?
- *Insurance.* What insurance amounts and types are required? Is the owner an "additional insured"? This is terribly important and needs to be conveyed to the insured's insurer!
- *Indemnity.* What indemnity and hold harmless promises are required of the contractor? Does coverage insure against or indemnify against liabilities caused by the owner's negligence or partial negligence? Is the clause one that is enforceable under state statutes or court precedents?
- *Subrogation.* Is subrogation by an insurer to a paid-party's rights waived?
- *Claim Restrictions.* What are the restrictions or time limits on making claims for lost time, added costs, and so on, and on change order submissions? See A.R.S. §§ 12-821.01, and 11-622 covers the timeliness for making claims against public entities. Other states have similar statutes regarding if, when, how, and the time frame in which a claim may be filed.
- *Warranties.* What warranties does the contractor give? And to whose work—just the contractor's own or for the subcontractors' work, materials?
- *Venue.* Is there a choice of law or choice of venue clause that runs afoul of the state laws?
- *Attorneys' Fees and Interest.* Are there clauses regarding attorneys' fees or interest on delayed payments or defective work claims?
- *Dispute Resolution.* Is there a good (fast) dispute resolution clause, such as arbitration by fairly selected and impartial arbitrators (with reasonable controls on discovery), in a locale and *on a schedule* that is reasonable? See A.R.S. §32-1129.05(i) requiring that the arbitration or mediation occurs in Arizona. Other states have similar statutes that should be reviewed to ensure that there is an avenue of good (fast) dispute resolution.[5]
- *Mediation.* This should always be incorporated as a condition precedent to arbitration. If mediation fails, then the next step would be arbitration.
- *Termination.* Can termination be done on short notice? What costs, damages, profits will be paid if there is a termination not "for cause."
- *Waiver of Consequential Damages.* This is a creature of the 1997 AIA document revisions. It is ordinarily harder on owners than on contractors. Most competent attorneys representing the owner would never let this be incorporated.

2.6 A MATRIX TO BETTER UNDERSTAND AIA CONTRACTS

The following matrix was developed to assist in understanding the relationships and responsibilities between parties when using AIA Contracts A101-B141—A201 1997 version. When references are made to Arizona Revised Statutes, it is recommended that a review of other states' statutes is conducted to determine if the interpretation is compatible.

Architect	Owner	Contractor
(a) Duties		
B141§1.2.1	A201 § 2.2.3	A101 § 2.0 & 3.1.2
Cooperate	Furnish all surveys. Pay for the work and, per A201 §2.2.2, pay for all permits prior to executing contract.	Do the work in the contract documents unless indicated to be work to be done by others.

---

[4]Taken in part from Arizona Revised Statutes & Rules; check the Revised Statutes in your state.

[5]Taken in part from Arizona Revised Statutes & Rules; check the Revised Statutes in your state.

Architect	Owner	Contractor
§1.2.3.2	B141§1.2.2.1	A101 § 3.3.1
Use professional skill in plans and prepare schedule to meet owner's time frame	Give information to architect and in a timely fashion, especially as needed for architect to put a lien for his or her services on the site. See A201 § 2.1.2. and 2.2.4.	Be responsible for choosing the "means and methods" of construction.
§1.2.3.4	B141§1.2.2.7	A202 § 3.4.1
Keep confidentiality	Tell architect promptly if owner becomes aware of fault or defect in plans or schedule.	Pay and provide for all labor and materials (plus transport), tools and equipment, utilities, permits to be obtained after execution of the contract, and other facilities and services to perform the work.
§1.2.3.5		A201 § 3.4.3 and 3.9
Avoid economic conflict of interest		Hire skilled subcontractors (per AIA 201 § 5.2) and employees and enforce good order and discipline among them. Employ a competent job superintendent.
§1.2.3.6		A201 § 3.11
Follow governmental codes		Keep a set of plans marked up "currently" with field changes so as to have an as-built set on-site.
§1.2.3.7		A201 § 3.10.2
Inform owner of errors in plans/schedule		Keep track of submitting products submittals and "shop drawings" per §3.12.1 so architect can timely review under § 3.12.5.
B141 § 1.4		A201 § 3.13 & 3.15 &10.2
Defines the scope and limits on performance by insertions. **AN EXTENSIVE EXHIBIT TO THE B141 LAYS OUT EVERYTHING THE OWNER GETS**, value engineering, budget estimates, evaluation of bids, schematic design, construction documents (plans and specs), bid valuations, construction work evaluation, administration, observation, submittal review, change order evaluation, punch list, etc. See A201 § 4.2.11 saying architect will interpret the contract documents.		Keep the site cleaned up and safe and clean it up totally at the end.
A201 § 4.2.2		A201 § 9.2
Visit the site from time to time to "observe" the work and tell the owner if he or she spots any defects in work.		Prepare a "schedule" of values to be used by architect in approving the payment applications.
A201 § 4.2.5 and 9.7		A201 § 3.2.1
Review and approve contractor pay applications, in a timely fashion.		Compare all contract documents on any portion of the work and field measure before starting each portion, reporting any errors discovered to the architect.
A201 §3.12.5		A201 § 3.3
Review shop drawings as submitted by contractor, but for conformance to design, not safety.		Direct and supervise the performance of the work.

(*continued*)

Architect	Owner	Contractor
(b) Ownership of Plans		
B141 § 1.3.2.1 and .2	A201 § 1.6.1	B141 § 1.3.2.3
Architect owns the plans. Owner only has a license to use the plans to build the project. See § 1.3.7.7. architect can use pictures to promote him- or herself	Architect owns the plans.	Contractor has a license to use plans to build the project.
(c) Compensation & Termination		
B141 § 1.3.8	B141 § 1.5.2	A101 § 4.5.0
Architect can terminate services on 7 days' written notice, if not paid. If there is an interruption (temporary suspension), the price will be renegotiated.	Defines the nature of the compensation and when payments are due.	Says how progress payments, final payment, and retention are to be handled. Payment is by percentage of accomplishing a schedule of values, minus retention.
	B141 § 1.3.9	A201 § 9.2
	Covers items included in compensation	Prepare at the start of the job a "schedule of values," so pay applications can be evaluated by the architect for percentage of completion.
	B141 § 1.3.8.4	
	Owner can terminate on 7 days' written notice if "substantial" failure of performance.	
(d) Mediation		
B141 § 1.3.4.1		A201 §4.5.1
Mediation is a condition precedent to the right to arbitrate a dispute with the owner. See A201 § 4.5.1.		Mediation is a precedent to the right to arbitrate a dispute between owner and contractor.
(e) Arbitration		
B141 § 1.3.4.2 and 1.3.5		A201 § 4.4.5
AAA arbitration is mandated (*after* a *quick* mediation). A201 §4.6.1.		If the architect decides and you want to "appeal," an arbitration request has to occur in 30 days after receipt of that decision. Consolidation is not allowed. See Mediation and Arbitration A201 § 4.5 and 4.6.
1.3.5.4		
Consolidation of parties other than owner and architect will not be allowed.		
(f) Liens for Services		
B141 § 1.3.4.1		Not mentioned, but A.R.S. §32-981 allows liens, if timely recorded after a 20-day notice is sent earlier.
The architect can lien and, per 1.2.2.1, the owner has to give him or her any needed information to permit a lien to be recorded.		

Architect	Owner	Contractor
(g) Consequential Damage Waivers		
B141 § 1.3.6	A201 § 4.3.10	A201 § 4.3.10
Each party *waives* consequential damages (lost rents, lost profits from another job that could not be taken on). (Duplicates A201 § 4.3.10.)	Contractor and owner waive them to each other.	Contractor and owner waive them to each other.
(h) Role of the AIA A201		
B141 § 1.3.7.2	Incorporated into the A101 Owner / Contractor Contract.	
Expressly incorporated for terms' "definitions."		
(i) Primacy of Insurance —Indemnification		
B141 § 1.3.7.4	B141 § 1.3.7.4	A201 § 11.3
Waiver of recovery for a loss that is outside the property insurance coverage, which is provided, if it is provided.	Waiver of recovery for a loss that is outside property insurance coverage provided, if it is provided.	Sets out guidelines for liability insurance, workers compensation insurance, and property insurance.
A.R.S. § 32-1159	A201 §11.4.7	A201 § 3.18.1
Any granting of indemnity for another's sole negligence is void, as it is against public policy.	All parties waive for their insurers the right to be "subrogated" to another's rights and to then chase the miscreant.	If insurance does not cover a loss, indemnify owner, architect, and architect's consultants from tort claims, for personal injury or damages to property to the extent it was caused by negligence of the general contractor or the subs.
(j) Construction Schedule		
B141 § 1.2.3.2	A201 § 8.3.1	A201 § 3.10.1
Check and follow the contractor's tracking of the schedule that was prepared for the work.	If owner or architect delay the contractor, the contractor will get days and dollars.	Promptly *after* contract award, prepare the schedule for the work, revise it during the job at appropriate intervals, and finish the work on time.
(k) Flow Down Clauses		
A201 § 5.3.1		A201 § 5.3.1
In subcontracts, per A201 § 5.3.1, all subs are to assume, toward the general, all obligations of the general toward owner.		In subcontracts, per A201 § 5.3.1, all subs are to assume, toward general, all obligations of general to Owner.
(l) Implied Warranties		
		A201 § 3.5.1 and 12.2.2.1
Kubby vs. Spearin cases	Per case law, the owner warrants that the plans are sufficient to do the job—and that the site will be available.	Materials are to be good quality and new, work to be free from defects and to conform to the contract documents.
(m) Express Warranty		
		A201 § 3.5.1 and 12.2.2.1
		Materials are to be good quality and new, work to be free from defects and to conform to the contract documents.

(continued)

Architect	Owner	Contractor
(n) Termination		
	A201 § 14.1 & 14.4.3	A201 § 14.1 & 1.4.4.3
	Watch for cost! If termination for convenience, lost profits are recovered by the contractor.	*Watch for cost! If termination for convenience, lost profits are recovered by the Contractor.*
(o) Change Orders		
		A201 § 7.1
		Contractor has to do change order work as ordered by the architect, even though the costs may not be agreed upon until later.
(p) Substantial Completion & Punch List		
A201 § 9.8.3		A201 § 9.8
Check if contractor's application for substantial completion is correct and issue a certificated substantial completion, proving usability for the intended purpose.		Upon substantial completion, give the owner a punch list and timely do the items thereon.
(q) Liquidated Damages		
		Can be provided by contract clauses, but they must be reasonable and not constitute a penalty
(r) Prompt Pay		
		A201 § 9.6.2
		Promptly pay the subs upon being paid.
(s) Claims		
		A201 § 4.4.5
		Have to be submitted first to the architect for decision. If one dislikes the decision, arbitrate the architect's decision by 30 days after receiving it.
		A201 § 4.3.2, 4.3.5 and 4.3.7
		Must be made within 21 days after the event took place or after first discovering that an earlier event may be grounds for a claim.
(t) Implied Duty to Not Hinder the Other Contracting Party		

Rawlings vs. Apodaca case in Arizona.

2.7 KEY CHANGES TO THE 2007 AIA DOCUMENTS

Since 1888, thus for over 120 years, the American Institute of Architects (AIA) has promulgated standard form documents for use in construction projects. The AIA General Conditions document (known as the A201) has gone through 16 revisions since 1911, and this latest revision of the form documents (released in December 2007) has altered just a few of the numbers in the numbering system by which we all refer to the AIA form documents. Here is a table to help you quickly find the key documents using the old numbering system compared to the new system for the AIA form contracts.

Available at http://www.aia.org/docs_free_paperdocuments are some helpful AIA-prepared "Comparatives and

Old Number	Contract	New Number
A101	Owner / Contractor Stipulated Sum Agreement	A101
A107	Abbreviated Owner / Contractor Agreement—Stipulated Sum (Limited Scope Projects)	A107
A111	Owner / Contractor Cost Plus Fee with GMP	A102
A114	Owner / Contractor Cost Plus Fee Without a GMP	A103
A201	General Conditions	A201
A401	Contractor / Subcontractor Agreement	A401
B151	Abbreviated Owner / Architect Agreement	B101
B141	Owner / Architect Without Predefined Scope (likely just design work)	B102
B141	Owner / Architect Agreement for a Large Complex Project	B103
B141	Owner / Architect Full Agreement with Scope of Services, Design through Contract Administration	B201
	Architect / Consultant Agreement	C401

Comparisons" that chart the new document and show particular clauses or changes from the former document. If you go to the AIA website, you will particularly want to look at two commentaries: A201-2007 regarding key clauses in the A201 and B101-2007 (formerly B141) Owner / Architect Agreement. Likewise, on the AIA website, you will find the new AIA A201 (2007) line by line comparison to the old AIA A201 (1997) and the new AIA B102 and B201 (Parts I and II) compared line by line to the old AIA B141 (1997).

The following material addresses the key changes that were made in the relationships on the job by the text of the 2007 AIA documents, particularly the A201 General Conditions. There are changes in the A101 or the B101, but they track what was done to the A201.

So, let's now consider the AIA A201 (2007) General Conditions, going article by article, to see the key changes from the 1997 version. These are comments as to the *important* alterations of the A201 General Conditions document, and a few of the important sections that were retained without alteration.

Article 1

A. Instruments of Service and Electronic Data:

In Article 1 (§1.17) in the new A201, the old term "Project Manual" is removed from use. A new "ample" definition of the "Instruments of Service" is inserted into the General Conditions. The definition is widened so as to include *any creative work* (tangible or intangible) prepared by the architect under a new B102 and B103 Agreement. New Sections 1.5.1 and 1.6 clarify that these may be *electronic* documents and refer the parties to a very extensive separate "protocol" document for transmitting information and contract documents electronically. That protocol can be filled out and attached to the A101 or the B102 contracts. In other words, the new 10-year revision by AIA has acknowledged what we owners of IPODs and MP3 players and digital PAD phones would be quick to say: Eighty percent of all data these days are electronic or stored electronically, and the architects wanted *that* their copyright for that data protected. Under new Sections 1.5.1 and 1.5.2, anyone other than the architect has a *limited license* to use the broadly defined instruments of service, and then only for the purpose of performing this project (and its remodel or repairs).

B. Duties *Inferable* From and Consistent with Contract Documents:

This very important clause remains in place and unchanged.

C. Initial Decision-Maker ("Idm") And Courthouse Litigation Instead of Arbitration:

In the old AIA (1997) A201, the architect was always and universally the first "station" to which a contractor / owner dispute had to be taken. *Only after an initial decision* made by the architect could a claim move on to mediation, and then to arbitration. *All* disputes ultimately would go to arbitration for resolution under the 1997 A201 General Conditions.

As we will see below in Article 15 of the 2007 documents, it is no longer necessarily true that the initial decision-maker will be the architect.

And, litigation *(instead of arbitration)* is offered in a "check the box" format, as was never the case before. In Section 1.1.8 of the new AIA document A201, the term "initial decision-maker" is defined to indicate that the decision-maker may *not* be the architect.

These are two huge changes in the long history of AIA form documents.

Article 2

A. Information on Finances that the Contracator Can Pry Out of the Owner:

It used to be, under the 1997 version, that *whenever* (before and during the job) a general contractor just requested in writing that the owner give the contractor documents to prove that the owner had funding available to do the work, those documents had to be supplied to the contractor. Owners used to alter that clause often, annoyed that they had to "stop and fetch it" whenever the contractor "yanked the chain."

Now, under the 2007 version of the A201 General Conditions (§2.2.1), after the commencement of the work, the contractor's right to demand proof of finances from the owner is *limited*. A request can only be made if (1) the owner fails to make payments when payments are due; (2) there is a material increase in the contract sum by change order; or (3) the contractor can show, in writing, that there is some reasonable concern that the owner is or will be unable to make a payment.

B. Notice of Contractor Default:

In the 1997 version the owner had to give a seven-day notice and then a three-day separate second notice to be able to have the right to jump in and perform work that the contractor had done wrong and was not removing or correcting. In the 2007 version, this has changed in § 2.4 to *one* 10-day notice by an owner.

Article 3

A. Caveat Regarding the Contractor's Obligation to Catch Code Non-Compliance in Plans:

Sections 3.2.3 and 3.7.3 have only a few words that have been altered, but the change slightly broadens the contractor's obligation to report any violation of *any* code that the contractor discovers in the plans, not just a "building code."

B. Burial Grounds:

New Section 3.7.5 deals at length with a problem that has interrupted projects in our Western states, the discovery of archeological sites or "burial grounds." It also deals with wetlands, which we rarely find in the West. The section requires the owner to act promptly to avoid a discovered burial ground from delaying the work and raises the possibility of the contractor being entitled to an equitable adjustment, that is, a change order for delay costs. See §3.3.1 for an added bit of text about such costs imposed on the owner.

C. Vetting the Contractor's Proposed Superintendent:

Sections 3.9.2 and 3.9.3 are changed so as to expressly require the contractor to notify the owner and architect of the name and qualifications of the proposed job superintendent. The owner has the right to make a reasonable *and timely* objection to the proposed superintendent.

Article 4

A. Architect:

Many pieces of Article 4 were "scissored out" and relocated into the new Article 15. The title, which used to be "Administration of the Contract" has been revised and broadened to simply read "Architect."

Section 4.2.1 appears to shorten the time during which the architect's contract administration duties continue so those duties no longer extend into the one-year warranty period.

The entire process for making claims is moved out of this article (Sections 4.3.1–4.6.6) and is relocated to a new Article 15 (specifically, Section 15.1). As mentioned before, the sections relocated to the new Article 15.1 no longer assume that the architect will necessarily serve as the person initially deciding the claims related to the project.

Articles 5, 6, 7, and 8

No changes of significance occur. Since Article 7 is about change orders, this is a good portion of the A201 to leave untouched. It should be noted that when Contract Change Directives (CCDs) are issued, revisions to Section 7.3.9 have the architect or Initial Decision Maker (IDM) rendering an opinion as to what portion of that CCD work the owner ought to be paying to the contractor *while* the work is being done and while the correct amount owed for the work is being sorted out through the claims process.

Article 9

A. Joint Checks When Subs are Unpaid:

This article is on "Payments and Completion" and a new Section 9.5.3 states that if an architect refuses to certify a contractor's payment application, under § 9.5.1.3 (failure of the contractor to pay subs properly), then the owner has the option (risky though it may be) to pay subs on the job *with joint checks*. If he or she does so, the owner has to inform the architect so that such is reflected in the next payment application.

This change gives owners a little more control when relations with a general contractor have broken down and the owner is attempting to prevent key subs from demobilizing and walking off the job, only to have to get them back later on.

B. Proof that the General Contractor is Timely Paying Subs:

A new Section 9.6.4 permits an owner to make a written demand on the general contractor for *proof* that the general contractor *has* properly and timely paid over to the subs sums that the owner has given to the general contractor for subcontracted work. If the general contractor does not reply within *seven days*, then the owner has the right to *approach the subs directly to ascertain their payment status*.

This means that the owner no longer has to wait an entire 30-day payment cycle (when lien waivers have to be provided) to learn if the last payment did or did not get to the subs properly. It gives the owner a tighter leash on a general contractor who seems to not be paying subs, "enjoying the float," and thereby slowing the subcontractors' motivation to perform on the job.

Article 10

In this article on Protection of Property and Hazardous Materials, Section 10.3.5 now says that when a contractor brings on-site a hazardous material and negligently handles it, the contractor has to pay the owner any remediation costs or damages.

Article 11

A. Renewals of Insurance Certifications:

Section 11.1.3 now requires the contractor to not only deliver to the owner the *initial* certificates of insurance, but also all *renewals* and all *replacements* of those original certificates.

B. Additional Insureds

It used to be that Section 11.3 in the A201 provided an option whereby the owner could force the contractor to purchase general liability coverage for the owner as primary coverage. This is stricken from the A201 and, instead, Section 11.1.4 requires the contractor to name the owner and the architect as "additional insureds" on the contractor's general liability insurance coverage during the job *and* requires the contractor to name just the owner as an "additional insured" during the completed operations coverage.

Article 12

There are no significant changes in this article.

Article 13

Federal Arbitration Act

Section 13.1 used to say that the law of the place of the project's performance was the governing law for the contract. However, as it is now rewritten, the following has been added: "except that, if the parties have selected arbitration as the method of binding dispute resolution, the Federal Arbitration Act (FAA) shall govern." This change has potentially significant importance because a construction contract's arbitrable disputes were handled under the old 1997 A201 by AAA Arbitration Rules. Now, a whole body of Federal Law about arbitration has suddenly been made applicable, *if* arbitration (instead of litigation) has been selected for dispute resolution purposes. Many contractors are striking this reference to the FAA in favor of AAA.

Article 14

There is no important change in the Termination for Cause or for Convenience sections. However, note that the old requirement that an owner had to prove that the contractor "persistently" failed to perform or follow the laws or codes no longer exists. Now, mere *repeated* violations by a contractor will be enough to allow termination to occur. It does not seem like a *real change*—just a minor clarification.

Article 15

In the old 1997 A201 General Conditions, there never had been any Article 15. Claims-processing requirements were included in the section on the architect's Administration of the Contract portion of the 1997 A201 (old Section 4.3.1–4.6.6). However, all of those claim-related provisions have been *moved into the new Article 15* and a wholesale revision has been made in (i) the question of whether the initial decisions on claims will be automatically made by *the architect* and (ii) whether a disappointed party will take that claim's initial decision to arbitration, or to litigation.

A. Initial Decision-Maker ("Idm") Role:

Section 15.1.2 repeats the old requirement that WRITTEN NOTICE must be given of all claims, whether for money or for time or for consequential damages or for other relief.

All the way through Article 15, reference is made to the initial decision-maker ("IDM") who was first mentioned in Section 1.1.8 as being a person to be selected and identified in the contract (in the A101 or the B102, for example). **BUT, IT IS NOTED THAT THE IDM IS NOT NECESSARILY THE ARCHITECT.** If no *other person* is chosen as the IDM, *then the architect is to serve as the IDM*.

THE IDM MUST STILL RECEIVE A CLAIM 21 DAYS AFTER OCCURRENCE OF EVENT(S) GIVING RISE TO THE CLAIM, OR 21 DAYS AFTER RECOGNITION OF THE EXISTENCE OF THE CLAIM. This very important notice requirement remains unchanged.

B. Time to Arbitrate or Litigate a Claim:

The rule used to be that *within 30 days* after an architect / IDM's initial decision, a disappointed party had to challenge the decision by demanding arbitration OR ELSE lose the right to ever go to arbitration (Old Section 4.4.6).

However, the *new* rule (under new Section 15.2.6.1) is that a happy or unhappy party may, within 30 days after the initial decision, demand that the other party (usually the disappointed party) file for mediation within 60 days of the initial decision. And, if the disappointed party then *fails to file for mediation* within those 60 days, the initial decision by the IDM *becomes final*. That means that any chance to arbitrate or litigate is then lost by the disappointed party.

C. Arbitration vs. Litigation:

In the new 2007 B102 or A101 contracts, the AIA has made a huge departure from the 1997 versions and all earlier versions. For some 80 years, AIA contract forms have been set up so that all disputes (after initial decision) simply **GO TO ARBITRATION.**

In the new 2007 B102 or A101 contracts, the parties check a box whereby, after mediation, the unresolved dispute goes to litigation OR to arbitration. The document forces the parties to select either some trial judge in the courthouse or an arbitrator as the final decision-maker; before, it *was always an arbitrator* who was the final decision-maker.

The new AIA A201 (2007) Section 15.4.1 says that *if* arbitration *has been selected* in the underlying contract, then an arbitration under the AIA Construction Industry Rules will occur and *ALL KNOWN* CLAIMS MUST BE INCLUDED IN THE DEMAND WHEN IT IS MADE. FAA and federal law are relevant if there is a question of whether or not a dispute is an arbitrable dispute.

D. Consequential Damages Waiver Clarified:

The new 2007 AIA A201, Section 15.1.6 repeats the old Section 4.3.10 waiver of consequential damages that owners don't like. But the new section explains in detail just which consequential damages are expressly included in the waiver. This is really a clarification.

The new Section 15.1.6 says:

"This mutual waiver [of consequential damages] includes [but is not limited to]:

1. damages incurred by the owner for rental expenses, loss of use, income, profit, financing [costs], business and reputation [losses], and for loss of management or employee productivity or of the services of such persons; and
2. damages incurred by the contractor for principal office expenses, including the compensation of personnel stationed there, for losses of financing, business or reputation [losses], and for loss of profit *except anticipated profit arising directly from the work*.

…nothing … shall be deemed [however] to preclude an award of liquidated damages when applicable, in accordance with the requirements of the Contract Documents." (Emphasis added.)

E. Consolidation of Arbitrations:

For nearly 60 years, the AIA documents have contained clauses that fought against one arbitration under an AIA contract being consolidated into some other arbitration under the same contract. The new Section 15.4.4.1

expressly allows consolidation *if* everyone consents and if the other arbitrations employ similar procedures for selecting arbitrators.

F. Statute of Limitations:

Under a new 2007 AIA A201, Section 13.7.1, there is a hard "end-stop" for the State Statutes of Limitations, which is made applicable. *No* action can occur *later than* 10 years after substantial completion. The State Statutes control, but this says that no claim can be made, in any event, after ten (10) years have elapsed from final completion.

Consensus Documents

Several organizations, not happy with the new AIA 2007 form contracts and the 2007 A201, have "bolted" and refused to endorse (for their members) the 2007 AIA documents. Instead, they have produced what they have called the "Consensus Documents." We have not attempted here to address the Consensus Documents and to point out how they are different from the AIA 2007 documents. However, one key feature that appears to be different is the Consequential Damages waiver clause. The Consensus Documents contain what is called a LIMITED waiver of consequential damages, which. "The Consensus Documents" have limited the waiver and have has thereby "made points" with owners.

2.8 INSURANCE AND INDEMNIFICATION

One of the most discussed issues surrounding the construction industry today is liability insurance, or for that matter any type of insurance. The unprecedented amount of litigation that has developed throughout the industry due to the failure to *do it right the first time* has resulted in many long-time insurance carriers denying or dropping coverage to insureds involved in construction defect litigation. Losses relating to litigation have resulted in soaring rates for insurance (with high deductibles) from the few remaining insurance carriers who will underwrite the construction industry. At the outset, we will distinguish among the three types of protection that play a major role in the construction industry: surety bonds, indemnity, and insurance.

Surety bonds

Surety bonds are purchased by the contractor from a third party, which will normally be a recognized insurance carrier, such as Fireman's Fund, Merchants Mutual, or Liberty Mutual and generally fall into four distinct categories: license bonds, bid bonds, payment bonds, or performance bonds. In essence, the surety becomes the guarantor to the owner (obligee), or party that solicited bids and/or retained the contractor to perform the work (1) that the work will be properly performed at the quoted price and (2) that all subcontractors and materialmen will be paid should the contractor fail to do so or become insolvent. The surety does not take on this role as guarantor lightly and, prior to issuing the required bond, will conduct a thorough investigation into the contractor's finances and ability to pay. A part of every surety bond agreement is a personal guarantee that must be signed by the contractor (principal). Under this personal guarantee, the contractor gives the surety a mortgage or trust deed, an assignment of job proceeds, a chattel mortgage, or a Uniform Commercial Code (UCC)-1 security interest and UCC-1 financing statement on the contractor's equipment, receivables, and so on in favor of the surety. By so doing, the surety has a means to recapture some or all of the monies that it may have to expend under the bond, should the contractor fail to perform. To better define the process, the surety is a third party contract agreement that is paid a fee by the contractor to "stand by the gates to the project site," ready to rush in to do what the contractor might fail to do under the construction contract. If forced to perform, the surety will go after the contractor to recoup all monies that had to be paid out for such performance by the surety. The following addresses in a little more specificity the various surety bonds, what they cover, and how they affect the contractor.

License bonds are normally a statutory requirement under state law requiring as a condition of being granted a contractor's license that the contractor post a bond based upon the amount of estimated gross receipts that can be attached by persons harmed by the contractor's failure to perform according to statute, regulation, or contract. Prior to releasing funds, a hearing will normally be granted, allowing the contractor the opportunity to make good on the issues within a set period of time. Failure on the part of the contractor to perform in accordance with the order handed down by the licensing board authority will then allow the injured party to recoup some amount of funds from the bond as established by the authority.

The **bid bond,** in essence, is a guarantee that the work as defined under the contract documents will be accomplished for the price as quoted by the lowest responsible bidder and will normally have a time established stipulating the latest award date. Should the lowest responsible bidder fail to enter into a contract for the amount stipulated in the bid prior to the award date, or for such other reason, as failing to be able to provide a performance and payment bond as required by the contract documents, then the owner may look to the surety to pay the difference between the low bidder and the next responsive bidder who will enter into an agreement. The surety will then look to the contractor to recover these funds.

In all cases of a surety relationship, three parties are involved: owner, contractor, and surety. On public jobs, the contractor cannot do the work without posting a surety bond. Under federal work, the contractor must post a "Miller Act Bond" in the total contract amount. Under state and local governmental projects, the contractor will be required to post a "Little Miller Act Bond." On private jobs, payment and performance bonds may be a requirement under the contract to ensure that the owner is not left "holding the bag" if the contractor becomes insolvent. Payment bonds are particularly important, as failure to pay a subcontractor or material supplier can result in a party recording a lien on the owner's property (the project) until the outstanding debt is satisfied. The existence of a lien can also lead to legal complications, particularly in the absence of properly prepared contractual agreements among all parties concerned.

Performance bonds come in a variety of types. A general performance bond between the owner and contractor will normally be based upon the contract documents, local codes, and ordinances that must be adhered to if an occupancy permit is to be granted by local governmental authority. Another example of a performance bond previously included a 10- or 15-year roofing materials bond, which in essence acted as an insurance policy for the owner guaranteeing that when properly applied, the materials would stand up to local weathering conditions for a specified length of time. If the materials failed to meet the criteria set up under the materials bond, they would be replaced usually on a declining scale over the life of the bond. These roofing bonds are no longer common, having been replaced by roofing guarantees. Performance bonds need to be reviewed carefully, as it is common for the maker to create a number of loopholes that will limit the bond maker's liability.

In virtually all cases, if the contractor does not have enough assets to back the personal guarantee or does not have enough "well-heeled brothers-in-law" or shareholders who will co-guarantee the amount of the bond, then the contractor simply cannot obtain and post the required bond and, therefore, cannot get or do the work.

Indemnity

Indemnity consists of one party's promise to "hold another party harmless" in the event that a certain event occurs on the job. The "indemnitor" is the person providing the indemnity, and the "indemnitee" is the person who gets the protection of the indemnification.

Indemnity is created by either a contract clause or by virtue of some common law precept, which is embedded in certain special relationships.

Indemnity is a "risk-shifting device." You will find in nearly all construction contracts that the owner wants to be indemnified by the contractor. Consistent with the flow down clause, which pushes all duties downstream to the subcontractors, the contractor wants to be indemnified by the subcontractor, who actually does the work. Indemnification is not covered in AIA 101 but can be found in A201 General Conditions, which explains all features of the job's legal relationships. In view of how directly indemnity affects a contractor, it is important to know and understand the various types of indemnity:

- **General Indemnity:** *Subcontractor will indemnify contractor from all losses (and expenses) resulting from the subcontractor's work.* This type of clause does not at all address the "social issue"— whether the subcontractor will pay for the general contractor's own negligence. Under an Arizona court decision, *Mardian Construction vs. Pioneer Roofing, 1981,* since no express language within this clause addresses the effect that indemnitee negligence will have on the subcontractor's duty to indemnify, the clause is only a "general" indemnity agreement. Under the *Mardian* case, the general contractor will not be paid or protected against his or her negligence if the clause is a general indemnity clause. Public policy, as declared by the *Mardian,* says that when the clause does not clearly and specifically force the subcontractor to pay for the general contractor's own negligence, the law *will not* protect the general contractor.

- **Broad form:** *Subcontractor will indemnify contractor from subcontractor's negligence and loss, even if caused in part or in whole by the contractor's negligence.* The case of *Washington Elementary School District 6 vs. Baglino Corporation, 1991,* an Arizona Supreme Court Decision says that if the language is sufficiently clear, this clause *can* hold the subcontractor (indemnitor) to pay for the contractor's (indemnitee) sole negligence, but an Arizona Statute, §32-1159, qualifies the court decision. The statute says that the general contractor's *sole* negligence will not be indemnified. In other words, if the clause is clear enough, the subcontractor can agree to indemnify a general contractor for the general contractor's own partial but not sole negligence.

  Concerning broad form clauses on public jobs, ARS §34-226 says that the general contractor will be protected only against the subcontractor's proportion of the negligence. So if clauses on public jobs are Broad Form or Intermediate Form, they will be "blue penciled" and revised so as to reduce the general contractor's protection to a "comparative negligence indemnity," an indemnity against only the indemnitor's portion of the negligence.

  Section 32-1159 states that a clause in a private (nonpublic) architectural or construction contract that requires indemnity against the sole negligence of the indemnitee is against public policy and void. This statement allows the enforcement of intermediate form clauses that we see in private job contracts but not the enforcement of broad form clauses. This arrangement is obviously not as good for contractors on private jobs as the §34-226 statute is only applicable to public jobs. Since some states may consider broad form clauses to be against public policy, it is recommended that one review applicable state statutes to determine if such a clause is acceptable.

- **Intermediate form:** *Subcontractor will indemnify contractor for subcontractor's negligence and loss caused in part by contractor's negligence.* In Arizona, this clause will be void only on public jobs, for which only comparative negligence (the indemnitor's negligence) is protected against under ARS §34-226.

- **Comparative form:** *See the language in AIA A201 §3.18, which states that the subcontractor will indemnify the* *contractor for subcontractor's negligence, but only to the extent caused by the subcontractor.* On public jobs in Arizona, this is the only clause that is enforceable.

- **Common law indemnity:** *This occurs when there is no contract clause and can best be defined as follows: When a person has discharged a duty owed by him or her, but which, as between himself or herself and another, should have been discharged by the other, he or she is entitled to indemnity unless the person is barred from receiving payment by the wrongful (negligent) nature of his or her own conduct.* In other words, the party who is seeking common law indemnity is going to have to be *completely free* of fault (or negligence) before asking for common law indemnity, an equitable relief claim. Because of this very high no-fault hurdle, successful common law indemnity claims are infrequent. Finally, if there is an express contract clause about indemnity, that clause's existence wipes out all possibility of having common law indemnity claims because the contract language governs over common law.

Insurance

When the first insurance enterprises appeared in the late 1600s, they were denounced as encouraging a lack of accountability for one's own actions. Starting out slowly through the 1700s, the insurance industry thereafter grew rapidly with the increased industrialization of Western society. Controversy concerning insurance seems to constantly have been an issue. In China Grove, Tennessee, after tornados destroyed the town in April 2006, teams of Mennonites showed up to assist those of their faith whose barns and houses had been flattened. "Most of the Mennonites here do not have insurance; they say it goes against their religious beliefs. They rely on each other to rebuild after disasters." "Buying insurance takes away from helping each other," said Oscar Yoder, age 45. "I don't condemn it, but we chose not to. It draws us closer together. We believe God will provide and we do not depend on insurance."

Today, most of us rely on insurance to cover us for health care costs, automobile accident costs, and liability exposure for slip and fall accidents on premises we own. In the construction industry, insurance is virtually a mandatory requirement, but what is insurance and how does it work? Unlike surety companies under surety bonds, this time, the insurance company (which has been paid a premium by the insured for the coverage) *will not have the right* to go back against the insured for reimbursement of the company's losses.

This business of contracting for someone to pay the insured's exposure to persons whom the insured harms by engaging in negligent activity is considered "okay." This risk transference and the actuarial and legal underbrush that surrounds this business of insurance constitute one of the largest industries in America. It is a complex and much regulated service industry. The following reflects some of the various types of insurance available to the construction industry:

- *Commercial general liability insurance (CGL),* protects the contractor from having to pay for injuries that the contractor, its employees or agents might cause to other parties' "persons" or to other parties' property. The following is a typical example of a claim associated with CGL: If a contractor constructed a steel entry framework leading to a parking garage and it collapsed killing a construction worker entering the parking garage and injured a passer-by, along with damaging a portion of the parking garage, the insurance company would pay for the loss by the injured party to the injured party in order to protect the insured.

  In the case of latent defects found on a project after completion, the CGL insurance carrier for the general contractor would be brought into the picture; following almost immediately thereafter, the subcontractors under contract to the general contractor would be brought in as third parties thus involving their CGL insurance carriers. It is not uncommon to have 20 or more subcontractors involved in any construction defect litigation with an equal number of insurance companies and, in some cases, more. The object of getting all parties involved in such litigation is to spread out the damages, requiring each of the third-party defendants to address and pay for the resultant damages that have been determined to be their responsibility.

AIA 201 § 3.18.1 reads in part "to the extent not covered by insurance, the Contractor agrees to indemnify the Owner and Architect." This clause was obviously written so as to "not get in the way" of the third-party insurers' first paying the owner for losses under CGL policies. Only *after* CGL insurance pays, does an indemnity by the contractor apply. Virtually all CGL policies contain an exclusion to avoid liability to the insurer to pay a contractor's *contractually* assumed obligations (indemnity, for example).

It must be remembered that all CGL coverage will have exclusions for repair or replacement of the contractor's work. The insurance will pay for the resultant damage caused by the contractor's work—but will not pay to repair or replace the work performed by the contractor. In other words, "Do It Right or Pay the Price," because your insurer is not going to pay to redo your work. You, the contractor, will have to pay to redo the work while the insurer pays only for the resulting damages that your work did to other people's work.

This can become a two-edged sword, particularly for an unsuspecting owner who has contracted work out to a poorly financed small building contractor and encounters major construction defects and scheduling problems. The owner believes that the CGL will protect from a contractors poor workmanship and serious construction defects; however, as noted above the coverage only relates to *the resultant damage* caused by the contractor's work but will not pay to repair or replace the work performed by the contractor.

This means that if the project is under construction and it was discovered that there were serious construction defects affecting the structural integrity of the structure and the contractor could not or would not properly remediate the problems the owner may be forced into paying another contractor to come in and remediate the defects, as the defects had not caused resultant damage. It is essential that the owner be listed as an additional insured under the GCL policy and fully understand what the insurance carrier defines as resultant damage.

Too often litigation is brought against a contractor for the express purpose of collecting damages for construction defects under the GCL policy, only to find that there was no coverage and the contractor has insufficient funds to correct the deficiencies, or declares bankruptcy, leaving the owner "holding the bag" along with unwarranted litigation expense.

- *Professional General Liability Insurance (PGL).* This insurance is normally purchased by architects and engineers and protects the professional from injuries which its employees or agents might cause to other parties' "persons," or to other parties' property. A typical example of a claim associated with PGL would be that a roof collapsed because the engineer failed to design the roof to handle the dead loads of the equipment placed on it, resulting in a collapse injuring two construction workers. Under such a scenario, the insurance company will pay for the loss by the injured party to the injured party in order to protect the insured.

- *Builders Risk Insurance.* This policy, normally purchased by the owner, will, in many cases, also be purchased by the contractor to guard against risk of loss to the project by fire, flood, vandalism, and so on. If there were a disaster, such as a heavy windstorm that occurred during the course of construction resulting in damage to work already started or in place, this insurance would jump in and pay the project's owner for job site–related loss.

- *Professional & Omissions Insurance (E&O).* Errors and omissions insurance basically covers the insured for any errors or omissions made in the preparation of design documents used to show the contractor the types of materials required and how a structure or parts of a structure should be built to meet the overall design intent. Should the structure fail due to an error or omission in the design, this policy kicks in to pay for what it will take to correct the failure This policy is normally considered one that is primarily for architects and engineers; however, since the Kansas City Hyatt Hotel catwalk's collapse in 1981, there has been a struggle among architects, engineers, and contractors over who is responsible for design elements contained in shop drawings. Since that disaster, architects give fewer and less complete details, simply indicating on the contract documents that the contractor should "build to code." The contractor doesn't understand the detail and submits an RFI (request for information) to the architect, receives an evasive response, and submits a sketch or shop drawing, which the architect *reviews*, but never *approves*. This "arm wrestling" contest between the architect and contractor, by which the architect tries to shift liability to the contractor for key interfaces or components of the job, continues to grow in intensity. Because of this battle, contractors have inserted certain clauses into the AIA documents in an attempt to turn the tables back on the architect, requiring the architects to expressly tell contractors if contractors are required to perform design work. For this reason, it is often the case that a contractor must look to E&O coverage, particularly if performing design/build work. In fact, many owners will now require the contractor to provide an E&O policy when performing design/build work.

- *Automobile Liability,* and self-insured motorist coverage will obviously be an insurance policy that all construction professionals will have and as it is not unique to the construction industry, further discussion is not required.

- *Subrogation* is generally a part of all CGL policies. Subrogation allows an insurer to step into the insured's shoes and go after other parties to recapture dollars expended by the insurer to cover the claim against their insured. Under AIA 201 § 11.3.2, there is a waiver of subrogation. If that prevention of subrogation is going to be a part of the contract, the contractor must get the insurer to agree to that feature of the contract. If the contractor fails to get the insurer to agree to a waiver of subrogation, the contractor may be in deep trouble with the insurer and be sued regarding work already performed. The insurer may reject coverage because the contractor gave away the insurer's right to obtain subrogation.

Apart from subrogation issues, it has become increasingly common for those obtaining CGL coverage to ask that there be "additional insureds" named on the policy. What that change in the policy does is reduce the number of persons to whom the insurer can go to for subrogation, and perhaps increase the insurer's exposure to pay out coverage for some incident.

- *Policy Limits, Deductibles, and Other Loop Holes.* It is not uncommon to have an insurance company set policy limits on any of the noted policies or establish hefty deductibles for their insured, particularly if the insured's actions have prompted a series of claims to be filed against the policy. Due to the increased numbers of construction defect litigations occurring, many insurance carriers are canceling their insureds, increasing their deductibles, reducing the amount of coverage, not providing coverage on various issues, or setting tight time limits involving when a claim can be filed.

There are two basic kinds of CGL policies. On CGL "claims made" policies, the only timing issue concerning whether or not there is coverage occurs when the claim was actually made. However, if we have an "occurrence" policy, then there are several accepted theories as to when coverage comes into play. "First exposure" versus "first manifestation" versus "continuous occurrence" shows that multiple results are possible on the question of whether coverage applies. What an insured or a claimant wants is a "stacking of triggers" resulting in multiple duplicative coverage (several policies applying to the claimed loss).

The issue of causation ("what was the cause" and "was it a covered cause") opens the door to the question of "who's at fault" or "how do we apportion the loss between various responsible parties?" Where there are concurrent causes for a loss, insurance may cover one cause, but not others. The courts have generally ceased trying to determine the "principal" cause or the "effective cause" and are moving to the apportionment of causation and, thus, the apportionment of coverage.

Many states, counties, municipalities, and larger contractors are self-insured, which in many ways is a misnomer, because they will normally purchase layers of CGL insurance from separate carriers. The only thing that is relevant to multiple policy (seriatim in time) coverage, where there are many different policies, or a stacked layer of policies, is that you must give *timely* notice of claims under policies to *all* insurers, or you may lose your coverage and face the claimant's demand without coverage that you could have had.

It is essential to thoroughly review any insurance policy, whether it's your own policy or the policies of subcontractors, sub-consultants, or vendors and understand their limitations, particularly on such issues as a vendor limited dollar amount policy, or "wasting" policy coverage. Under such coverage, every dollar expended by the insured's carrier for defending its position against a claim you have made leaves fewer dollars of coverage to satisfy the claim. Such "wasting" policies are becoming common. Not only do they cap the insurer's loss, but the "wasting" feature obviously totally changes the dynamics of the claimant's fighting with the insurer, trying to get the loss settled.

There is no question that insurance coverage can be quite complicated; however, most courts have ruled in favor of the insured when the language contained within the policy is less than explicit; therefore, the language must be such that the uneducated insured cannot be "done out" of reasonable expectations by the insurance industries' "sharpie" lawyers and their "wordsmithing" of arcane clauses. In other words, the policy had better say what it means or the courts will hold the insurer to what the insured's reasonable expectations of coverage were.

2.9 ALTERNATE DISPUTE RESOLUTION

When disputes arise on the typical construction project, the relevant contracts will ordinarily have addressed the issue of where and when those disputes will be resolved. Usually, the contracts require that the design professional on the job act as an initial "arbitrator" of disputes as a condition precedent to the disputes being submitted to mediation, arbitration, or litigation. Failure to timely submit a dispute to the project architect or project engineer can be raised as a complete defense to the later filing of an arbitration request or lawsuit. Assuming the architect/engineer has determined a dispute, the party not satisfied by the determination set forth by the architect/engineer may (within a stated period, usually 30 days) proceed to the next stage of dispute resolution permitted under the relevant contract clause. On large claims (or on all claims, if the contract so states), the next stage may be the filing of a lawsuit in the trial court of the county designated by a contract clause for jurisdiction and venue, or, in the absence of such a clause, wherever statutes permit. In lieu of courthouse litigation, most contracts in the field of construction call for arbitration of disputes.

The courts in almost all states are overburdened by civil litigation and plagued by a shortage of judges. The time from filing a complaint to trial court resolution by trial is eight months to six years, depending on the state and the county that have been selected. It is rare that judges in the trial courts have had significant experience in the field of construction. Frequently, litigants attempt to bypass the jury system and have a construction case heard by a judge only (this is termed a "bench trial"). That strategic decision somewhat improves the chance of obtaining the correct result, but few judges have great knowledge of the field. Placing matters involving construction before a jury is not only an expensive approach (because of the need to have a large number of experts on many aspects of the case), but it also injects uncertainty into the resolution of a complicated set of facts and issues. Juries often misunderstand some of the conventions, terms, standards of care, and related matters that are known to persons active in the field of construction.

Litigation is most appropriate for large claims for which exhaustive (expensive) preparation for combat in the courtroom is required. In the trial of a $2 million claim regarding the structural failure and partial collapse of a university student union building with delay/impact damage due to faulty plans provided by the owner, it might be economical and advisable to spend $300,000 on the plaintiff contractor's behalf, taking 10 depositions and working up to a 12-day jury trial. If the claim were a $100,000 matter involving similar bad plans, it would not be economical to spend large resources on the pretrial discovery and the resolution of the dispute. The past 25 years have seen an incredible escalation in the hourly rates charged by litigation lawyers, in the number of billable hours devoted to pretrial discovery, and in delays caused by crowded court dockets. With the price of resolving a dispute in the courthouse escalating out of sight, and with the increasing complexity of construction disputes, the attractiveness of alternative dispute resolution (ADR) has increased. ADR can be addressed in a number of ways. The three most popular ones found in the construction industry are discussed next.

2.10 MEDIATION

Despite all of the cost-saving features of arbitration, it is still an expensive process to reconstruct events of a yearlong job a year or two after they occurred. Paying a platoon of lawyers, spending one's own time preparing the case without a chance of being paid for that time, and waiting for certainty as to the outcome of the case can take a heavy toll on the disputants.

For that reason, in the past 10 years, there has been an increase in the use of mediation, often as a required precursor (condition precedent) to the occurrence of arbitration or a courthouse trial. In mediation, usually before any significant monies are spent on discovery or trial preparation, the parties use an outside party to act as a facilitator to resolve the dispute. The mediation requires the principals from all parties involved in the mediation (plus their lawyers, if there are any) to physically meet in a neutral place. Usually, there is a "general session" with all hands present, followed by the mediator's meeting with first one side and then the other separately (caucuses). The mediator attempts to focus the parties on what the dispute hinges upon, the expenses and delay of "trying" the case, the possibility that something could derail or devalue one or another claim or defense, and the advantages of a fair settlement. No settlement is reached if the parties do not agree on one. Sometimes, some issues are settled and the remaining issues set over for trial. Most mediators report approximately 70 percent of mediated cases settled in whole or in part.

Some states and most dispute resolution associations have rules providing that (as with settlement negotiations) the exchanges of information and comments in mediation are "privileged" and cannot be used as evidence in arbitration or courthouse litigation.

2.11 ARBITRATION

The American Arbitration Association (AAA) has published a study showing that claims under $75,000 are averaging 3.5 months from filing to arbitral award. The same study states that claims from $75,000 to $500,000 have been averaging 10 months from filing to arbitral resolution. Apart from speedy resolution, arbitration offers an informed "dispute-resolver." The arbitrator acts like a judge and is ordinarily knowledgeable in the field in which the dispute has arisen. In filing and processing a Demand for Arbitration, the AAA will permit the parties to specify the fields of experience desired for the arbitrator who is to be selected.

For five reasons, the expense of arbitrating a dispute is ordinarily far less than a courthouse resolution of the same dispute. First, the "trial" before the arbitrator will be shorter because the arbitrator knows and understands the field. Second, pretrial discovery is ordinarily strictly limited and controlled. Third, arbitrators don't have 2,000 active cases on their caseload, as a trial judge may. Arbitrators give nearly uncontested attention to the few cases before them, and they accelerate the schedule to get the dispute heard and resolved. Fourth, except in rare instances, there is no "appeal" of the arbitrator's rule or errors in an arbitrator's view of the law or the facts. This is a risk for the users of arbitration, but this finality is an essential component of the economical nature of the dispute resolution process. Fifth, the rules of evidence are relaxed in arbitration to speed the hearing and save expense.

2.12 SPECIAL MASTER

On rare occasions, when some complicated fact-finding work needs to be done, a judge may decide that he or she does not want to listen to two opposing expert witnesses with radically different accounts of what each saw or sees at the construction site. On occasion, under Rule 53 of the Federal Rules of Civil Procedure, the judge will appoint a Special Master to investigate the situation as an officer of the court and to report to the court on the factual situation so the court can then take action under the law. The court actually has the power (per Rule 53 (h)) to decide whether to accept, reject, or accept in part the findings of a Special Master but will defer greatly to the expertise of the Special Master and will normally endorse his or her reported findings.

The appointment of a Special Master may also occur when sound minds prevail in a dispute or in formal litigation when the parties involved believe the entire issue or portions of the issue can be resolved by appointing a third party (Special Master) who has special knowledge or abilities to work with the parties to resolve or attempt to resolve the issue. If agreeable by opposing counsels, the Special Master can be vested with the authority to render direction to the parties involved to arrive at resolution.

SECTION THREE

GOVERNMENT & PUBLIC ISSUES

or

Working with Rules, Regulations, & Uninformed People

3.0 JUMP-STARTING SECTION 3

Dealing with governmental and public issues is an everyday occurrence within the construction industry. In reviewing this section, the reader should take the following points into consideration:

1. Failing to comprehend the role of governmental agencies regarding the construction industry may result in misunderstandings, disputes, disappointment, and failure to ensure that the work is done properly.
2. Compliance with governmental codes does not guarantee a durable or a quality project because codes generally address only safety issues.
3. Some states have minimum workmanship standards that may be enforced by the Registrar of Contractors and industry standards may exist—but the operative word here is "minimum."
4. Governmental agencies license and inspect, but many inspectors are faced with case overloads, lack of technical knowledge, and inadequate training. The owner or contractor who assumes that the presence of government inspectors will guarantee that a quality project results is mistaken.
5. Governmental agencies that make mistakes in inspections and in plan reviews ordinarily cannot be sued because they have governmental immunity and do not have any duties to the property owner or the contractor.

The purpose of this section is to develop a broader understanding of various governmental and public related issues, many of which affect architects, engineers, construction professionals, developers, owners, and members of the public on a daily basis. Some issues come to light only after a project has been completed, while other issues influence a project from its inception. What all projects have in common are the public's perceptions and expectations during the project's entire lifetime.

Factors common to most issues are rules and regulations in the form of federal acts, state statutes, county and city ordinances, building codes, and other such documents that have been adopted by jurisdictional governmental agencies. These rules and regulations are normally monitored by federal, state, county, and municipal agencies charged with the responsibility of interpreting them to ensure that the public's interest is protected and that work being performed is in compliance with these mandates. In most instances, these rules and regulations are necessary and have improved the quality, safety, and performance of construction.

Public perceptions, on the other hand, can be influenced by many factors unrelated to such rules and regulations. Such perceptions result from emotions, bias, and personal or corporate trauma that may have given rise to demanding action from a governing authority. Unlike trade practices, professional ethics, industry standards, and other such requirements developed by industry professionals, governmental and public regulations came about through demands from the general public to governing bodies; by public forum; litigation and the establishment of legal precedents; and through public committees, community leaders, home owners associations, and other such entities. As a result of these perceptions, the burden of proof that the job was done right becomes an ever-growing concern for every architect, engineer, contractor, and owner. Unfortunately, governmental requirements often result in increasing the time and cost of the work being performed. In some instances, these requirements improve the project; in other instances, these requirements result in delays and can influence the overall cost of the project while providing no added benefits. Having to provide reams of unnecessary or unjustifiable paperwork to some uninformed, unqualified, and/or irresponsible personnel will negatively affect the timely progression of the work. In the residential sector, unnecessary regulations increase the cost of homes, which limits their affordability to potential buyers. It is hoped that bringing these conflicting or unnecessary demands to the forefront prior to start of a project can result in the issues being properly addressed and resolved before they become "real problems." Proactive identification, discussion, and resolution of perceived project issues result in a betterment of the project for all concerned.

Public Perception

The process of a project design submittal and approval by jurisdictional entities (town, city, county, state, federal, etc.) and compliance with applicable codes and mandates do not guarantee the long-term performance or quality of a project. The general public and the construction industry fail to recognize that building codes generally address only safety issues and that these codes do not consider quality or performance. The issuance of a building permit and later a certificate of occupancy does not assure long-term performance of a building.

The property owner who believes that any part of the design, construction, and post-construction effort does not meet rules and regulations has the right to file a claim with the state's regulatory agency against entities responsible for that portion of the work. This section points out both the good and the bad aspects of dealing with many of those governmental and public issues that confront construction professionals on a daily basis.

Since a number of claims turn from a simple warranty issue into a major construction defect litigation case, it is appropriate to consider how public issues can and do get out of control. To understand how all these laws, rules, regulations, codes, and so on, which were set up to protect the public, have led to misunderstandings and, in some cases, a total lack of awareness of the facts, one needs to have a general knowledge of how and why they were developed. By bringing these issues to the fore, professionals throughout the industry can work together with their state, county, and local governmental officials to help improve the system: Educational seminars, addressing deficiencies in the current system can be put on by local architects, engineers, and construction experts. Those who deal with the litigation of such decisions on a continual basis can have a major influence on governmental agencies, making positive changes and/or updates in antiquated or inappropriate laws, rules, regulations, and codes. "Good rules, when properly and knowledgably applied, are good for everybody."

3.1 FEDERAL LAW & ITS INFLUENCE

Prior to 1892, all governmental involvement related to the control of construction was at the local level. An early example of this local-level activity is the Plymouth Colony, which in 1626 passed a law that new houses should not be thatched but had to be roofed with board or pale. With the slum conditions in many of the leading American cities getting out of hand, the federal government stepped in to investigate in 1892. Much like the factors influencing the development of effective programs today, political interference, lack of funds, and lack of properly trained investigators limited these investigations. The first significant legislation governing the construction industry was passed in 1929, when the New York State legislature passed the "Multiple Dwelling Law." Although many states followed New York's lead, nothing of national substance transpired until the first federal housing law was passed in 1934, after which the Federal Housing Administration (FHA) was created to carry out the objectives of this act. This act spawned the Home Loan Bank Board, Federal National Mortgage Association, Communities Facilities Administration, and the Public Works Administration. By 1937, the federal government was heavily involved in the housing industry, clearing out slums and replacing them with new homes and living quarters. The Federal Housing Act of 1959 made federal funds available to cities that wished to clean up their slum areas and spurred one city after another to begin to develop community renewal plans. To access these funds under Section 301 of the act, communities had to develop a comprehensive inspection program. These inspection programs launched the building inspection industry into what it is today. In fact many of the requirements currently found in the various building codes were adopted from those standards set up by the FHA in the 1950s. The major cities started to develop inspection codes to meet the requirements set forth in the act. Regional nonprofit groups began to form and develop regulations governing building requirements in specific areas of the country. Examples are the Building Officials and Code Administrators International, Inc. (BOCA), International Conference of Building Officials (ICBO), and the Southern Building Code Conference International, Inc. (SBCCI), which banded together in 1994 and formed the International Code Council (ICC), the official publisher of most of the country's building codes, which have become the dominant codes that we will be using in this book.

The complexities and responsibilities of trying to manage the various federal agencies charged to administer and oversee these federal programs became almost insurmountable, and, in 1966, the Department of Housing and Urban Development (HUD) was created and given prime responsibility for the federal government's involvement in the field of housing. On certain projects, federal law takes precedence over state and local law, particularly when federal funding is used in the construction financing of a project. Federal funding to state and local agencies will also have a major effect on how state and local laws are formulated. For instance, state statutes developed for the State Industrial Commission, or other such responsible agency, on safety issues affecting construction or the work place will normally follow Ocupational Safety & Health Administration (OSHA) standards in order to secure federal funding to help monitor and support the states' enforcement programs. It is therefore essential that the contractor (and sometimes the architect and engineer) be familiar with such programs and how they will affect their businesses. The fines imposed against a business found to be noncompliant with a federal law can be quite substantial, particularly if someone were injured or killed in the performance of the work. The federal acts that would most likely affect architects, engineers, and particularly contractors in the performance of their duties would include the following:

- Davis-Bacon Act of 1931
- Federal-Aid Highway Act of 1952
- Wilderness Act of 1964
- Wild and Scenic Rivers Act of 1968
- National Environmental Policy Act of 1969
- Occupational Safety and Health Act of 1970
- Water Pollution Control Act of 1970
- Clean Air Act of 1970

- Safe Drinking Water Act of 1974
- Clean Water Act of 1977
- Surface Transportation Assistance Act of 1978
- Water Resources Development Act of 1986
- Pollution Prevention Act of 1990
- Americans with Disabilities Act of 1990

Revisions and additions to these acts continue to be made as each year passes and administrations and public needs change.

The federal government has, from almost its initial conception, been a major, if not the leading, entity for the enunciation of public needs. In 1824, the Supreme Court ruled that federal authority covered interstate commerce, which prompted the General Survey Act, authorizing the president to make surveys of roads and waterways of national importance that would allow for the transport of commercial, military, and public mail. This survey responsibility was assigned to the Corps of Engineers, which to this day remains an integral part of the construction, maintenance, and repair of these roads and waterway systems. Since its inception in 1802, the Corps of Engineers has probably contributed more to the fields of architecture, engineering, and construction in the way of design; specifications; job site safety; and the means, methods, and management of programs than any other federal agency. In 1968, the Corps established the Construction Engineering Research Laboratory to study construction materials, design, energy, utility systems, housing habitability, and maintenance. The Corps also has a Hydraulics Laboratory that has made major contributions in the areas of sedimentation, turbulence, and river meandering.

In 1970, the Environmental Protection Agency (EPA) was formed. Prior to this time, the federal government was not structured to have any type of significant impact on those pollutants that might harm health and degrade the environment. The EPA has the primary task of repairing the damage already done to the natural environment and establishing criteria that will guide the public and industry in making a cleaner environment a reality. Assisting the EPA in this challenge is the Corps of Engineers.

Likewise, the Federal Highway Administration (FHWA) has been a continuing influence in developing many of the standards that architects, engineers, and contractors use today. Since the Federal-Aid Highway Act called for uniform geometric and construction standards for the interstate system, the various state highway agencies banded together to form the American Association of State Highway and Transportation Officials (AASHTO), which today publishes design standards and other information available to all architects, engineers, and contractors working on our intrastate highway systems. Many other federal agencies play a part in our everyday activities even though the general public is unaware of these activities. Unless we are working on federally funded programs, we never realize just how influential the federal government has been in developing better design, engineering, and construction on roads, waterways, and structures.

It is not uncommon to run into "conflicts" when dealing with issues affected by federal legislation, particularly when such legislation takes precedence over other governing codes. The Americans with Disabilities Act (ADA) requires wheelchair accessibility to most commercial facilities and to a certain number or percentage of apartments or units in multifamily residential developments. The act requires that sidewalks, stoops, porches, or patios be at the same elevation as the floor of the facility in order to provide a level landing for wheelchair access. This results in a conflict with building code requirements for vertical clearance below the weep screed when a stucco finish is applied to the exterior of a building. The exterior concrete slab placed at finished floor elevation at the entry may cover the weep screed, thus preventing proper drainage of the stucco system. Moisture infiltration results in mold and mildew. Allowing water to enter the interior of a structure poses a danger for "slip and fall" accidents. Placing an exterior concrete slab too close to or over the weep screed has been declared a construction defect in a number of residential and commercial lawsuits. Currently, no provisions in the ADA regulations or the building codes resolve such conflicts. Increasing the height of the stem wall by 4 inches, except at primary entry/exit points, and installation of a weather-resistant threshold or trench drain at the entry could satisfy this requirement but has yet to be addressed by the ADA regulations. It is essential that professionals within the industry bring these issues to the forefront so they can be resolved without the need for costly litigation.

Many in the design and construction industry may not be aware that working on federal or Native American land does not normally require a plan check or building permit. However, this does not mean that the professional has "carte blanche" on the work to be performed. Federal or Native American projects are normally beset with numerous regulations that may affect the performance, such as the following:

- Mandated hiring practices requiring employment of a percentage of Native American labor in the workforce.
- Mandated hiring practices requiring pre-apprentice or other type of on-job training programs that will be monitored by a federal agency, such as Housing and Urban Development (HUD).
- Mandated Quality Control–Quality Assurance programs that monitor the work being performed with the power to shut the work down if it does not comply with a federal or Native American specification. Many of these specifications may include clauses that attach penalties if the product does not meet the quality assurance standards defined in the specification.
- Further, since the provisions of a building code, such as special inspections, do not apply on Native American projects, it may not be clear to either the designer or

contractor exactly what regulations do apply. Without a clear understanding as to the applicable standards and with a lack of adequate inspection, the potential for errors and omissions is increased.

Failure to follow such requirements can delay payment for work performed until the matter has been resolved. Because such rules and regulations may be different from those found outside the boundaries of the Native American reservation, architects, engineers, and contractors are at risk when working on Native American projects.

3.2 STATE LAW & ITS INFLUENCE

The control of state building practices will generally be found in the form of state statutes. In many cases, a statute may be general in form, relating to the whole community, or it may be directed to a specific person or entity. Federal laws that affect state governments will normally be adopted by a statute delineating any additions, deletions, and/or modifications in order to fit the specific needs of the state. Many states have adopted their own specific building codes that take precedence over local city and county codes. In some states, the state building code may apply only to certain types of structures, such as manufactured housing, state colleges and universities, hospitals, and nursing homes. In virtually all instances, major civil projects such as state highways, watersheds, and rivers will be controlled by state and/or federal codes, rules, and regulations. Some states, including California, Florida, Michigan, New Jersey, and New York, have developed their own specific building codes utilizing excerpts from such organizations as BOCA/ICBO/SBCCI/ICC and modifying them to fit the conditions unique to their state, such as earthquakes, hurricanes, and tornadoes. It is essential when working on any state-sponsored project to determine whether or not the state may have an applicable code, statute, or regulation that may influence the work to avoid possible problems or disputes.

Virtually every state has a statute that establishes agencies such as a state Board of Technical Registration and a Registrar of Contractors for the purpose of testing, qualifying, licensing, and overseeing the work of architects, engineers, and contractors. In general the people heading up these agencies are appointees of the governor to serve a term in office and are tasked to uphold the adopted laws, rules, and regulations. The statutes that create these agencies normally state that the purpose in forming the agency is to provide for the safety, health, and welfare of the public through the promulgation of rules, regulations, standards, and qualifications for those seeking registration as an architect, engineer, or contractor, along with imposing such penalties as may be mandated for the violation of the rules and regulations. The following reflect some of the basic duties and responsibilities of these agencies along with some concerns that have affected their effectiveness.

State Board of Technical Registration

Normally, a board would be composed of three or more members, as required by statute, appointed by the governor and/or the legislature, such as an architect, an engineer, and a member from the general public. Some states have separate boards for architects and engineers. The board would have the responsibility for carrying out the powers and duties as mandated by state statute, which may include the following:

- Consider and pass upon applications for registration or certification.
- Conduct examinations for in-training and professional registration.
- Hear and pass upon complaints or direct an administrative law judge to hear and pass on complaints and charges.
- Specify the proficiency designation for the various branches of engineering.
- Secure the services of others to administer tests to applicants, investigate complaints, and otherwise provide support in carrying out its duties and responsibilities.
- Investigate any unregistered person practicing in a board-regulated profession or occupation and issue such civil penalties as may be established by law.
- Take disciplinary action against the holder of a certificate or license of registration who is charged with the commission of certain acts, such as
  - Fraud or misrepresentation in obtaining registration
  - Gross negligence, incompetence, bribery, and so on
  - Aiding or abetting and unregistered or uncertified person
- File injunctions through the superior court against any person practicing in a board-regulated profession or occupation without proper registration.
- Members, agents, and employees of these boards are immune from personal liability with respect to acts done and actions taken in good faith within the scope of their authority.

Unfortunately most of the responsibilities of administering the duties of the board fall on people who have little technical education or training, which impedes some of the functions the board was established to perform. One of the key issues that seem to come up is the definition of "engineer." Most state boards insist that to practice engineering, you must be registered by the state. However, in fact there are more graduate engineers working in various fields than there are those who are registered under state law. To claim that a mechanical engineer with a Ph.D. in thermodynamics cannot advertise that he or she is an engineer seems ludicrous. To be investigated by a retired police officer or other person without a technical background because you are not a state registered engineer performing value engineering on a construction project and then attempt to explain to a similar investigator why a registered engineer was negligent in the design of a post-tensioned slab on grade falls into the same category. Because of these anomalies, it is essential that the registered professionals within the industry take it upon themselves to clarify the scope and definition of those professionals requiring registration versus those professionals who are working in or as consulting engineers for industrial, manufacturing, processing, utilities, and other areas that also may be serving the needs of the public. Additionally, investigators need to be qualified to investigate the issues that require investigating. It is totally inappropriate to have an unqualified person interpret an issue concerning which he or she has no training, education, and/or experience.

Contractors State License Board—Registrar of Contractors

Each state may have a different name for the agency that controls the requirements of licensing and monitoring of contractors. For simplification, "Registrar of Contractors" is the term used here to define this agency. Like other regulatory agencies, the sole purpose of the Registrar of Contractors is to provide protection for the health and welfare of the public. In concept, this is to ensure that a contractor offering service to the public is qualified to do the work and has adequate resources to ensure that the work will be completed. Each state determines the various classifications of contractor licenses and the requirements an individual must meet in order to obtain the license. In general, most states include three basic classifications: general engineering contractor, general building contractor, and specialty contractor.

The general engineering contractor classifications will normally cover roads, bridges, dams, and large earth-moving public works projects.

The general building contractors will normally cover the construction of industrial, institutional, governmental, commercial, and residential structures.

Specialty contractors will normally include single-trade contractor such as electricians, plumbers, air conditioning contractors, painters, plasterers, landscapers, and environmental contractors. Many trade contractors may work as subcontractors to the general contractor on a project.

In order to secure a license, each contractor must have a responsible managing employee and qualifying party to take and pass an examination to confirm knowledge in a selected classification, and knowledge of the state laws governing contractors. The licensee may also be required to show proof of financial soundness, agree to maintain workers' compensation insurance, provide a surety bond or cash deposit to cover possible claims against the contractor, and perform other requirements as required by the state. Once the qualifying party has passed the examination and become licensed, most states require that the party remain an employee of the contracting company and not take other employment that would conflict with his or her ability to adequately supervise the work performed by the licensee. This responsibility presumes that the qualifying party is competent and capable of performing the work expected of a trained tradesperson and has sufficient knowledge to know that the work being performed meets industry standards, required codes, and rules and regulations as may be mandated by the governing authority. Unfortunately, numerous "schools," for a nominal fee, will provide classes to those wishing to become contractors, providing answers to virtually every question that may be asked in the examination, and guaranteeing that the qualifying party will be able to pass the examination. This circumstance has allowed many applicants to enter the industry who are not qualified and lack the competency and experience to know or understand requirements in their respective field of contracting. Additionally, the qualifying party may have little or no authority over the work that is being performed.

These issues, along with the lack of contractor training, are primary factors in the construction dispute equation. It is not uncommon to find that the general contractor and/or his or her staff, as well as the subcontractors, lack adequate training and experience for the work they perform. In some states, the required examination will not truly measure the competence of the applicant. Programs are currently underway for a national registration of contractors. It is hoped that those responsible for developing this program will take the time to address the importance of having the qualifying party be truly competent in the classification being sought. The need to ensure the continued proficiency of the qualifying party is of paramount importance and should be addressed by requiring mandatory educational credits in respective classification as a prerequisite for license renewal. Trade associations, manufacturers, and governmental assessment training all offer an excellent means of acquiring these educational credits. It is essential to any successful program to keep on top of new and developing industry standards. Requiring continuing educational credits as a prerequisite for license renewal will be a step in the right direction.

Field inspectors employed by the state are normally required to be certified through such organizations as BOCA/ICBO/SBCCI/ICC, and for the most part they are dedicated and want to do a good job of evaluating contractors' workmanship. However, many state field inspectors are faced with overload conditions, lack of technical knowledge, and/or inadequate training, which often lead to field evaluations that are not complete or accurate. For example, an inspector judged craze cracking on the surface of a concrete slab to be "compound cracking." First, there is no term or definition of "compound cracking" in the concrete industry. Second, craze cracking is a surface condition primarily resulting from premature finishing while the surface of the concrete is too wet and has no effect on the strength of the overall concrete.

Too often the field inspector's findings are arbitrary and do not reflect the true facts of the issue. In many instances, this is due to the field inspector being a tradesperson with excellent experience in one trade but charged with the responsibility of evaluating all trades. Being unable to spend sufficient time to properly inspect a job, coupled with a lack of knowledge and/or inadequate training, often results in unwarranted findings regarding disputes. Continuing education programs and/or required special inspection by a registered professional on complicated assignments could be implemented to address these issues. Many of these conditions are the result of inadequate funding for the responsible agency, which can be an insurmountable problem if the state is having budget problems. To circumvent such issues, many states, counties, and cities have reverted to contracting out plan review and inspection services. Other possibilities for budget-strapped agencies include securing assistance through outside sources by offering tax incentives to professional firms if they furnish qualified professionals on a pro bono publico basis.

Any property owner or member of the general public may file a complaint against a contractor through the office of the Registrar of Contractors. The Registrar has a duty to investigate the complaint and render a decision. If the contractor's work is found to be deficient, the state will issue a work order requiring the contractor to correct the deficiency within a certain time period or risk suspension of his or her license. If the contractor fails to comply, the damaged party or the state may request a hearing in front of an administrative law judge (trier of fact), or other such official. The Registrar of Contractors hearing process is, in some instances, flawed. Some of the administrative law judges have extensive experience in construction issues, but others do not. To compound the problem, the hearing officers depend heavily on the findings and testimony of the untrained field inspector, who often misinterprets facts. This places a heavy burden on both the complainant and respondent to not only present their case but also to inform and educate the hearing officer, in order to obtain a fair and just decision. It is not uncommon in some states for the trier of fact to issue civil penalties and/or require the case to be turned over to the superior court as a civil action, particularly if the investigative findings uncover fraud, gross negligence, or other unlawful acts.

Minimum Workmanship Standards

Virtually every state and territory of the United States of America has a requirement that any person who intends to build and/or install something for the general public in exchange for money must become a licensed contractor in accordance with the applicable laws of that state or territory. Under the applicable law, many states have included a section defining the minimum acceptable workmanship standards each licensed contractor must meet to maintain a license in good standing. These standards fall under the control of the Registrar of Contractors and are published by the Registrar and made available to contractors and the general public. Many of these standards were initially a part of the Manual of Acceptable Practices developed by the Federal Housing Authority in 1979. While some of these practices are as good today as they were when they were initially developed, many others are outdated and do not apply in today's world. These minimum acceptable workmanship standards are for the most part based on building codes developed by BOCA/ICBO/SBCCI/ICC, and other regulatory agencies. As previously noted, many of these codes and regulations do cover life safety and general engineering issues but fail to address the quality of workmanship, which in most instances is paramount from the viewpoint of the property owner. For example, building codes specify a minimum strength for concrete but do not address other important properties such as water tightness, permeability, wear resistance, or durability. In addition, many issues have come about through the advent of new materials and installation procedures that are not mentioned in the codes or standards, yet constitute a very large portion of today's construction industry. Many of these new materials and procedures affect those standards incorporated in these manuals, rendering a number of the existing standards incomplete and inaccurate. Coupled with this is the simple fact that many of these standards do not adequately define workmanship.

The workmanship standards will often conflict with nationally recognized industry standards and practice. Which standard takes precedence can become an issue in trying to resolve a conflict between the property owner and the contractor. If the workmanship does not comply with the state standard, the directions for remediation are often vague and inadequate, leading to further conflict rather than resolution. If conflict and litigation are to be reduced in the industry, the way in which the minimum workmanship standards are to be defined and applied needs to become a priority of the state Registrar of Contractors and those in the construction profession. To accomplish this may require higher qualifications for Registrar of Contractors field staff and providing better training at every level to properly address both the needs of the public and the lack of oversight on the part of contractors. It may also require a committee of experienced and knowledgeable professionals to aid in determining the minimal standards. Because members of most state boards and registrars are political appointees, they may have little if any knowledge about the problems affecting the construction industry today. For this reason, it is essential that the person in charge be not only a politician and a knowledgeable building official, but also an advocate for improving the industry by making contractors accountable for their actions.

3.3 COUNTY ORDINANCES & THEIR INFLUENCE

In most instances, the building safety department, highway department, and so on for any county was established by county ordinance and voted upon by an elected Board of Supervisors. As with state and municipal governments, normally a law, usually a county ordinance, will indicate that the county has a responsibility to protect the health, welfare, and safety of the public within the county. County jurisdiction normally includes the unincorporated areas within the county, including county roads, flood plains, water resources, and general building construction. County ordinances will normally adopt a recognized code, such as that of BOCA/ ICBO/SBCCI/ICC and other regulatory agencies. What is needed, however, is a committee of registered professionals to interpret these national/international codes for local implementation. This is not to deny that some counties do have an effective program that does take the time to properly address code issues. As with state and municipal agencies, the county may modify or add to these codes to suit the needs of the county. It is interesting that county government in many locations is somewhat lax in its responsibilities regarding plan check and building inspection services. It is not unusual to travel the rural county roads and find nonconforming buildings or work being performed without proper permit. Whether this circumstance is the result of insufficient funding, "out of sight out of mind," or a conscious decision is unclear. An example of this condition follows.

A residence was constructed in an unincorporated community, approximately 70 miles from the county seat. According to the county, the home was constructed in a "Rural No-Service Zone," and a building permit and inspections were not required. As such, no inspections were ever made by the county through the entire course of construction. This issue does not relieve the contractor of his or her responsibility under state law, which will normally require that no home can be constructed that does not comply with the Minimum Property Standards, the Manual of Acceptable Practices of the Federal Housing Authority, and/or the latest edition of the International Building Code, or other such rules and regulations. As can be seen in Figure 3.1, the home looked to be an ideal residence with wonderful panoramic views of the surrounding countryside.

Closer observation of this home showed that the deception on the part of the contractor who built the home and the realtor who sold it was extreme and broke numerous state statutes and county ordinances. Had the county exercised its responsibility to protect the public interest and provided on-site inspections, many of the defects would have been prevented and the resulting trauma experienced by the new home owner would have been avoided as well.

FIGURE 3.1 A pleasant looking home in an attractive desert setting until one takes a closer look.

FIGURE 3.2 Exterior wall light gauge steel stud has been placed directly into and through the concrete footing allowing it to be exposed to exterior soils, storm water runoff etc.

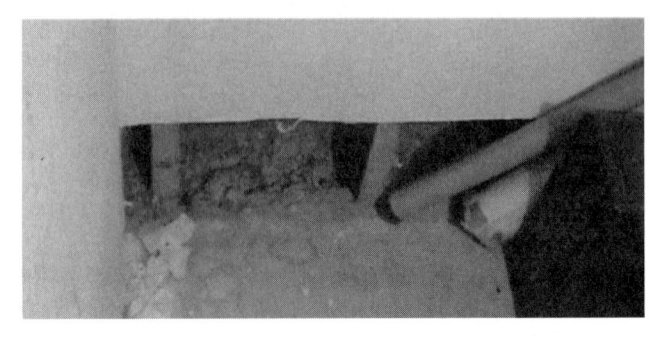

FIGURE 3.3 Interior view of exterior wall showing the light gauge steel framing with no base plate and inserted directly into the concrete stem.

The time-consuming and costly litigation that developed would never have happened.

A few of the many examples of shoddy and grossly negligent construction found in this home are provided in Figures 3.2 and 3.3.

The exterior of the entire home was framed with steel studs that were set on grade, after which concrete was "poured" around the studs as depicted in the photos. Little was done to properly terrace, retain, and stabilize fill material for a home that was constructed on a hillside lot. Plywood was used on the back of the metal studs in the lower-level garage to act as a retaining wall to hold back the loose fill holding the slab on-grade in place under the kitchen. As the plywood rotted out, the loose fill entered the cavity between the studs, leaving voids under the slab and allowing rats to enter the garage ceiling space under the living room and eat the insulation off the electrical wiring. This, coupled with total disregard of recognized industry standards, compounded the defects to such a point that trying to develop a reasonable remediation plan was virtually impossible, as it would have resulted in the new home owner having to pay more for the remediation than what it cost to purchase the home. See Figures 3.4 to 3.10.

The preceding is a representative example of what takes place when there is no monitoring of construction activities

FIGURE 3.4 Here we find the transition between the lower garage plywood retaining wall and the upper slab on-grade. Of key interest is that there was no real footing or stem at the upper level of the home. The contractor had simply placed a thickened edge to the concrete slab over uncompacted fill.

FIGURE 3.5 The underside of the first-floor deck shows evidence of serious failure problems requiring that it be shored up and new joists and shims be added. Note the serious water intrusion from above, indicating that the deck has not been properly waterproofed.

FIGURE 3.6 Floor joists had split from the earlier failure of the deck framing and new 2x8 s were scabbed onto the existing joists; however, none of the new joists, installed by the original contractor, were run all the way to the ledger, which was shot onto the concrete slab on-grade or held in place with a $1/2$ inch anchor bolt.

FIGURE 3.7 There was no slope built into the front porch; this, coupled with no elevation change from the exterior to the interior of the home and settlement of the retaining wall in back of the garage, allowed storm water runoff to drain back into the home.

FIGURE 3.8 Pulling back the carpet in front of the sliding glass door reveals how the OSB deck sheathing is already starting to deteriorate and encourage mildew and mold growth from the vast amount of water entering the home during rainstorms.

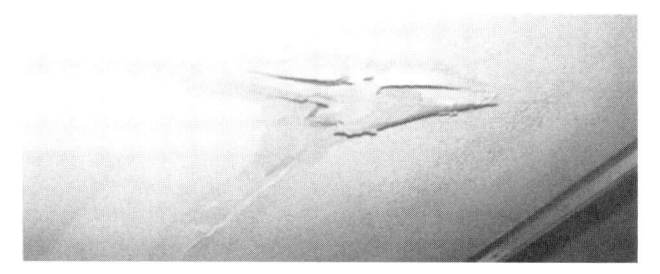

FIGURE 3.9 Storm water entering the home from the front porch deck into the living room eventually ended up destroying the ceiling of the garage, resulting in the formation of mold and mildew throughout.

irrespective of the work's being performed by a licensed contractor (which was the case with this home), a local odd jobber, or the home owner. This type of situation does not speak well of county government and its responsibility to protect the health, welfare, and safety of the public within the county.

On the other hand, there are counties that take these responsibilities very seriously and do an excellent job in plan check and field inspection services.

FIGURE 3.10 The garage floor also has a reverse slope, allowing storm water entering the garage totally destroying much of the homeowner's belongings stored on the floor of the garage.

3.4 LOCAL GOVERNMENT/ MUNICIPAL LAW & ITS INFLUENCE

By city ordinance, local municipal building departments are charged with the responsibility of protecting the health, welfare, and safety of the public and enforcing such building and life safety codes as may have been established under state statutes and local ordinances. Each municipality may modify and/or expand such requirements to fit the needs of its city as determined by its elected officials. In most cases, the city ordinance will adopt a recognized code such as those prepared by BOCA/ICBO/SBCCI/ICC and other regulatory agencies. In most cases, these ordinances do not address the quality of construction and/or workmanship standards; therefore, code compliance will not ensure quality, long-term performance or workmanship.

It is interesting, and often surprising, to discover that factors that contribute to a given construction problem or dispute may be rooted in a code provision or a local interpretation of a code provision that actually requires a faulty condition or procedure to take place. An example is a building inspector who required the stucco lath to be stretched tight, which prevented the embedment of the lath in the stucco, resulting in major cracks, water intrusion, and microbial growth. Another is a jurisdiction that required expansive clay soil under sidewalks of a new school be compacted to 95 percent of the Proctor value, resulting in major distortion and breaking up of the sidewalk after the first major rainfall or the misconception that a single density test taken on the finished building pad will be representative of the condition of the entire subgrade fill mass that exceeded 12 feet and ended up settling more than 11 inches within the second year, resulting in major structural damage.

It cannot be overemphasized that just because a building has been approved by the local building official and a certificate of occupancy issued stating that the building meets code, it does not mean that it in fact does meet code.

The following illustrative examples are provided to show the results of one project after the framing inspection had been approved by the local city building inspector. When reviewing these photos, keep in mind that by not bringing these deficiencies to the attention of the contractor at the time of inspection, the only person being damaged is the contractor. The city will in most cases be left out of any litigation unless gross negligence can be proven. Therefore, any deficiency that would prompt litigation and a subsequent investigation would drop these deficiencies right at the doorstep of the contractor. See Figures 3.11 to 3.24.

Courtesy of Western Technologies, Inc.

FIGURE 3.11 The ledger at the Simpson Strong Wall has been cropped to accommodate the hip beam. In the process, the hold down strap has been cut short.

FIGURE 3.12 Shear wall does not extend to the roof diaphragm as required by Detail 9 of Sheet SD3. Significant ductwork is now in the way of repairing the shear wall.

FIGURE 3.13 Close up of top of shear wall shown in photo above showing ductwork and lack of plywood/OSB in attic space.

FIGURE 3.14 Shear panel not present in attic for gypsum board sheathed shear wall.

FIGURE 3.15 No shear transfer has been provided at the perimeter. Note the gap between the roof sheathing and the top of the wood blocking.

FIGURE 3.16 No H-clips were provided on low roof TJIs.

Another example of issues not recognized by local building officials is that the deterioration of concrete may be caused by a phenomenon known as "salt crystallization." It is the result of salt crystals deposited in the surface pores of the concrete. Concrete foundation stem walls are exposed to moisture from the adjacent soil. Water may migrate up the exterior surface or may enter the stem wall below grade (ground line). In either case, the water contains natural, soluble salts that are carried upward. When the water rises above the adjacent grade (ground level), it will migrate toward the exposed surface and evaporate into the atmosphere. The salts will be deposited in the surface pores of the concrete where they will cause a disruption of the cement paste or binder (glue). Research, performed by the Portland Cement Association, has shown that this phenomenon can occur even in relatively dense concrete. However, lower strength or less dense concrete is more at risk of this deterioration.

FIGURE 3.17 No blocking or connection straps were provided around the opening as required by Detail 3 on Sheet SD4.

FIGURE 3.19 No blocking or straps provided around openings.

Most building codes allow the use of 2500 psi concrete, which is relatively porous and has a high permeability. This allows a greater flow or migration of water wicking upward in the stem. It also allows water to evaporate deeper in the surface of the stem and to accelerate the deterioration. A contractor or builder may be sued for a perceived defect even though he or she is in compliance with the building code.

Although it would not ensure that salt crystallization would not occur, raising the minimum compressive strength to at least 3000 psi would reduce the permeability of the concrete and lower the risk of salt crystallization. This is generally not a structural issue. It is a durability issue.

In some cases, special structural inspection is required for certain building systems. This requires either the engineer

FIGURE 3.18 Hold down strap is inset from end of shear wall.

FIGURE 3.20 No edge nailing was provided around opening in shear wall.

FIGURE 3.21 No truss hangers were provided on the short hip trusses.

FIGURE 3.23 At the exterior wall, few studs are full height. Should check calculations for wind on the exterior wall.

of record (who designed the building) or a certified special structural inspector to inspect the structural elements. Generally, that inspection is more complete, but, again, it is limited to the structural elements. Too often, the whole review and inspection process only results in a false sense of security. In view of these issues, an owner cannot rely on the public inspection process. To make things even more difficult, building officials are immune from personal liability with respect to acts done and actions taken in good faith within the scope of their authority. This makes it quite difficult to pursue any type of litigation, unless one can prove gross negligence.

FIGURE 3.22 No LPT-4 shear connectors were visible as required per the "Plate Attachment Schedule" on Sheet S-11.

FIGURE 3.24 The stair stringer is split on both sides of the intermediate support wall.

3.5 WHO CATCHES DEFECTS BEFORE, DURING, & AFTER?

Prior to the commencement of construction, the design must be submitted to the local authority for a plan review and a construction permit. This review is primarily conducted to ensure compliance with local codes and ordinances and is merely a cursory search for gross errors. This review covers only minimum safety issues as reflected in the codes and is unrelated to the overall quality of the design being reviewed.

During construction, local governments require inspections. In many jurisdictions, the inspectors are overloaded, having 20 to 30 different projects per day to inspect. During the course of a job, it often occurs that the first two or three inspectors come on the job site and pass (green tag) various systems. However, a week or two later a different inspector arrives and rejects (red tags) one of the systems that had been previously approved.

The government inspectors are required to act as they believe appropriate under the applicable code. The last inspector to view the system "owns it" and will red tag it if he or she thinks it does not meet code—even though other inspectors previously stated that the system was code compliant *and* the owner and architect had relied upon the earlier inspectors having green tagged the system.

The public does not understand that many of the required inspections are not performed by government inspectors. On complex projects, the International Building Code (IBC) requires "special inspections" to be performed by private inspection companies. If these special inspectors are selected based upon "low-bid" cost concerns, the inspections that they perform may be flawed or insufficient.

The usual manner in which defects are discovered after construction involves a home owner complaint, one that rises to the level of a Registrar of Contractors' proceeding. A complaint procedure exists whereby a home owner may raise issues with the construction as performed by the contractor and obtain a governmental Registrar investigation; work letter; and, in some cases, a hearing before a hearing officer. The Registrar may issue an order suspending a contractor's license in the event that he or she fails to repair the defects that have been found to exist during the investigation. It is also possible for an owner to proceed against the architect or engineer before the state Board of Technical Registration in the event that defects in design (which constitute professional negligence) are discovered after construction.

Before, during, and after construction, the architect and engineer have a legal and ethical duty to the public/government to disclose life safety issues to the authorities if an owner ignores the professional's advice on design or installation. See paragraphs 7 and 11 of R4-30-301 (Rules of Professional Conduct).

3.6 WARRANTY ISSUES VERSUS NORMAL MAINTENANCE

One of the key issues that come up in nearly every construction defect case is that during the warranty period, the contractor claims that the issue complained about is really an owner maintenance issue and does not fall under the normal two-year materials and workmanship warranty as required under most state statutes. It is therefore important that every contractor clearly define the difference between what is covered under the warranty and what is considered to be an owner maintenance requirement as the home owner can lose his or her rights if he or she fails to perform maintenance. Under most state laws, the implied warranty means that the contractor warrants that the work performed not only meets all established applicable codes, rules, and regulations but will pass without objection in the trade under the contract description; is fit for ordinary purposes for which such work is intended; is properly built in accordance with applicable design, specifications, manufacturers, and industry standards and written instructions; and conforms to the promises and affirmation of fact as contracted for by the owner. It is essential that every contractor thoroughly understand that by not properly defining the warranty issue and not having the owner read, sign, and understand what the warranty covers, he or she leaves the door open to litigation being filed to determine whether the defect was the result of contractor negligence/performance or prompted by the home owner's lack of proper and timely maintenance.

Many of the larger contractors, home builders, and developers have created warranty performance standards that are provided, in full or in part, to the owner at the time of substantial completion, along with maintenance manuals, operating instructions, list of responsible subcontractors, and other such documents. In the case of home builders, such warranty documents are available for a prospective buyer to review prior to purchase with a condensed version provided at the time of sale. The warranty durations may vary and will often depend somewhat on state statutory requirements. Construction warranties will generally extend from 1 to 2 years on materials and workmanship, while equipment warranties may extend from 5 to 20 years or better. In any case, the owner should review what the warranty performance standards cover and for how long such coverage lasts. Many times, the performance standards may have clauses that were not taken into consideration by the owner prior to accepting the work. Some examples of issues involving warranty performance standards follow:

- Repair of drywall cracks during the first year will be performed one time and then are considered an owner maintenance issue with the caveat that the contractor does not guarantee an exact match.
- Coverage by the contractor for subterranean termite infestation damage to the structure may only cover

damages sustained in the first year even though the termite service warranty says protection is good for five years.
- Shade changes or discoloration in grout is not normally covered under the warranty as most contractors do not endorse or apply any grout sealer.
- While the warranty may repair plumbing leaks "as needed," there may be an additive clause that the contractor is not responsible for consequential damages resulting from those leaks.

The question is often asked, Does the contractor's implied warranty of good workmanship (*Kubby vs. Crescent Steel Co.*, 105Ariz.459, 466 P.2d 753 (1970) extend to anyone other than the owner who hired him or her to build the residence? Yes, the warranty extends to second owners of residences, but they are considered to stand "in the shoes" of their predecessor in title, so that if the first owner waived a defect, the second owner cannot sue on it (*Richards vs. Powercraft Homes, Inc.*, 139 Ariz. 242,678 P.2d 427 (1984). This does not apply to second owners of commercial construction (*Hayden Business Center Condominiums Assoc. vs. Pegasus Development Corp.*, 209 Ariz. 511, 105 P.3rd 157(A3 App 2005). In commercial projects, the builder's implied warranty of workmanship runs only to the first owner. The warranty is limited to those "in privity" of contract to the builder.

In virtually every state, the new home owner is protected by an implied warranty created by the laws of the state for a designated period of time after taking possession of a newly constructed home, provided that the home was constructed by a licensed contractor. Generally, this time period is two years after occupancy and/or the date on which legal occupancy occurred. Legal occupancy, in most jurisdictions, occurs when the local governmental agency responsible for inspecting the home during the course of construction issues a final occupancy permit or certificate of occupancy. The occupancy permit generally declares that the structure has been built in accordance with local codes and ordinances and is ready to be occupied.

Once the contracted work has been completed and the structure occupied, the owner has the option to file a complaint with the governing authority within the warranty period, as set forth by state statute, after discovering a construction defect or finding that the building contractor's workmanship falls below the minimum workmanship standards or governing codes and regulations. This does not preclude the filing of a complaint against a contractor prior to the completion of the work. In most states, complaints may be filed by others, including but not limited to claims by other contractors, owners, co-owners, joint tenants, their successors in interest, or one who has direct contact with licensee in the performance of his or her work. If the contractor or respective subcontractor fails to correct the deficiency, the governing agency will then investigate the matter in the presence of the claimant and responsible contractor or subcontractor. Should deficiencies in the construction be found, the governing authority will normally issue an order to the responsible contractor to correct the deficiency within a fixed period of time. The claimant must provide access for the contractor to correct the deficiency. If the contractor does not correct the deficiency, he or she will suffer the consequences as dictated by the laws of the governing authority. Failure to correct the deficiency may open the way for the claimant to secure relief through the state's recovery fund or other such avenues as prescribed by state statute. In essence, the claimant maintains an action against the contractor's surety or cash deposit, as may be required by the state, until the complaint has been satisfied or dismissed. In most cases, the premise under which the governing authority will make its ruling will be the minimum workmanship standards for licensed contactors as adopted by the state and testimony from the state's field inspector.

At times, construction defects may not show up until after the normal two-year warranty period prescribed by state statute. Under this circumstance, the owner may be forced to accept the issue and repair it unless it can be proven to be a latent defect. Under these conditions, most states have laws concerning what constitutes a latent defect and have established a longer period of time to report and/or address such defects through the legal system. Apart from actions by the Registrar of Contractors, the owner may have contract provisions under which the owner can sue or seek relief through arbitration.

3.7 STATUTE OF LIMITATIONS— WHAT'S INCLUDED?

In every state, statutes prescribe limitations to the rights of action, declaring that no suit shall be maintained nor charges be made unless brought within a specified period of time after the cause of action accrued. Statutes of limitation are legislative acts prescribing periods of time within which certain rights may be enforced. There is a difference between a statute of limitations, which can be extended by a delay in discovery of a defect, and a statute of repose, which cannot be extended by late discovery. In most states there is a statute of limitations and/or statute of repose on "latent" defects, defects that could not be discovered by reasonable and customary inspection or defects unknown to the property owner, of which, in exercise of reasonable care, he or she should have no knowledge. Defects include such issues as

- Underlying soil conditions that when subjected to uncontrolled moisture infiltration compress and/or swell resulting in damage to the structure.
- Lack of proper tie-downs, hurricane straps, and other such devices that are required to secure and stabilize the structural frame in a stable condition when the structure is experiencing wind forces and/or seismic movement.

- Lack of proper weatherproofing, exposing the structure to environmental conditions that attack its structural integrity.
- Improperly sized members that fail when subjected to normal dead and/or live loads for the class and type of structure.
- Improperly prepared and/or assembled duct or pipe fittings that result in injecting water, sewage, gas, air, and so on into, under, or around a structure that cause and/or promote injury or damage to the public and/or structure.

3.8 HOME OWNERS & HOME OWNER ASSOCIATIONS

Going beyond the workmanship standards, the trend lately has been toward the formation of home owners associations (HOAs) that establish their own restrictive covenants, rules, and regulations (CC&Rs), legally filed under applicable state laws. These HOA requirements govern how and what type of dwelling can be erected, including the type of architecture the dwelling must meet, what colors can be used, what size and type of window can be used, to landscaping requirements, and so on. These CC&Rs do not take the place of general laws, but supplement them. They are normally established and formalized at the commencement of a multihome development when all of the property is owned and under the control of a developer. Some HOAs have been formed after some of the dwellings were constructed and occupied, but it becomes much more difficult to form standards that would be agreeable to all of the individual home owners within a given development as the number of individual property owners increases.

The HOAs experience of real and/or alleged problems with their dwellings has opened up a new litigation industry sweeping the country. Coupled with class action suits filed on behalf of a series of home owners in any given development, construction defect litigation has become a full-time job for many attorneys and construction defect consultants. On the other hand, construction defect litigation has become a financial burden on architects, engineers, and contractors and to their respective liability insurance carriers and sureties. While there is no question that much of this litigation may be needed, one must be cognizant of the fact that litigation is expensive and time consuming for all parties involved. It would be much better to develop a proactive approach in resolving the issues or eliminating the causes of litigation by doing it right the first time.

3.9 THE HOME INSPECTOR, PUBLIC HERO OR TROUBLE MAKER?

One line of defense for the general public against construction defects found in homes today is the growth in the number of home inspectors. The home inspector is a relatively new profession that was created in the late 1970s, primarily through real estate industry representation orders to provide investigative reports on possible defects found in a home prior to sale. Additionally, these inspectors might be retained by a new home buyer to verify that the disclosure statement provided by the seller is correct.

These home inspectors examine everything from the roof down, including electrical, plumbing, heating, ventilating, and air conditioning. Our experience has proven that some of these inspectors are quite competent in their inspection process, while others are incompetent and uninformed on many of the issues affecting a home.

A number of home inspection organizations currently exist, such as the American Society of Home Inspectors (ASHI). ASHI has created a professional standard of practice and code of ethics for home inspectors, requiring that applicants submit a written examination and provide a minimum of 250 fee-paid reports that must comply with the established ASHI standards of practice before an applicant can become a member. Various training schools have picked up on the services provided by such organization as ASHI and developed standard forms that have been accepted by such organizations as ASHI for use by those, who for a nominal fee, can obtain these forms and be recognized as a home inspector. Just having copies of these inspection forms does not guarantee that the person using the form is, in fact, a qualified home inspector. However, with many such inspectors being charged with negligence and misrepresentation, it has become increasingly important for the public to verify credentials of the inspection agency, as well as the inspector, before retaining the agency. Currently, some training schools are both registered with the state education authority and licensed by a state governing authority. Training developed in such schools will usually be approved by various agencies and organizations that govern and/or promote inspection services, such as the following:

- California Real Estate Inspection Association (CREIA)
- American Society of Home Inspectors (ASHI)
- Office of Real Estate Appraisers (OREA)
- Foundation of Real Estate Appraisers (FREA)
- National Association of Home Inspectors (NAHI)

Continuing education requirements are mandated in the bylaws of ASHI, CREIA and NAHI. It is essential to retaining credibility as a certified home inspector that one continues to improve one's education through on-the-job experience, as well as through the training programs and courses offered by the continuing education programs of these organizations. It is also a great comfort for the home owner who is depending on the information provided by these home inspectors before committing to the purchase of a new home.

Caution should be exercised by the public when selecting a home inspector as there have been cases in which a non-qualified home inspector renders opinions on issues that he or she has no business being involved in. A certified home inspector who has been properly trained will always recommend that serious issues be referred to qualified professionals.

3.10 INDEMNIFICATION

Is indemnification limited by state statutes? In most states the answer is, yes. In Arizona, A.R.S.§ 34-226 and § 32-1159 each say that a clause in a construction contract by which a contractor, architect, or engineer has indemnified a person or entity from that entity's sole negligence is void against public policy and unenforceable. In the case of public works, the statute says that the indemnitor cannot be made to pay for *any* of the negligence committed by the indemnitee. In the case of the private job, the clause protects the indemnitor only from having to indemnify against the *sole* negligence of the indemnitee. See also West's annotated California Code § 27-82 and Colorado Code § 13-50.5 -102(8).

3.11 THE PUBLIC ADJUSTER VERSUS THE INSURANCE ADJUSTER

It is not uncommon in times of disaster (fires, floods, earthquakes, etc.) for home owners to look to a public adjuster to evaluate property damage. Although their insurance company provides the home owners an adjuster at no cost, many home owners seek the services of an outside adjuster who has no relationship with the insurance carrier to help settle the claim. Public adjusters are professionals who are available to the insured party to inspect a reported loss site, determine the damages, prepare supporting documents, review the insured's coverage, and determine current replacement costs, while exclusively serving the client rather than the insurance carrier. Public adjusters will normally charge a fee of as much as 15 percent of the total value of the settlement for their services, plus expenses. Many times after a disaster, the State Department of Insurance will set the percentage that a public adjuster may charge. Unfortunately, a number of public adjusters are not qualified to perform a proper analysis of the damages sustained and/or have a history of inflating the cost of such damages in hopes of reaping a larger fee. When such an issue occurs, the insurance carrier has the right to challenge the findings and retain an independent adjuster to evaluate the damages after which an umpire is agreed to by the two adjusters to hear their respective findings. After hearing the arguments from each adjuster, and reviewing the damage, the umpire makes the final award, which may or may not be agreeable to both parties. The umpire's fee is normally split between the parties involved. The public adjuster's fee is not ordinarily covered by the home owner's insurance policy and therefore comes directly off the top of any settlement figure agreed to by the home owner's insurance carrier. Additionally, when a dispute occurs between the home owner's insurance adjuster and the public adjuster as to the cost of repair, the resolution can take months and may often lead to litigation. Because of these unfortunate circumstances, it is always in the best interests of the property owner to verify that the public adjuster is a member in good standing with such organizations as the National Association of Public Insurance Adjusters (NAPIA), as well as checking qualifications through the State Department of Insurance, an insurance agent, a lawyer, or a friend. It is important to avoid adjusters who go from door-to-door after a major disaster, unless they are properly qualified and can provide reputable references.

Normally, during major disasters, the home owner's insurance company will retain the services of qualified architects, engineers, and contractors to assist in evaluating and pricing the cost of repairs. Any home owner has the right to review the cost of repair with the insurance company's adjuster. Most important, every home owner policy should be reviewed with the agent periodically to fully understand what's covered and what's not. It is particularly important if improvements and/or additions were made to the home or personal property was added that would require additional coverage. One of the challenging issues to affect home owners over the past few years is the occurrence of mold.

As a result of the considerable amount of litigation that has transpired over the past few years, and the difficulty in estimating potential losses associated with mold remediation, the insurance industry has modified its standard policy language to exclude mold, rot, and mildew damage from most contractor general liability, commercial, and home owner policies. This change has not precluded mold claims from being paid through judicial interpretation of such policy clauses, resulting in distorted and unpredictable jury verdicts. These issues are currently being addressed by insurance regulators and legislators. If changes are not made, the economic consequences to all concerned can be severe. It cannot be expressed too forcibly that it is important to understand how to properly seal and waterproof not only the exterior of a structure but interior facilities such as showers, steam room, indoor swimming pools, and so on. Failure on the part of the architect, engineer, or contractor to take this warning seriously will lead to severe economic exposure. Too often, insufficient detail on how to waterproof the building envelope is shown on the plans or is improperly addressed in the specifications. This exterior envelope includes roofing, siding, stucco, decks, stem-walls, and foundations.

3.12 POOR, BIASED, & MISLEADING REPORTING TO THE NEWS MEDIA

It is not uncommon when problems to a structure occur that unknowledgeable people, in their haste to cast blame and demand action, make poor, biased, and misleading statements to the media or others. Unfortunately, in a rush to make news and create public awareness, the media often do not take the time to determine if the statements are accurate and representative of the facts. This has, at times, resulted in embarrassing and financially troubling consequences to the construction professionals involved. This often occurs when there is a major disaster such as a fire, earthquake, landslide, and so on, and the media concentrate on the individuals who have suffered from the disaster before knowing all the facts. Many times, the investigation discovers that the primary damage was the result of negligence on the part of the individual affected by the disaster rather than the design professional. Because of such possibilities, it is essential for all construction professionals, when confronted with a significant issue that may attract media attention, to ensure that they immediately and professionally remain on top of that issue, directing it toward a satisfactory resolution. Trying to hide from such an issue or down play it only makes an insignificant problem look bigger than it really is.

SECTION FOUR

MYTHS & FACTS

or

Issues That Lead to Misunderstandings & Confrontation

4.0 JUMP-STARTING SECTION 4

Unfortunately, there is an abundance of myths when it comes to the field of construction. In this section, we have attempted to separate these myths from facts and clear up some of these misconceptions.

Starting Out With Simple Facts

1. It is a good idea to employ a reputable home inspector to examine a home prior to making a purchase. A competent inspector can find dangerous and/or expensive-to-correct defects in the home that would not be noticed by the average home buyer. But, be aware that there are vast differences in the abilities of inspectors and very limited oversight of the home inspection business.
2. A home may be successfully constructed on almost any soil type or geological feature if the proper preparation of the site is accomplished. However, the cost of site preparation may easily exceed the cost of the structure if exotic or otherwise difficult locales are selected.
3. Proper soils testing at a construction site can be costly, but shortcutting these important tests is often more costly.
4. Controlling the flow of water onto as well as off of a building site is extremely important to the long-term stability of the building.
5. Especially during the early drying phase, concrete slabs on grade will allow migration of some water (this is called breathing), and this fact should be taken into consideration when planning floor coverings (e.g., porous versus nonporous) and even type and placement of furniture. This is a serious and complex problem, and most designers don't understand it. Unfortunately, there is no perfect solution to deal with the condition created when an impervious floor covering is installed on a concrete slab-on-grade.

The layperson is usually not familiar with what, in fact, differentiates a construction defect from acceptable construction practice. Such people have heard numerous stories related to construction issues ("*My brother-in-law works for a contractor.*") but never had a chance to meet and discuss their concerns with a competent and experienced construction professional.

Unfortunately, construction defect lawsuits involve both real as well as imagined defects. There is no doubt that the glut of construction defect litigation has exposed some very shoddy work, particularly in the residential construction industry. Section 7, Construction Issues, addresses many of the underlying causes leading to construction defects. However, it is not uncommon to have normal construction variants added to real defects by predator lawyers and their "experts" to inflate the claimed damages.

As with most general statements, there will be exceptions; therefore, it is recommended when purchasing a home that a competent certified home inspector be retained to look for problems that would not normally be detected by the home buyer. Be aware that not all home inspectors are certified and they could be odd jobbers or tradespeople of relatively narrow experience who are trying to capitalize on a problem in the home-building market. As well qualified as some home inspectors may be, they will not normally have the qualifications and experience of a professional forensic construction investigator. To better differentiate between the two, refer to Section 9, Specialized & Investigative Services.

There is an understandable resistance on the part of the prospective home owner to paying a fee to someone to examine either a new or used home to search for defective construction. After all, "wasn't it built to code?" It is vital to understand that "code" has little if anything to do with quality. Building codes relate only to safety and short-term function of basic services such as electrical and plumbing. Add to this the unfortunate fact that, especially in "hot" housing markets where homes are being built in large numbers, the public code inspectors often simply cannot do the job.

To assist both current and prospective owners, we have compiled a list of conditions, some of which are real defects and some of which are not, with appropriate comments. We hope that these are both directly helpful and generally interesting to the reader.

The following myths and facts address many of the actual conditions observed in the field and are presented to put some of these questionable issues in proper perspective.

4.1 SOIL MYTHS

Myths and Facts for the Layperson: Many times the layperson (home owner, realtor, reporter, lawyer, tradesperson, business person, or others) is not familiar with what in fact differentiates a structural problem or a construction defect from acceptable construction tolerance and practice. They have heard numerous stories related to construction issues but have never discussed issues and concerns with an experienced construction professional. The following information may assist in addressing many of their concerns.

Construction defect lawsuits claim damages for real or imagined defects. The glut of construction defect lawsuits has exposed shoddy construction. Some of the underlying causes of this problem are a lack of trained and experienced supervision, a shortage of qualified labor, and rushed schedules to meet consumer demands. Some of the problems are actually consumer driven. Once the owner makes a decision to build, he or she often wants the project delivered on an unrealistic schedule.

Potential buyers do shop price, which pressures builders to cut corners to compete. There are three variables in any financial dealings the size, the cost, and the quality. You can select only two and the third will be fixed by the first two selected.

Many issues raised in lawsuits are actually normal conditions and are not defects. Many claimed defects are without merit but are added to legitimate issues by predator lawyers and their "experts" to inflate the claimed damages. This section is included to put some of the issues in perspective.

As with all general statements, there will be exceptions. An experienced construction investigator can readily identify those conditions that are exceptions to these basic facts listed here. Based on years of experience, the myths and facts presented here apply to an overwhelming majority of actual conditions observed in the field. However, when in doubt, contact an experienced professional in the specific field in question, that is, a geotechnical engineer for soil issues, a structural engineer for structural issues, a civil engineer for drainage or paving issues, or a roofing/waterproofing expert on roofing and water leakage issues. Many home inspectors are not qualified to evaluate perceived problems.

MYTH: Potentially swelling clay soils are unsuitable for housing development projects.
FACT: Many housing developments throughout the West and Midwest are built in areas affected by expansive clay soils. These soils will normally remain stable as long as the moisture content remains relatively stable. Development and construction on soils with a high swell potential can be successfully accomplished, providing steps are taken to mitigate the risks, including the following:

- Providing proper drainage to ensure that storm water runoff and all forms of water, that is, irrigation, planters, and down spout discharge, are directed away from the home's foundations.
- Providing a drainage collection system that captures and disposes of roof runoff and subgrade water before it affects the soils adjacent to and/or under the home.
- The use of piers and grade beams to carry an elevated floor system, thus allowing the soils to shrink and swell without affecting the structure.
- The use of a reinforced waffle slab on-grade, thus allowing the soils to shrink and swell within the waffle voids without affecting the structure.
- Elevating and maintaining the soil moisture contents at or above optimum moisture conditions before placing the concrete slab on-grade, whether the slab is post-tensioned, heavily reinforced, or conventional. Compaction of the soil should be limited to a lower degree of compaction while maintaining moisture contents at or above optimum moisture.
- The soils report prepared for a development will normally address potential swell values and probable heave magnitude of the soils where the test samples were taken and tested. Moderately swelling soils present a much lower risk than those with a high swell potential. Generally, soils with a 1.5 to 3 percent swell present a relatively low risk of significant heaving.
- Irrespective of the methodology that may be used, the key to controlling the shrink/swell soil problem is providing adequate positive drainage to keep water away from the foundations and slabs-on-grade.

MYTH: Dumped or spread fill zones will self-compact with time and after a few years will not be subject to settlement under structures.
FACT: Dumped or uncompacted fill zones are in a loose, low-density state. They may remain that way until new structural loads are placed and the completed site is now subjected to roof runoff, landscaping irrigation, utility service line leaks, or other surface water sources. The combination of higher structural loads, low soil densities, and infiltrating water results in nonuniform settlements and distress conditions.

MYTH: Hillside slope benching and preparation of native surface soils are not necessary when located under new fill zones.
FACT: Slopes steeper than about 5 to 1 (horizontal to vertical) should be benched to form horizontal platforms on which the new fill soils are placed in horizontal compacted lifts. These benches form notches in the original ground surface and are keys for preventing the down slope movement of the fill mass.

MYTH: Compacted fill zones need just a few density tests taken when the testing company is called out to take a test at the direction of the contractor.

FACT: This type of "on-call" density testing only gives the contractor and the owner information about a single 4″ x 6″ prism of soil (sand cone) or an 8″ or 10″ density test (nuclear gauge). This test is valid and appropriate only for the soil at the location and depth tested. It says nothing about any adjacent soil since the tester did not see if the soil layer was compacted using the same equipment and procedures. To be effective, and to achieve a "certified" or "engineered" fill, the testing company must have its eyes on the fill pad construction full time. This way, the tester knows that the proper uniform procedures and equipment use resulted in passing tests that are representative of the entire fill lift.

MYTH: The harder (denser) a soil is, the better it will be for support of structures.
FACT: This is true, in general, for granular, non-expansive soils. The denser a soil is, the more internal shear strength and less settlement potential the soil mass will exhibit. However, for clay soils, this is totally *untrue.* The higher a clay soil density is and/or the drier the clay soil is, the more expansion potential the in-place soil will have. Clay soils must be placed, compacted, and maintained (prior to and after construction) within narrow ridges for both density and moisture content.

MYTH: Site drainage, particularly around the perimeter of a structure, is not very important to the successful long-term performance of a building.
FACT: One of the most important aspects of soil/structural interaction is the provision and maintenance of positive site graded slopes away from a structure. Many times the final grading and lot drainage are left to an equipment operator who may or may not have any idea of what the drainage slopes and conditions are to be for the project.

MYTH: Loosely placed or dumped fill zones as backfill behind retaining walls are acceptable and will result in lower lateral stresses on the wall.
FACT: The statement is not only false, but it is also an incorrect means of filling behind a retaining wall.

MYTH: Sloping soil surfaces up and away from the back of a retaining wall or sloping soil surfaces down and away from the bottom of a retaining wall will have no effect on the wall's lateral stability.
FACT: Sloping surface behind the top of a wall will increase the lateral wall pressures and tend to push the wall out or rotate the wall more. Downward sloping surfaces at the base of the wall result in a significantly lower passive soil pressure at the wall toe. The reduced toe pressure can result in lateral translation and sliding of the wall or rotational movement of the footing/wall.

MYTH: Depth of footing embedment below the native or finished soil surface at the base of a retaining wall will have no effect on wall stability.
FACT: The allowable bearing pressure is reduced as the footing depth is reduced.

MYTH: Positive and adequate drainage systems behind and through a retaining wall are not necessary.
FACT: Inadequate or lack of drainage behind a retaining wall results in increasing soil moisture behind the wall and a significant increase in lateral pressure against the wall.

MYTH: Placing landscape materials that require watering adjacent to the foundations of a building presents no problem.
FACT: Water from any source may cause foundation movement regardless of the type of soil and therefore should be avoided. Planters next to a building often result in serious foundation problems and unsupported claims of poor construction. Placing a faucet on the exterior wall above the point where utilities pass under the foundation may result in serious problems. Any leaks from the faucet can penetrate the utility trench and migrate under the home affecting soil equilibrium.

4.2 CONCRETE MYTHS

MYTH: A concrete slab on-grade is a structural element and part of the foundation system.
FACT: A concrete slab on-grade floor in building construction, where the slab is contained within the interior face of the exterior stem (bearing walls), supports only the nonbearing (nonstructural) partitions. The upper floors and roof are supported on the bearing walls or structural systems, which are supported on independent footings. When an interior bearing wall is supported on the slab on-grade, that part of the floor is considered a part of the foundation. Otherwise the concrete slab on-grade is simply a floor "floating" within the stem walls, which form the perimeter.

MYTH: Post-tensioned concrete slabs on-grade are a part of the structural foundation.
FACT: That portion of the post-tensioned slab supporting the gravity loads, that is, bearing walls supporting the upper floors and roof, is the foundation system; however, the remainder of the post-tensioned slab is simply a floor.

MYTH: Post-tensioning a concrete slab on-grade will eliminate all cracking.
FACT: Post-tensioned slabs on-grade will have fewer cracks, and those cracks will be smaller (narrower); however, post-tensioning will not guarantee that no

cracks will occur. Minor hairline cracking in a post-tensioned slab is not an indication of a problem and is not considered a construction defect.

MYTH: Post-tensioned concrete slabs on-grade will prevent movement due to shrink/swell characteristics exhibited by expansive soils.

FACT: The post-tensioned concrete slabs on grade will not prevent movement due to shrink/swell movement of expansive soils. The purpose of the post-tensioning is to hold the slab together when movement does occur and to reduce the abruptness of the movement. The effect of heaving and deflections within the superstructure *may be reduced slightly* due to the distribution of the dead (weight of the structure) load; however, this benefit will be limited, particularly in areas of the country that have large deposits of montmorillonite (class of highly expansive) clays.

MYTH: A crack in a concrete slab on-grade or foundation stem wall is an indication of some type of soil movement or a structural problem.

FACT: First, the concrete slab on-grade is not a structural element. Second, a crack does not necessarily indicate soil settlement or heaving. Unless the crack is significantly large (wide) and each side of the crack shows signs of differential movement (up, down, sideways, or a change in slope), most likely the crack is the result of normal drying shrinkage and/or thermal stress. Most cracks found in concrete stem walls are caused by these stresses and are relatively uniform in width. Cracks with a significant variation in width may indicate differential movement.

MYTH: Any crack in a concrete slab on-grade is a defect.
FACT: All concrete is subject to drying, shrinkage, and cracking. This is a normal part of concrete and is not necessarily a defect; however, by adding excessive amounts of water to the mix prior to placement, excessive and enlarged cracking may occur and require remedial work. This type of condition is considered a defect.

MYTH: Concrete slab floors are perfectly level and flat; therefore, variations in the levelness of the floor reflect a structural problem, foundation movement, or other such maladies.

FACT: In normal concrete construction, it is virtually impossible to place, finish, and cure a concrete slab on-grade that is either perfectly level or flat. Tolerances for levelness and flatness are published by the American Concrete Institute and recognized throughout the building industry. Various jurisdictions will also have minimum standards adopted that specify the allowable tolerances for flatness of concrete slabs on-grade. Generally, the tolerance for deviation from level for a slab on-grade is 1 inch from one side of a slab to the other. Studies indicate the average deviation from level for residential construction is approximately 1.25 inches. The flatness tolerance for floors is approximately $\frac{1}{4}$ to $\frac{5}{16}$ inches departure from a 10-foot straightedge placed randomly on the floor according to ACI 117.

MYTH: Moisture accumulation on a concrete slab on-grade indicates defective concrete.

FACT: All concrete and soils contain moisture. Moisture flowing in and out of concrete is a normal process called breathing. Moisture problems develop when, and because, coatings, membranes, or floor coverings are installed on or under concrete that disrupt the normal flow of moisture. The concrete is not the problem. It is the application of an impervious membrane that alters the normal process. There are steps that can be taken to decrease the permeability and minimize the amount of water vapor passing in and out of the concrete. Simply using concrete with a lower water/cement ratio will increase the density of the paste and reduce permeability; however, reducing the water/cement ratio too much may increase shrinkage, cracking, and/or curling as a side effect in the overall slab performance.

It is important to understand that any change made to the concrete mix, placement, finishing, or curing procedures will have side effects. There are no exceptions to this statement. In many cases, the resulting side effect is more serious than the initial problem being considered.

- There is a risk of a moisture problem every time an impervious membrane or moisture-sensitive floor covering is installed on a concrete slab on-grade, which would include installation of composition tiles, wood flooring, seamless vinyl, epoxy, urethane, acrylic, porcelain tile, and rubber-backed carpet.
- Carpet, unfired tile (ceramic), and natural stone generally breathe, making them relatively compatible coverings for a concrete slab on-grade; however, placing an impervious covering on top of these materials, such as plastic runners, chair mats, or flat-bottomed furniture, may very well result in condensation and water accumulation.
- The preceding examples are not the result of defective concrete, but the result of placing an incompatible material or system on the concrete slab on-grade.

MYTH: Concrete spalling and deterioration (such as sulfate attack) are the result of poorly mixed concrete.
FACT: Sulfates of various concentrations are found in different materials, including soil. If suspected to be

present in soil in certain locales, sulfates should be identified during the soil investigation process. If the sulfate content of the soils exceeds an established threshold listed in the International Building Code, special requirements are specified for the concrete that will be in contact with the soil to minimize the risk of sulfate attack.

- It should be noted that all cement, and thus all concrete, contains small amounts of sulfates. About 2 percent gypsum (calcium sulfate) is added to the raw feed in the grinding process to control flash set of the cement. Therefore, identifying the presence of sulfate in concrete does not by itself indicate sulfate attack.
- Sophisticated testing methods can quantify the amount of sulfate present in concrete in place, and its specific form. Observed deterioration may actually be due to "salt crystallization," which is a physical attack as opposed to a chemical attack.
- Severity of the problem and remedial action depend on which condition caused it. The problem may be the result of existing soil conditions and has nothing to do with the concrete mix. It should be noted that some fertilizers and other landscape materials contain sulfates and when placed next to stem walls can cause deterioration.

MYTH: Cracks in brittle floor coverings, such as ceramic tile, are an indication of defective concrete, a structural problem, or a foundation problem.
  FACT: Most cracks in brittle floor coverings placed over a concrete slab on-grade can be attributed to a number of factors, the least of which is related to defective concrete and/or a structural problem:

- Most cracks in brittle floor covering installed on concrete slabs-on-grade are reflections of cracks or joints in the slab. Normal thermal expansion and contraction cracks cause lateral movement of the slab and opening and closing of the joint or crack. Due to the different coefficients of thermal expansion, a reflective crack may develop.
- Cracks or control joints in the concrete slab on-grade should be covered with a quality grade slip-sheet to minimize reflective cracking. Proper installation of this material allows the slab to move independently from the floor covering. Reflective cracking in brittle floor covering installed on concrete slabs is, for the most part, an installation problem as opposed to a concrete problem.
- When brittle floor covering is to be installed over a concrete slab on-grade that has been placed upon moderate shrink/swell soils, the procedure of covering the area with sheet vinyl prior to installing the finish floor covering will normally avoid much of the cracking that may develop due to soil movement.

- In the Southwest, Saltillo tile is quite popular; however, this tile is generally cured in the sun and prone to becoming concave or bowl like. Improperly installed, it can result in a series of hollow spots under the tile, which may fail under foot traffic or other loads.

MYTH: Adding water to concrete at the job site to make it easier to "pour" will not affect the strength or quality of the concrete.
  FACT: Adding quantities of water (above the design water/cement ratio) to a load of concrete will reduce the quality, reduce the strength, and adversely affect all of the properties of the concrete. Excessive water in the concrete mix increases shrinkage, cracking, and curling. The water/cement ratio controls all of the properties of concrete. In addition to the strength, the water/cement ratio controls the permeability, wear resistance, resistance to freeze thaw, and resistance to chemical attack. Because excess water is detrimental, concrete should be "placed" rather than "poured."

MYTH: Allowing concrete to remain in the truck for extended periods of time will not affect the quality as long as mixing continues.
  FACT: Leaving concrete in a truck for an extended period will reduce the quality and affect all of the properties of the concrete, particularly in hot, dry weather. Hydration of the cement begins immediately after adding the water at the batch plant. ASTM C 94 requires the concrete to be discharged from the truck within 90 minutes after the water is added at the batch plant. In extremely hot, dry weather, even 90 minutes may be too long.

4.3 FRAMING MYTHS

MYTH: Using galvanized steel connection plates ensures against rust.
  FACT: Just because metal connector plates for wood trusses are hot dip galvanized doesn't mean they won't rust. The galvanizing provides protection against rust due to occasional exposure to moisture; however, where chronic exposure to moisture exists, such as roof framing directly below roof drains, the sacrificial coating is more quickly used up, exposing the now-unprotected bare steel.

MYTH: Using framing hangers has greatly improved the quality and ease with which a carpenter can frame a structure. If they fit the size of the wood member, they must be okay.
  FACT: Hangers are manufactured in many sizes and capacities. Capacity is the key word. A framer should not use a hanger that is on hand just because it fits the member. The hanger has to provide adequate support for the carried member and

FIGURE 4.1 Galvanized Steel Does Not Rust?

FIGURE 4.3 Hangers Are Adjustable?

FIGURE 4.2 If the Hanger Fits, It Must Be OK?

FIGURE 4.4 The First Truss Can Support All the Overframing?

adequate transfer of load to the supporting member to avoid connection failure.

MYTH: Framing hangers can be easily adjusted to fit most conditions.
FACT: Light-gauge steel hangers are manufactured in various widths, lengths, and seat slopes and with various capacities. Bending or otherwise modifying a hanger, although easy to do, is not an acceptable alternative to using the correct hanger. Bent hangers are significantly weakened at the bends and cause excessive connection movement when they straighten back out under load. The hanger bearing seat has to properly fit the carried member to provide adequate bearing area and avoid hanger deformation, damage to the carried member, or possible failure.

MYTH: The first truss can support all of the overframing.
FACT: On wood truss roof systems with wings, the wing roof extends back and connects back to the main roof trusses. These planes are often conventionally framed. It is not acceptable to attach the intermediate and ridge supports for the overframing rafters to the first common truss for the wing. This truss was designed for 2-foot on center loading only. The overframing loads have to be distributed over the roof trusses below. It is also necessary to laterally brace the top cords of the truss below the overframing if the roof sheathing was not installed.

4.4 MASONRY MYTHS

MYTH: Grout joint failure and voids are an indication of structural problems in masonry walls.
FACT: Voids and cracks in mortar joints of masonry do not indicate failure or a structural problem. Such conditions will normally point to poor workmanship on the part of the mason who may have added too much water to the mortar mix. If the wall has been properly

reinforced and grouted, simple tuck-pointing will resolve the problem.

MYTH: All cracks in masonry walls indicate either foundation failure or a structural defect.

FACT: Hairline cracks in mortar joints in masonry are usually caused by shrinkage and thermal stress; they are normal and do not indicate either failure or a structural problem. Small cracks may also develop at the corners of window and door openings. An experienced investigator can differentiate between types of cracks and determine the cause. Minor tuck-pointing may be required to fill the cracks. Paint will generally cover the hairline cracks.

MYTH: Panelized masonry block yard fences do not need foundations.

FACT: Panelized fences do require a foundation. The panelized fence consists of 8″ by 16″ pilasters spaced about 13 feet on centers with 4-inch thick block units in between. The pilasters are reinforced vertically and anchored into a small foundation, normally augered into the soil with a posthole auger. The panel blocks fit into a vertical channel in the pilaster blocks. The panels have joint reinforcing in the horizontal mortar joints. The vertical joints in the panel blocks are not mortared. They simply have a small tongue and groove on the vertical edge. Contractors often place a mortar or grout layer only in a shallow trench to provide a level base for the panel blocks in lieu of a proper foundation, as shown on the approved standard details.

Because of this lack of foundation, the panel is subject to settlement, particularly if the yard is heavily watered. This will show up as step cracks in the infill panels. If the cracks open to the point of being objectionable, the panel should be taken down and a proper foundation installed and relaid to eliminate the cracks.

MYTH: Cracks in stucco applied over a panelized block fence indicate failure.

FACT: Reflective cracks will develop in stucco over both the horizontal and vertical joints in all masonry. Normal thermal movement of the block system causes reflective cracking. Generally, applying stucco to a masonry fence is not recommended. Flaking of stucco on the panel blocks where they fit into the pilasters is also normal because of normal thermal movement. These are not defects and are simply the nature of the system. Of course, if the panelized fence was not properly constructed (see the previous myth), there will be cracking in the stucco relating to that movement. But, cracking *per se* does not prove faulty construction of the wall.

4.5 STUCCO MYTHS

MYTH: A crack on stucco indicates a failure or defective material.

FACT: Stucco is made with Portland cement, which means it is subject to normal drying shrinkage. The majority of cracks in stucco are caused by drying shrinkage and/or restrained thermal movement. Failure to protect the stucco from premature drying during the first few hours immediately after finishing will significantly increase the number of cracks. Although cracks are unsightly, they generally do not adversely affect the long-term performance of the stucco. Larger cracks may be filled with an elastomeric caulking prior to painting. Paint should bridge smaller cracks. The cracks may reappear over time, which is an indication that it is time to repaint the wall.

MYTH: The mix proportions and sand content of stucco are critical.

FACT: Manufacturers and suppliers of stucco materials provide mix proportions for their materials. Generally, these proportions are given as so many pounds of sand to a bag (usually 80 pounds) of the stucco material. In the real world, the stucco crew will add a certain number of shovels of sand per bag of the mix in the mixer. The workers should gage the number of shovels to the prescribed weight in the manufacturer's directions, but it will not be exact. Although different manufacturers usually bag similar products, their recommendations for mix proportions (cementitious/sand ratios) may vary significantly. In effect, this shows that the quality of the stucco is not sensitive to the sand content. In fact, higher sand contents will theoretically reduce shrinkage. Experience has shown that if the stucco is relatively hard and is not easily scratched with a sharp instrument, it will probably perform well in the long term.

MYTH: Stucco is waterproof.

FACT: Although stucco sheds most of the water from rain or other sources, it is not "waterproof." It is recognized that water can penetrate the stucco. In some cases, water can condense behind the stucco. A moisture barrier should be placed behind the stucco with a weep at its base to prevent the water from entering the building. When properly installed, this moisture barrier is effective.

MYTH: Expansion joints and weep screeds are not required for the "one-coat" stucco system.

FACT: All stuccos are cementitious materials that crack from shrinkage. Galvanized or fiber reinforcing mesh will distribute the cracks and limit the crack widths to some extent, but cracks will occur. A rough texture finish will reduce the visibility of cracks.

Conversely, a smooth finish will increase the visibility of cracks. Additionally, cracks will be transmitted at the joints of improperly fitted rigid expanded polymer panels (such as Styrofoam blocks).

In most applications, control joints are not used in the one-coat stucco system. However, it must be understood that random cracks will develop as a result of omitting the joints. As noted, stucco is not waterproof and, therefore, weep screeds are required to allow water, from any source, that collects behind the stucco to escape. Expansion joints should be used to separate stucco from other materials such as masonry or changes in framing such as shear walls.

Evaluation of the stucco finish should be made from a distance of at least 10 feet.

MYTH: Cracks in stucco are an indication of settlement or heaving due to expansive soil and are a structural problem.
FACT: As noted in previous myths/facts, most cracks in stucco are caused by normal drying shrinkage and/or restrained thermal movement and do not indicate abnormal movement of the structure, but settlement will also cause cracks. It is important to be able to tell which is which. Cracks caused by abnormal movement of the structure will have very different patterns and characteristics. An experienced investigator can recognize the different crack types and identify their sources.

MYTH: Using an elastomeric coating over one-coat stucco systems will result in a crack-free finish.
FACT: While an elastomeric coating may hide cracks in stucco, applying it over any cement-based material may create much more serious problems, including blistering, water bubbles, and delamination of the elastomeric coating. As previously noted, all concrete-based materials should be allowed to breathe, allowing water vapor to flow in and out on the material.

Unfortunately, "elastomeric" a descriptive term which is now applied to almost anything that has even a minimum elasticity, be it ordinary house paint or caulking; therefore, one must keep in mind there is a wide variation between elastomerics and the many products on the market with some elastomeric characteristics.

As a result of the number of coating materials on the market that can be applied over severely cracked one-coat stucco systems, the cause of the cracking must be determined and mitigated, and a thorough power washing of the exterior face of the stucco must be conducted before proceeding with any such remediation. If the system has been poorly installed and lacks proper bonding, or there is trapped moisture behind the stucco finish, applying an elastomeric coating will only compound the problem, as the latent moisture will cause blistering. The ideal application of elastomerics occurs when leaking or water migration from the exterior needs to be stopped on substrates that don't need venting out or are vented by other means. It is therefore recommended that a qualified expert in synthetic stucco systems be brought in to evaluate the condition before rushing to a "quick fix" solution to crack issues.

MYTH: Stucco is too alkaline and causes paint to fail.
FACT: Stucco, as all Portland cement–based products, has a high pH, which means it is highly alkaline. Calcium hydroxide is a byproduct of cement hydration. Calcium hydroxide makes up approximately 20 percent of the cement paste, even after hydration is essentially complete. The normal pH of concrete is approximately 12.5. The high pH inhibits corrosion of embedded steel and is therefore beneficial. The calcium hydroxide near the surface will combine with carbon dioxide in the air to form calcium carbonate, which is essentially neutral (pH near 7). However, if moisture from any source migrates through the stucco, it will carry calcium hydroxide back to the surface and restore the high pH. If the paint is not compatible with the high pH, a suitable primer should be applied first.

MYTH: Expansion joints and weep screeds are not required when applying a one-coat stucco system, due to its elasticity and impermeability.
FACT: One-coat stucco systems do have excellent elasticity, but only to a point. The product is still primarily a cementitious material and will, therefore, breath, expand, and contract, particularly in climates with extreme temperature swings. These characteristics will lead to uncontrollable cracking if control joints are not considered in certain configurations. Long walls that are subjected to extreme temperature swings should be broken up into 10- to 12-foot modules separated with an expansion joint to limit, as well as control, cracking. Any wall area that has different materials, such as part masonry and part wood frame, should also be separated with an expansion joint to handle the difference in thermal movement between materials.

One-coat stucco systems are not impermeable. Being a cementitious product, they will take on water in a similar manner as any other concrete-based product. By not providing an avenue for trapped moisture to escape, such as via the weep screed, pressure builds up, forcing the moisture back through the exterior surface or into the substrate, resulting in additional damage.

4.6 ROOFING MYTHS

MYTH: Tile roofs are lifetime roofs, requiring little or no maintenance.
FACT: A tile roof is only as good as the underlayment. For many years, tile manufacturers recommended No. 15 or No. 30 roofing felts for underlayment. Such felts

in a desert climate, where surface temperatures under the tile can exceed 150 degrees Fahrenheit, didn't last a "lifetime," and many were seen to fail in five to eight years. Underlayment has now become a specialty product that is much more durable. The new underlayment membrane materials are synthetic, many with polymer reinforcing and an extended warranty.

MYTH: Full-length standing seam metal roofing can't leak.
FACT: Standing seam roofing has full-length joining with other identical shapes as well as joining metal at both ends. In addition, there are penetrations for pipes, ducts, skylights, and so on. Metal-to-metal connections are always difficult to make and need a skilled worker to make these connections, who can be difficult to come by. Because of these concerns, it is not uncommon to find leaks in a standing seam roof. It is, therefore, recommended to forestall metal roof leaks that are both difficult and expensive to repair by the use of a good peal and seal membrane to seal the substrate under the metal roofing.

MYTH: Ell metal flashing is a proper transition from balcony deck waterproofing under the slab to top of concrete.
FACT: The ell metal has problems because of its inability to fill and seal corners and joints and between the lengths of the metal strips. Waterproofing materials become affected by the thermal cycling of the metal and the sheet metal deterioration due to rust, thus resulting in a leak. Most waterproofing system manufacturers show their proprietary materials filling the corner and turning up to the top of slab for continuity and for successful long-term operation.

4.7 ROADS & PAVING MYTHS

MYTH: If the asphalt pavement mix used for freeways, highways, and major arterials will withstand high volumes of heavy trucks, it should provide super performance when used for parking lots, driveways, and residential streets where only light traffic exists.
FACT: In order to withstand heavy traffic, asphalt mixes are modified by decreasing the asphalt content and increasing air voids to resist rutting and shoving, both of which reduce the pavement's durability. If these modifications are used for light-traffic pavement, where these measures are unnecessary, a less durable pavement with increased cracking and raveling will result. The high loading is actually necessary to make the modified asphalt perform properly.

MYTH: A 1-inch reduction in thickness for a flexible pavement design of 2 inches of asphalt concrete (AC) over 6 inches of aggregate base (AB) should not significantly affect the finished pavement's performance under traffic.
FACT: If the entire 1 inch of deficient thickness is in the asphalt concrete layer (1″ AC and 6″ AB), the number of axle loads to failure for the pavement is reduced approximately 80 percent. If the entire 1 inch of deficient thickness is in the aggregate base (2″ AC and 5″ AB), the number of axle loads to failure is reduced by nearly 50 percent. Combinations of deficiencies in both asphalt concrete and aggregate base totaling 1 inch would result in loss of pavement performance between those percentages.

MYTH: This recently placed pavement is in good condition. There is no need to do anything for it while it is in this condition.
FACT: Placement of a seal coat while a pavement surface is still in a good condition will protect the pavement surface and assist in maintaining that condition while the effects of weather and traffic will be absorbed by the seal coat. The seal coat will need to be replaced every few years as it deteriorates in order to maintain protection for the pavement surface below it. If the pavement surface proper is allowed to deteriorate, restoration of that surface will be difficult and will generally exceed the cost for applying and maintaining a seal coat.

MYTH: Seal coats applied to asphalt pavements are simply a paint job that temporarily improves the pavement's appearance but really does nothing to enhance pavement performance.
FACT: Pavement seal coats are often referred to as sacrificial surfaces that deteriorate due to traffic and weather while protecting the underlying pavement from these effects. Seal coats should be applied before the pavement has undergone significant surface deterioration to maintain the pavement surface. Seal coats need to be reapplied after they have deteriorated to a point where the pavement surface is subjected to traffic and weather conditions. They generally should be reapplied on a 3- or 4-year cycle. Squeegee-applied sealer that includes a finely graded mineral aggregate, such as ground slate or slag, will have a longer life than a spray-applied liquid bituminous sealer. Heavier duty seal coats, such as slurry seal, microseal, or chip seal, apply a significantly greater thickness and have a considerably longer life than the lighter seal coats. Cape seal, which is a combination of chip seal, often utilizing a rubber-modified binder, and slurry seal actually exceeds the intent of a seal coat and enhances the structural property of the pavement, in addition to providing surface protection.

4.8 GENERAL MYTHS

MYTH: The successful long-term performance of a residential structure (and other lightly loaded commercial structures) is "assured" as long as everything is designed and built to the local building code requirements and has been inspected by the governing municipality.

FACT: The code has been developed as a minimum code that is applicable around the world (hence the name International Residential or Commercial Building Code). Stop and think about this: How can an international code cover every type and kind of soil characteristic, material, and building practice for every local site situation? The code is a minimum and does not address long-term structural performance when it comes to soil/building interactions. The building code establishes minimum standards for safety but does not ensure quality construction.

Add minimum inspection by municipal personnel with limited knowledge and the end result may well be limited performance of the structure.

MYTH: Value engineering provides the client equal value for a reduced price.

FACT: Unfortunately, this much-used term applied to reducing the price of a project does reduce the price but rarely provides equal value in the way of quality or function. Examples of some value-engineering issues that are frequently used to reduce the cost of a project and reduce the quality, durability, and so on follow:

- One-coat stucco systems substituted for a specified exterior insulation finish system (EIFS).
- Reduce the weight of the roofing felts under tile roofs because the code does not require a heavier felt.
- Installing damproofing materials when waterproofing materials are needed.
- Reducing concrete strength from 3500 psi to 2500 psi in a slab on-grade because it's approved by the code.
- Cutting quality assurance testing of soils in mass fill because it disrupts the workflow.
- A three-ply built-up roofing system in lieu of a four-ply system.

MYTH: Termites will destroy the structural integrity of my house.

FACT: While termites do eat wood, they like the visible wood inside the house (trim, baseboards, hardwood floors, etc.) more than they like the structural timber hidden in the walls. When they do eat the structural lumber, they usually do not damage enough at one location to reduce the effectiveness of the structural members but continue on a line. It usually takes 10 or more years of infestation to cause possible structural damage. And then it will usually be isolated and can be corrected when found.

Therefore, when you discover termites, do not panic but have a reputable contractor treat them immediately and continue the treatment. The chemicals today are not as effective as the banned chlordane; thus, constant vigilance is usually necessary once discovered.

New treatment systems are becoming effective but only a reputable contractor can be trusted to properly apply the correct concentrations of the needed chemical. Also keep in mind that proper application of the right amounts of chemical is a health issue. Placing the chemicals in the wrong places, such as into below-grade air returns, can create a serious health risk for the home occupants.

MYTH: Our plans have been checked by the local building department; therefore, they are correct.

FACT: The plan review fees, although high in many cases, do not begin to cover the cost of a thorough review. One only has to look at the most commonly used code, the International Building Code, to understand how complex the requirements are to design a building. Add to that the zoning code, the fire code, the electrical code, civil engineering, and many others to understand why any single governmental agency cannot give a thorough review of all these sections for the fees paid.

SECTION FIVE

ARCHITECTURAL ISSUES & CONSTRUCTION DEFECTS

or

Is It Always a Contractor Responsibility?

5.0 JUMP-STARTING SECTION 5

In reviewing the architect's role, duties, and responsibilities, a few points need to be kept in mind while working through this section. Points that will help a reader fully understand the issues presented include the following:

1. These days, the architect is often not given ample authority within the construction command structure. If you are the owner, you need to understand who has control of which issues so as to assert your influence on the proper party. It is also unrealistic to expect the architect to accept liability without having effective control of the entire project.

2. If, as the owner, you wish to have construction liabilities clearly defined, you need to understand and influence the drafting of the construction contracts. You must also be willing to pay reasonable rates to design professionals when they take on appropriate responsibilities.

3. The importance of a good construction contract cannot be overstated. A good contract makes the responsibilities of all involved parties clear, is fair to all (and, therefore, enforceable), defines the objectives both in concept and structure, and is itself legal and enforceable. A competent attorney with significant experience in construction contract law can be an asset, as it is very easy, even with all good intentions, to write a contract that turns out to be counter to existing law.

4. As in so many fields, computers offer almost unimaginable potential to assist the design professional, from simply holding vast amounts of data and engineering drawings to allowing product information access on the Internet that can greatly increase search efficiency. However, the same potential can be used to cover up laziness and lack of logical thought. In competent hands, any tool is an improvement. In irresponsible hands, many bad things can happen. Mindless copying of specification sections or contract clauses is a curse.

5. Be very wary of the term "Value Engineering." The intended purpose is a complete review of the building plans by competent engineers and/or architects to search out costly procedures or materials that are not optimal.

This process is intended to result in a better project for less money. The process may also be used to highlight special needs, such as handicap accessibility, by providing input from professionals familiar with both the special needs and also the availability of suitable equipment and the application of fast-changing codes. However, in the worst-case example, the process simply inserts the cheapest components available, resulting in reduced cost but at the expense of function, quality, and long-term reliability.

The architect, in times past, was the "master builder," responsible for a project from its initial conception to its ultimate completion. The architect not only developed the ambience of the structure, but also the total environment surrounding the project, including even the foundations upon which the structure was constructed. Interestingly enough, structures even thousands of years old still stand today throughout the world as evidence of the brilliance of their day. However, those ancient structures did not have electrical, mechanical, or air-conditioning systems; elevators; and other amenities. Over the years, structures became taller, bridges became longer, and the complexities of the environment (i.e., wind loads, seismic reactions, geotechnical consideration, innovations in buildings systems and materials, and human factors) became more important in the development of monumental projects, requiring more in-depth scientific analysis. Enter the civil, structural, electrical, and mechanical engineers, who aided the architect in responding to those more complex issues. As time moved on, and population increased, and so did contributing factors that would influence the project, including governmental regulations covering how and where the project could be built, how tall it could be, what amenities it must have, what color it should be, and what type of vegetation could be planted around it. As such projects grew in size and complexity, the budgeting of available dollars became a primary issue, determining whether or not a project would actually be built. To help control cost, a new concept of "value engineering" became the buzzword. Cost control brought the advent of construction management, whereby the architect is selected and reports to a construction manager. As if these issues didn't complicate the architect's role enough on commercial work, enter the developer/home owners association whose representatives on multi-unit residential projects may not have a clue as to what impact their demands have on project design.

55

Needless to say, with all of this "help," the architect's role in and responsibilities to a project *should* have become limited and thus somewhat easier. On the contrary, not only did the architect's role and responsibilities become more difficult, but the chance of becoming involved in lengthy and expensive litigation also increased tremendously. The liability for design rests with the architect; therefore, the architect must be totally cognizant of any outside issue that may influence and/or affect the design. A commissioned architect must always remember that, as the leader, success is not only based on the project's design, but also on its ultimate cost and performance; therefore, if the architect's leadership of the project is somewhat warped, the entire project may not run well for all disciplines. In other words, the architect as "leader" is responsible for the sub-consultants and in-house disciplines.

The case law of Arizona and of a few other states allows that all persons on a construction project who have reasonably relied on the architect's documents may sue the architect if there are defects in the design (*Donnelly Construction vs. Oberg Hunt & Gilliland,* 139 Ariz. 184,677 P.2d 1292 [1984]). This principle was first stated in the 1958 decision in *United States vs. Los Angeles Testing Laboratory vs. Rogers & Rogers,* 161 F. Supp. 132, 136 (S.D. Cal. 1958). In this case, the general contractor sued the architect to recover costs incurred when the concrete structure being constructed was rejected. The contractor claimed the test reports provided by the architect were negligently reviewed throughout the course of the work, thus allowing the structure to be completed with under-strength concrete. The architect argued that having no direct contract with the general contractor, no duty was owed to the contractor, and, therefore, there was no liability for negligently performing the work. The court held that a duty independent of contract could be owed to third parties who rely upon the architect's performance.

Based on these two decisions, every architect needs to be aware that lack of contractual privity (Privity is defined as a mutual or successive relationships to the same right of property, or such an identification of interest of one person to another as to represent the same legal right.) does not necessarily bar an action to recover economic loss should the architect be found negligent. See Lordan "Arizona's Economic Loss Rule", March 2011, Arizona Attorney Magazine, page 17 and pages 23–24. While many states have some form of protection for third parties when the architect is negligent, the rule in most states is that the only persons who can sue the design professionals for design errors are those who are in privity of contract with the design professional (i.e., have a direct contract with the design professional). Because of the considerable number of construction negligence cases taking place throughout the United States, contractors are arguing that it is valid and reasonable to transfer liability to the architect because the contractor is compelled to perform the work per plans and specifications prepared by the architect. The argument continues that the contractor's personnel who perform the work do not have the knowledge or expertise to determine or challenge the design requirements and must, therefore, rely upon those plans and specifications to be accurate. This debate (as to who is responsible for the effects of a design feature's nonperformance) will continue to be a costly risk adventure for all parties involved, when it would have been much simpler to "do it right" the first time.

In view of the significant increase in litigation against professionals taking place throughout the country, this section will present a number of situations that have resulted in finding the architect negligent in the preparation of the design, which includes the preparation of plans and specifications. It should be realized that, in most jurisdictions, the architect has a duty to perform to the accepted standards of peers in the same geographic area and under the same circumstances. Failure to do so is commonly known as "professional malpractice." A failure to conform to the prevailing standards of care violates a general duty and is considered a tort, and, in some tort cases, the plaintiff may recover punitive damages for an egregious failure to perform according to the standards of care.

5.1 EARLY PROBLEMS

Many times, one gets wrapped up in developing a "memorable" design that depicts the architect's specialty, a design that is different from the work of peers. After all, a distinctive design is what most notable architects strive for and are commissioned to provide. This is all well and good, if the client has commissioned a unique design with an unlimited budget. It is essential for professional survival that the architect determines the client's ability to pay, then set a budget that is complete and accurate.

A good amount of background information on any prospective client is available through the Internet. One can investigate real estate holdings, the existence and ownership of partnerships and corporations, the finances of those entities, and (via various search engines) extensive information on the activities of the owner. The failures by architects to set adequate budgets for projects have diminished the profession's standing in the eyes of the owners and the construction industry as a whole. The increase in construction manager at risk (CMR) and design-build contracts reflects the growing desire by owners, and especially governmental agencies, to obtain adequate budgets at that initiation of a project.

The American Institute of Architects (AIA) has numerous contract agreement forms, such as the B141, that define the development and control of project budgets. It is important to carefully evaluate these AIA documents, as many have undergone major modification from the 1997 version. In the absence of AIA, AGC, or other similar standard contract agreements used today, agreements may be prepared by the client and/or the client's legal counsel.

Risk Management Is More Than Contract Language

Whenever possible the architect should utilize a contract agreement that is familiar, one that has specific guidelines clearly spelling out the professional duties and budgeted amounts, as the design moves forward. Knowing the financial limitations of the project up front is essential to the architect's relationship with and knowledge of the client. Whenever possible, and prior to their first contact, the architect would be wise to do some homework on the prospective client. Is the client legally stable and financially sound? What type of reputation does the client have in the community where the project is to be constructed? If a known entity, does the client have a reputation for constructing a quality project or one of lesser quality and performance to save on cost? Knowing these and other influencing factors that are going to affect the cost of the work is extremely important when preparing to meet with a prospective client.

It is essential that the architect maintain detailed notes of the first meeting with the prospective client. During that initial meeting, the architect must not only determine the risks of doing business with the client, but also the client's needs, wants, wishes, and budget constraints, as related to the project. It will be most essential in the overall evaluation process to know exactly where the project will be constructed. The project's location can, in and of itself, be the determining factor as to whether or not the architect can fulfill the client's needs, wants, and wishes and meet the budget constraints. Knowing the location of the project up front can save a world of embarrassment later. In today's business climate, it is not uncommon to spend a large portion of or the entire project construction budget in just obtaining governmental approval to build on a client's proposed site. How many times have budgets been agreed to, only to discover (before the schematic design phase of the work is completed) that the site approval costs exceed the cost of design? Additionally, it should be noted that a number of architects are prone to consume so much of the design fee in creating an aesthetic project that too little is left for consultants or their own in-house disciplines in terms of time and dollars for good comprehensive detailing. To properly manage risk, the architect must be cognizant that failing to provide a set of comprehensive contract documents can lead to costly litigation, not only from the client, but also from all parties who can be, or are, affected by the design end product. It is, therefore, essential for today's architect to not only become educated in risk management techniques, such as project budgeting and scheduling, but also to put them into use on a daily basis. By implementing a good risk management, budgeting, and scheduling system, the architect can better assess the goals set forth in the agreement with the owner, which leads to a more coordinated work effort. The importance of coordinating the work (client to architect and architect to the design team) cannot be overemphasized.

5.2 PROFESSIONAL COMMUNICATION & COORDINATION

It is normally the responsibility of both the architect and the client in the development of commercial and governmental projects to retain other professionals with the necessary competence and qualifications to assist in developing the overall design of the project. These other professionals and their respective responsibilities may include any or all of the following.

Client Responsibility

- Geotechnical engineering report and recommendations covering, but not limited to
  - Soil conditions and allowable bearing pressures
  - Migrant and on-site water conditions.
  - Foundation recommendations
- **Client Responsibility**
  - Land surveyors responsible for field measuring and verifying the boundaries:
    - Discover, locate, and report possible encroachments
    - Topographical mapping of the project.

Architect Responsibility

- Civil engineering design calculations, including preparation of plans, specifications, and details covering
  - Building location and property lines
  - Grading and drainage requirements of the project
  - Primary underground utilities
- **Architect Responsibility**
  - Structural engineering design calculations including preparation of plans, specifications, and details covering
    - Wind, seismic, and lateral forces acting on the structure
    - Live and dead loads affecting the structure
    - Foundations
    - Structural framing

- **Architect Responsibility**
  - Mechanical Engineering design calculations including preparation of plans, specifications and details covering:
    - Plumbing
    - Heating, ventilating and air conditioning (HVAC)
    - Boilers, steam fitting and process piping.
    - Mechanical processing, conveying, and manufacturing equipment.
- **Architect Responsibility**
  - Electrical engineering design calculations, including preparation of plans, specifications, and details covering
    - Primary power and service requirements
    - Electrical power distribution and controls
    - Decorative, accent, and emergency lighting
    - Lightning and ground fault protection
- **Architect Responsibility**
  - Specialty professionals may include
    - Interior decorators
    - Kitchen designers
    - Acoustical engineers
    - Hydrologists
    - Landscape architects.
    - Other specialists

In the classic configuration, once a team of support professionals (sub-consultants) is retained, the architect becomes the manager of "the team" and takes on the responsibility of coordinating the efforts to bring about a complete set of documents that meets the contracted design intent and applicable national, state, and local building codes. Once the design team is formed, it is the duty of the architect as "team leader" to see to the efficiency of the effort to be performed. If project efficiency is to be attained and/or improved, it is essential to maximize standardization. As an example, all of the design professionals should ensure that details and notes on the plans are structured in a similar manner.

As noted above, it is not uncommon for geotechnical, land surveying, and at times civil engineering professionals to be engaged by the client rather than the architect. When this occurs, it is essential that an open line of communication between these professionals and the architect and sub-consultants be created and maintained throughout the course of the work. Contracts for services between the architect and sub-consultants, such as the AIA C141, should clearly indicate the scope of work, allotted time, and fees. Documents such as these can be viewed on the AIA website, www.aia.org/docs_default.

As the work of one professional may affect the work of other design professionals, proper coordination is mandatory and must be well documented through meeting minutes and/or memoranda.

The increasing use of computers to do design work is creating a major problem in the industry. Designers are working on a 19- or 21-inch monitor and are unable to see "the context" in which they are working. Many of today's professional design offices utilize small cubicles that don't permit the professional to spread out the hard copy of the plans and thereby appreciate the context of the work This requirement of communication and coordination relates not only to contacts among the sub-consultant disciplines, but it also requires design professionals working within the same office to relate to the work of their team members.

Sub-consultants must be given the opportunity and responsibility to review the final design and specifications to assure themselves and the architect that their design elements and recommendations have been correctly incorporated into the final documents. All such reviews should be memorialized in writing.

Professionals are often retained directly by the client. It is essential that the agreement between the architect and the client clearly define what role these professionals will play and who will take on the responsibility of ensuring that the work product of the client's professionals will be properly integrated into the completed design, as well as the liability for any irregularity that may develop from the use of the work product.

The architect must be assured that the sub-consultants will continue to be involved with the project as it develops and reaches fruition. Design changes, client change orders, differing site conditions, weather, and other developments during construction must be evaluated for their effect on the original design, the project, and the project's performance. Such activities and related decisions must be well documented during construction for reference when questions regarding construction and performance are raised, in order that they are handled correctly and do not become claims.

The architect may attempt to limit the client's options regarding the bringing of claims by the owner by using AIA Document B-141, Standard Form of Agreement between Owner and Architect, or B151, Abbreviated Standard Form of Agreement between Owner and Architect. These documents include provisions noting that claims must be filed by the time of substantial completion, if known. Unfortunately most client claims are filed well after the time of substantial completion, due to an alleged lack of knowledge about the existence of the issues surrounding the claim. In order to better address the risk, an architect may consider inserting a waiver of client claims, except where it is determined that the cost, loss, or damage is solely caused by the gross negligence or willful misconduct of the architect. In many cases, such waivers have been found invalid by the courts. Such waivers should be reviewed with counsel before being considered.

5.3 ENVIRONMENTAL DURABILITY & MAINTENANCE

The architect and sub-consultants must provide for project performance in their design and that performance must take account of the environmental conditions at the project site.

The proposed project performance expectations must be developed in accordance with the severity of the surrounding environment so as to ensure lasting durability. This study must take into consideration natural, physical, and human environmental elements that will affect the project's performance. A good example of the failure to consider the environment is the mechanical engineer who designed a humidity control system for coastal conditions on an inland project that was at an elevation of over 7000 feet.

It is essential that these factors are determined prior to design and then the appropriate means of addressing these environmental elements must be incorporated into the design. The performance expectations must be incorporated into the specifications and maintenance requirement manuals before they are turned over to the client at completion of the work. This is extremely important when dealing with the design of multifamily units, such as condominiums, or any type of commercial facility in which people reside or work, for example, hotels or office buildings. In today's litigious society, it is not uncommon for a condominium home owners association or an office building's owner to seek damages for not being informed about how to address environmental issues, when in reality, those issues stem from improper maintenance on the part of the home owners association or the owner's lack of establishing a proper maintenance program.

The Importance of Maintenance Manuals

A thorough understanding of any required preventive maintenance program should be provided to the client together with the consequences of the lack of such maintenance. Generally, the development of maintenance manuals is left to the discretion of the developer/contractor and includes a package of technical data provided by others with no real thought as to completeness or the ramifications of not providing explicit instructions. The architect needs to be sure that the data contained therein properly address the entire building envelope, along with the HVAC system and appliances. Normally, these maintenance factors are covered in technical documents prepared by the manufacturer of a product or system and become a part of the operation manuals supplied to the client at the completion of the work. Unfortunately, this is not true with many products used in the design; often particular design elements require clarification to avoid future problems. Examples of such issues include the following:

- Chlorination and regular cleaning of condensate and storm water drain lines to prevent algae buildup and blockages resulting in severe water intrusion into the home or business.
- The ongoing need for surface cleaning and replacement of caulked joints, particularly in hot desert or extremely cold climates where caulking materials break down quickly, resulting in water intrusion and numerous other related problems.

It is important for the architect's risk management program to addresses the proper preparation and transfer of maintenance instructions to be used by the client, property owner, or governing body responsible for project maintenance after substantial completion. The transfer documents should include the architect's waiver of his or her responsibility should the client, property owner, or governing body fail to maintain and/or correct maintenance requirements as recommended in the maintenance manuals.

During the design development phase, the architect must look ahead to a project's maintenance program by taking the natural, physical, and human factors into consideration when addressing environmental durability during the design.

Natural Factors

Natural conditions likely to occur should be defined and incorporated into the design. These may include wind, seismic activity, water and water intrusion, elevation, humidity, freezing and thawing, extreme heat or cold, chemical attack, expansive or collapsing soils, and fire resistance. These elements are often project specific and must be defined by the architect and sub-consultants for the project and its site or location. These natural site conditions must be uniformly and consistently addressed and appropriate solutions applied by all parties involved during design and construction. It is not enough to provide for natural factors in one's design; the provisions for maintenance of the constructed design must ensure that the design once "built out" will be maintained and protected.

A classic example of the inattention to environmental factors is the use of "canned specifications" that were not modified for a specific project. In developing notes and specifications for a project that was built in north-central Arizona, the specifier failed to consider the environment. The elevation at the site ranged from about 2900 to 3200 feet above sea level, which is considered a relatively mild climate. The notes on the plan called for 2500 psi concrete. Air-entrainment was not specified. The plan also called for exposed aggregate for the walks and exterior stair treads. Figure 5.1 demonstrates what can transpire in less than three years when the person creating the specs does not understand the freeze/thaw ramifications for low-strength concrete.

Physical Factors

Physical factors incorporated into the design will include live load and dead load, and how these loads are applied, along with consideration of natural factors as previously noted. Strengths of materials to be incorporated into the design must be well defined and consistent with project specifications and the facility's intended operations. Appropriate factors of safety should be discussed by members of the team. Inclusion of factors of safety by one party involved in design development without consideration for factors of

FIGURE 5.1 Photos a and b reflect mild climate freeze/thaw exposed aggregate 2500 psi concrete deterioration.

safety utilized by other parties will generally result in an undefined final factor of safety that no one can evaluate. Mechanistic and probabilistic procedures have replaced factor of safety designs in some instances. This is particularly true when dealing with machinery that has been scrutinized under various testing procedures to determine the action and reaction at points of failure. This information is then made available to others for incorporation into structures that will house and/or support such machinery. In so doing, all elements of the design can be evaluated for both value and variability before being incorporated. This procedure will have the benefit of providing an overall value for the probable success of the design and a definition of the risk of failure to those sub-consultants charged with the responsibility of developing the engineering and design.

Human Factors

Human factors most often result in the greatest number of complaints being filed against the architect. Too often the architect, in the preparation of the design documents, will brush off human factors by stating that the work will be constructed in accordance with local building codes and "in compliance with Americans with Disabilities Act (ADA)." There is no question that many human factors have been addressed in various building codes and spelled out in ADA documents; however, numerous influential factors in every project affect ergonomics, from the chairs we sit in, to the floors we walk on, to the light we work by and the air we breathe. The architect and owner must be thoroughly familiar with these factors and assure themselves that the design and, most important, the specifications address the human factors as they relate to the commission in question. Leaving these issues to chance opens the door to costly litigation and the possibility of losing one's professional license. This is particularly true in public works, which allow federal government as well as state and local intervention. In view of the importance that governmental officials have placed on human factors, it is essential that the architect become thoroughly familiar with the operations and use of the client's project and their effect on the client's employees who will be working there and on any outsider who may come onto or into the project. An individual home, like an office building, will require an understanding of human factors such as the rise and run of stairs, handrail placement, ramp locations, ingress/egress in case of fire, and so on.

5.4 LACK OF KNOWLEDGE ON SPECIFIED PRODUCTS

A lack of knowledge of specified products often finds specific expression on the drawings and in the specifications. On architectural drawings, we often see examples of materials misapplied, incompatible materials mixed together, and application methodology not properly detailed (or spelled out in the specifications). This lack of knowledge has become more frequent with the advent of computer technology. Architects once developed hand-drawn details and formulated text on a product's performance. This process permitted a better understanding of how the product worked and how it related to other parts of the overall project design. With the advent of computers and the Internet, the entire thought process has changed. Today a technician can request information for a project through the Internet. After reviewing the sales pitch and nominal product data, it is all too easy to download product information and details directly from the Internet and insert them into the projects design documents, in a fraction of the time it took to do the job manually. This results in no regard for how the imported information may impact the overall project performance. This, coupled with "standard specifications" makes it almost a "slam dunk" to rapidly develop a set of construction documents, drawing upon the multitude of construction products now on the market. Does this streamlined process make for a better and more complete set of construction documents? Not necessarily. The computer-age adage of "Garbage in—Garbage Out" is quite true. Therefore, it is essential that the architect, owner, and sub-consultants have a thorough grasp on how smart and familiar this new generation of computer-literate technicians is with the products they are incorporating into the project's design. The architect must judge whether what is being produced is compatible with the overall design intent.

The architect and sub-consultant team must have a thorough knowledge and understanding of the requirements and performance of any specified product and how the product must be applied and/or installed, in order to maintain its warranty. However, it is also essential to be "cost conscious" and allow provisions in the specifications for alternates, provided they meet the required performance needs of the project. When value engineering is provided by the contractor, it must involve the architect and the sub-consultants to ensure that any such changes do not affect the integrity and performance of the finished product.

5.5 SPECIFICATIONS

The use of specifications has long been one of the primary responsibilities of the architect in developing a complete set of contract documents for the project. Unfortunately, for many years, specifications seem to have been relegated to a secondary role with many architectural firms. Much of this is due to the advent of canned specifications, such as those developed by the Construction Specifications Institute, (CSI), manufacturers' printed specifications, building codes, American Society of Testing Materials (ASTM) referenced standards, and so on. There is no question that these excellent resources have contributed greatly, helping the architect with the time it takes to prepare a set of specifications. The primary problem rests with the assumption that some of these very good organizations are developing clear and precise specifications and that it is no longer necessary for the architect to be concerned with "having to reinvent the wheel," so to speak. Nothing could be further from the truth. Specifications are the primary source of information regarding the construction of the finished product, ordinarily taking precedence over the architect's design details and plans. The legal system will always reference the specifications to determine whether the architect has provided the client and the building contractor the correct information from which to build. In the case of an accident, injury, or other misfortune resulting from a latent defect in the construction, the courts will ask whether the specifications may have caused the incident. With this in mind, the architect needs to reevaluate how to approach the specifications on each project.

Specifications General

The Construction Specification Institute (CSI) merits mention at this point for its years of effort at bringing a semblance of order to specifications and for fostering and encouraging excellence in specification practices. All facets of construction thrive on uniformity and accuracy, and CSI has been dedicated to those objectives. This does not preclude the need for the architect to be wary of the use of standard specifications, whether developed through CSI, building code associations, trade association standards, manufacturer specifications, or other sources. These form specifications are written to cover all conceivable elements that may appear on any project. However, each project is site specific and only a few of the requirements given within such form specifications may be applicable on the project for which they are being referenced. It may be very difficult to determine if a particular specification has application when its terms cover conditions different from those relevant to the project site. The architect needs to first be totally familiar with the site conditions and the overall design concept, including relevant details, before picking out an appropriate specification from a packaged set of specifications.

Specifications "Boilerplate"

In addition to the specification sections that describe the products and processes, which combine with the plans and details to form a building to be constructed, assorted other documents need to be addressed. These documents are known in most circles as the "boilerplate" of the contract documents and cover everything from pre-bid instructions, general conditions, preconstruction meetings, progress payments, and insurance to written provisions for on-site toilet facilities. Here again, wise choices must be made by the architect in cooperation with the client to ensure that all projects requirements, apart from materials and installation, are comprehensibly covered, to avoid being challenged should litigation occur. It is not uncommon to find that many owners and/or developers have created their own boilerplate, based on past experience. Irrespective of who prepares the boilerplate, the architect must carefully examine the documents as to applicability to the project and, most important, from the liability standpoint.

Standard Specifications

In most instances, when one refers to standard specifications, one immediately envisions those provided by manufacturers and published as a part of their promotional literature. Such information typically provides a general and overall description of the product or process and some installation or application information to assist the specification writer. Standard specifications that have been developed and published for application to a wide variety of projects may be effectively used. However, bidding and contract documents must be clear on which portions of the standard specifications apply to the particular project and any amendments to those standard specifications that need to be incorporated into the project. Standard specifications that are properly incorporated into the project documents may be very effective, and their use may be preferred or even required by some clients. The architect's specification writer must be knowledgeable enough to sort out the extraneous and include only the information that applies to the current project, in order to achieve the desired results. It is not unusual to find that the manufacturer's literature frequently requires additional and more elaborate information from the specification writer in order to properly mate the product or process with the particular project. The importance of the specification writer's understanding of the project cannot be overemphasized.

Specifications addressing a wide range of project conditions that are stored in a computer as a word-processing document under "standard specifications" must be used with caution. These documents are often inserted into project specifications as if they were project specific. They often contain irrelevant requirements or address features that do not appear on the project. In some instances, they may be totally incorrect for the site or the particular design being developed. This type of document must be carefully edited by a senior-level professional to remove irrelevant parts and to ensure that those parts incorporated into the project do have direct application to the project. Specifications "borrowed" from another project should be carefully reviewed for correct application for the project being developed.

Master Specifications

Whether or not the architect and the sub-consultants have evolved their own, purchased, or subscribed to a specification system that is meant to "cover everything," such a document is considered a "master specification." The idea, of course, is to have a comprehensive descriptive document that can be easily edited and, it is hoped, is accurate and correct. Like most other human endeavors, such a master specification is only as good as the knowledge of the individual assigned to the task of producing the project specifications or, as more commonly known, the project manual.

Insufficient Specification Detail

Too often, much is left to the contractors' and subcontractors' imaginations because the architect hasn't provided the needed and sometimes critical information in the form of an adequate set of specifications and/or details. If the selection of a product, detail, and/or assembly is left to the devices of a contractor, designers often are presented with something other than what they had in mind. And what the owner gets may be improperly done, incomplete, or simply defective because the installer did it wrong. Poor documents often result in the parties on the job having to rely on RFIs (requests for information), which are processed during construction and often used as a claim for delaying the job.

That is not meant to say that the architect is the source of all construction problems, just a fair share. Jobs on fast track and/or with small design budgets impact the architect's ability to prepare complete and accurate documents, thereby resulting in more RFIs; therefore, the importance of having complete and accurate specifications cannot be overemphasized. Too often, production of the specifications is delegated to a junior drafter or the project manager. Office staff size may not allow a full-time specifications specialist who can bring to bear knowledge, history, curiosity, and pride that result in a quality project manual that supports and enhances the other documents. If, however, compiling the specifications simply becomes an unpleasant task for one member of the design team, the end result may be chaos, numerous questions throughout the construction, added cost to the owner in the form of change orders, delay claims, and embarrassment and even censure for the firm's principals.

5.6 COVER SHEETS

Cover sheets or title sheets are normally the first sheet on a set of construction drawings or the project manual. In some jurisdictions, the information required on the cover sheet is dictated by the governmental agency that has the approval responsibility over the design. This is particularly true on public works or other such projects in which public funds are used. The cover sheet will generally provide the following information:

- The name and location of the project, accompanied by a vicinity map pinpointing the main crossroads and north/south coordinates.

- General property legal description, zoning, occupancy, and use of the project to be constructed.

- In a publicly funded project, it's not unusual to see the names of the governor/mayor, council members/commissioners, and possibly the contractor who was awarded the work listed on the cover sheet.

- The local building codes in force at the time the contract documents were approved for construction, including any special inspections required or governmental issues that need to be addressed during the course of construction.

- The names, addresses, and phone numbers of the architect and engineers responsible for the preparation of the contract documents as well as those who may have been retained by the owner, that is, the geotechnical and civil engineers.

- A list of applicable drawings by classification.

5.7 GENERAL NOTES

General notes are normally included as part of the civil, structural, plumbing, mechanical, and electrical plans. Architectural general notes may be included on custom designs but, for the most part, all of the architectural data are incorporated into the project manual or the specifications sheet. Notes on the contract drawings are an integral part of the drawings themselves. These notes should also be considered a part of the contract specifications, as they provide information that cannot be advantageously described in the drawings but, when used in conjunction with the line drawings, properly define the work. All notes on the drawings must be clear, informative, and relate directly to the work for which the contract documents were created. Too often, the same notes are used for more than one project and found to have no specific relevance to the work as shown on the contract plans. General notes will normally apply to many things found throughout the contract drawings, while notes that are applicable to a specific detail or issue that may be shown on one sheet of the contract drawings should be placed as close as possible to the detail or issue to which the note applies. Notes on the contract drawings are intended to be just as effective as if they were a part of the specification. It is essential that the architect review the general notes of all these sub-consultants to ensure there is not a conflict between the requirements defined by the sub-consultant and the applicable specification section.

This question of the precedence of one document over another document ought to be treated by the contract clauses relating to the project. It is sometimes thought that the printed word (e.g., in specifications) takes precedence over items shown on a plan, but such is not reflected in the AIA General Conditions. Owners often prefer to have a document precedence defined to eliminate conflict. A careful professional will not leave this important question without a specific contract clause dealing with document precedence and will spell it out in unquestionably clear language. Of course, the importance of this question of document precedence only serves to accent the importance of not using canned general notes together with canned specifications from different jobs, resulting in documents that are then inconsistent and in conflict.

Depending upon the size and complexity of the project, the use of general notes may be good or bad. Too many projects include too many general notes. On a simple project, general notes are often adequate and may serve as a substitute for the project manual.

A good project manual should include everything that would be in the general notes, and, thus, there will be no need for general notes. The detailed information in the project manual should be read and absorbed but, if there are general notes, the project manual often is not read enough or at all. Too many times, one goes on a job only to find that the project manual is not even on the job site and has not been read!

General notes have their place. The question is: When are such notes a valuable and needed adjunct to the project manual?

General notes should be consistent with the project manual, which provides more detailed information. If not properly coordinated, the inclusion of notes on the drawings in addition to the project manual information may create confusion. The better practice is to list items only once; however, in the event that information is listed twice, it must be consistent and correct in both locations.

One of the serious problems in the construction industry today is that the specifications are often written by a specification writer on a consulting basis, instead of by the architect of record. The consultant writing the specs may have little or no familiarity with the specific site, or the job in general. Engineering notes may well serve as an attempt by the engineer to cover certain required issues, because of an assumption that there may be little proper thought given to what the architect's spec writer may include in the project manual.

Often an engineer's notes become a way of superseding the project manual, with the expectation that the notes will become the basic reference for engineering concepts and the project manual will be ignored. When discussing this subject, most engineers become defensive because they have been burned on projects where the project manual, when finally published, did not properly address issues required by the engineer. Therefore, many engineers may insist on their general notes being left on the drawings, regardless of what the architect may or may not include in the project manual. It is useful for the engineering notes to be placed on the drawings so the spec writer is informed of the issues that the engineer wants to see included in the project manual when the manual is finally produced.

Regrettably, the engineers' insistence on having notes on the drawings helps to fragment the project, while the architect's failure to properly include the information in the engineer's notes in the project manual may leave the engineer's requirements unsupported. This has been an ongoing issue that needs a coordinated effort at resolution, followed by coordinated inclusion of information.

5.8 CODE REFERENCES

Code references must be current, accurate, and applicable. Code references are too often ignored but are brought to the forefront when problems and/or disputes erupt on the project. More often than not, the listed code references are obsolete or ignored, because the contractors and subcontractors don't normally have copies on their bookshelves or the copies they have are out of date and never used.

5.9 DETAILS

The contract drawings must be prepared by the architect in a clear and concise manner. This is essential, as the contract drawings must provide the contractor with enough information to properly bid on and construct the project. While some of the contract drawings show the overall design in plan, elevation, and section, the key to a successful set of contract drawings is the clarity and accuracy of the details. Excellent judgment on the part of the architect is required to determine what to show and how to show it. Apart from specifications, the architect's details define how the intricate parts of the project will be constructed. Seldom will the contract drawings alone be sufficient to provide all the information needed by the contractor, so the contract drawings are expected to be supplemented by sub-contractors' shop drawings or other detailed drawings. These drawings, together with the details on the contract drawings, provide a precise image to the contractor of everything needed to execute the work. This section is presented to show some issues surrounding architectural details, using actual examples and discussing how they affected the work.

Waterproofing Details

Waterproofing details remain one of the major issues in construction defect litigation and need to be properly addressed by the architect's design staff. See Figures 5.2 to 5.4.

The architect assumed that notations calling for "hardrock concrete with waterproof system over deck sheathing" would protect the buildings. Although the contract specifications called for a quality waterproofing material, the manufacturer's specifications did not address what happens when nails or staples penetrate the waterproof barrier. Nor did either specification address how to prevent storm water runoff from entering the ceiling system via the wood-framed columns or rough sawn beams. It is essential that the architect thoroughly understands how runoff water travels over the surface of a structure, and that the details, specifications, and/or the material manufacturer's specifications will properly protect the substrate from water intrusion and thus microbial growth. To do less opens the door for having to "pay the price."

Flashing Details

Roofing and flashing details, like waterproofing details, are essential in the design process and most often overlooked, leaving it to the contractor or the material supplier to come up with the details of how best to do the work.

Windows & Flashing Details

Figures 5.5 to 5.10 show the importance of providing accurate details that address not only the type of window being installed but also how it is to be installed to meet recognized industry standards and to ensure that the water tightness and structural integrity of the building is maintained. In this case, the health and welfare of the students had to be taken into consideration, due to the heavy microbial growth inside the substrate, requiring that the rooms affected be tented by industrial hygenests to prevent microbial spores from entering the classrooms.

The flashing installed over the sill block was installed per the architect's Detail 13, which did not correspond to window flashing requirements recognized by the following agencies:

- American Architectural Manufacturers Association—
Storm-Driven Rain Penetration of Windows and Doors
- California Association of Window Manufacturers—
Standard Practice for Installation of Windows with Integral Mounting Flange in Wood Frame Construction

By preparing a detail with no reference to applicable standards, the architect is the only designated standard. If the architect requires that the work be performed by applicable standards, then the contractor has a responsibility to request clarification of the architect's detail via an RFI.

The preceding illustrations demonstrate how insufficient details can lead to major water intrusion issues. The contractor, for the most part, followed the details; the architect approved the work; and the client in an effort to stop the water from entering the structure sealed the weep

FIGURE 5.2 Detail of trellis alignment at framed column shows wood beam resting on framed wood column. There is no indication of how the substrate of the framed column will be protected from the elements. The contractor constructed the assembly exactly as shown in this detail (a); the results are noted in photo (b) showing how the water ran down through the column into the deck sheathing and framing, creating serious microbial growth.

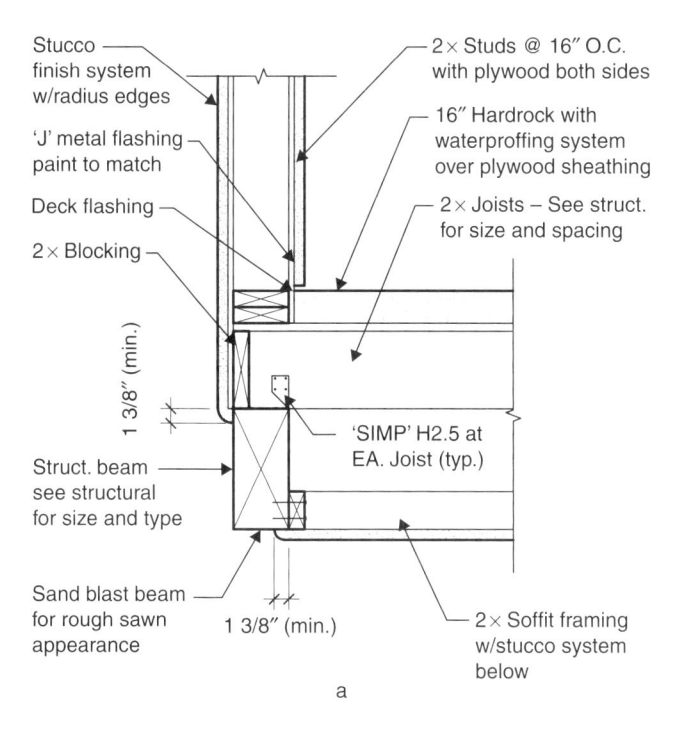

Architectural Issues & Construction Defects 65

FIGURE 5.3 Detail a of rough sawn header supporting patio deck. Note that there is no drip to stop storm water runoff from attacking the stucco ceiling below or is there any indication of how to vent the dead air space to prevent microbial growth. In addition what is meant by "16" Hardwork with waterproofing system over plywood sheathing. The contractor constructed the assembly exactly as shown in detail (a); the results are noted in the adjacent photos (b,c and d).

screed compounding the problem. Both architect and contractor had to "pay the price" for not "doing it right."

5.10 DO'S IN THE PRACTICE OF ARCHITECTURE

For the most part, the do's in the practice of architecture and engineering parallel the don'ts, as both are striving to attain the same goal: a well-designed project that meets or exceeds the expectations of the client and should result in bigger and more challenging commissions.

Provide enough fees in your contract to be able to perform and monitor the following "to do list":

1. Do monitor the efforts of the design team to ensure a well-coordinated project. This includes providing information to the owner's separately retained consultants (e.g., project geotechnical engineer) as the project develops.

2. Do understand the interrelationship of the different disciplines before starting design.

3. Do obtain, as early as possible, full program information. Be sure to consider the time required for plan review.

4. Do emphasize that the aesthetic design of the project should be completed by the time the schematic design phase is concluded.

5. Do document all of the input provided during the schematic phase.

6. Do create construction details considering the whole context of the project, not just the one detail. Recognize

66 SECTION FIVE

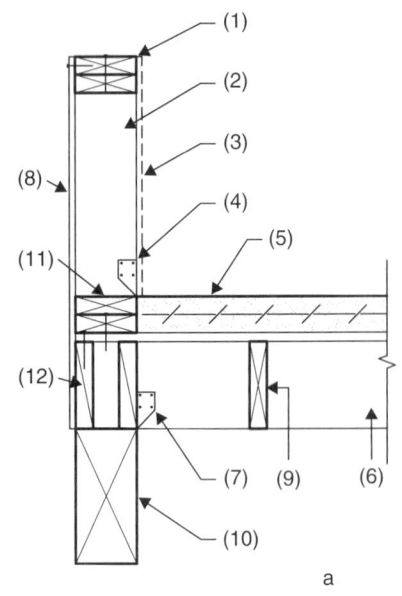

NOTES:

1. Double 2 × top plate w/16d at 12″ O.C.
2. Wood post
3. Sheathing material as occurs
4. Simpson H6 at 32″ O.C.
5. Topping over plywood shearhing
6. Wood joist at 215.
7. Simpson H3 at each joist
8. Plywood sheathing
9. 215A – wood joist runs parallel to wood beam.
10. Wood beam.
11. 2 – 2 × bottom plate w/16d at 12″ O.C.
12. 2 – 2 × Rim joists w/ 3 – 16d nails per rim joist.

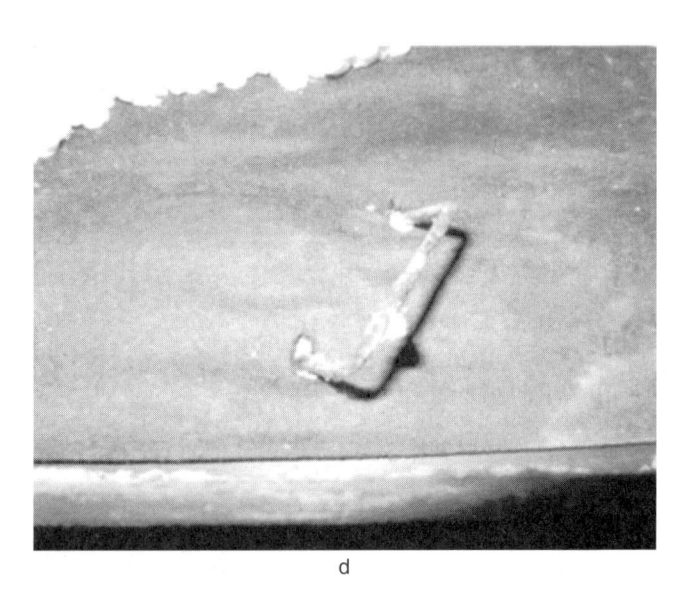

FIGURE 5.4 The structural detail (a) that shows the framing as depicted in Figure 5.3 shows the hard rock concrete defined as "Topping" but fails to show the flashing or expansion joint requirements that should have been addressed on the architectural detail in Figure 5.3. Not only are both details lacking in proper information they both fail to properly address how the flashing should have been attached to the framing and exterior finish. This lack of proper detailing resulted in photo b mold forming on the underside of the deck, c nailing the expansion joint allowing water to enter the substrate via the nails and d via the staples used to secure the waterproofing.

that computer screens limit the amount of detail that can be considered at any one time.

7. Do allow an appropriate internal and external coordination review before the construction documents are released for construction.
8. Do stay on top of the many different agencies performing the plan review and approval.
9. Do use experienced people for contract administration instead of the most junior person in the office.
10. Do make the effort to properly and timely review shop drawings.
11. Do coordinate with the contractor on the importance and timing of any RFIs submitted to the office.
12. Do consider all relevant disciplines when reviewing proposed change orders.

FIGURE 5.5 Typical water intrusion at window sill found in a number of classrooms throughout the school.

Architectural Issues & Construction Defects 67

FIGURE 5.6 Typical water intrusion at sill and the inside face of the exterior wall sheathing.

Notice that the bituminous flashing is located on the sill of the window and does not wrap up the jamb side of the framing. The sill metal flashing is just laying flat on the sill block and allowing the water to track into the substrate between the sill metal flashing and bituminous flashing material. There is no turned up edge to prevent or direct water to the outside of the building.

FIGURE 5.8 Note all of the weeps have been sealed by the owner to try and stop the water intrusion into the substrate.

FIGURE 5.9 Typical exterior flashing at sill; note how the flashing slopes back to the window sill, not away from the window. The CMU wainscot does not agree with the detail and slope down and away from the window.

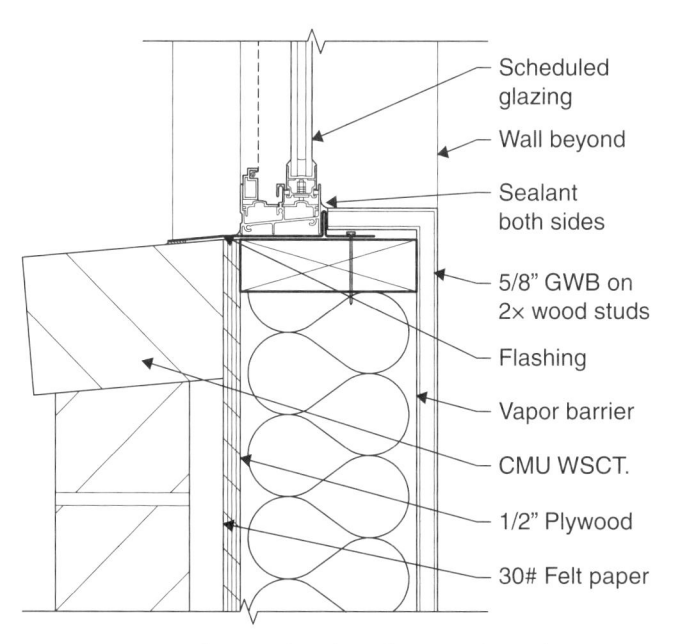

FIGURE 5.7 Detail 13 on the plans does not show the proper flashing for this type of assembly.

5.11 DON'TS IN THE PRACTICE OF ARCHITECTURE

1. Don't be isolated from the rest of the design team.
2. Don't let the code minimums control the design.
3. Don't let outside salespeople control product selection or dictate specifications.
4. Don't use canned documents without adjusting them to project-specific conditions.
5. Don't ignore the client's input.
6. Don't hire the cheapest consultants that you can find. They are often the most expensive.
7. Don't allow the client or yourself to change the project after the design development drawings without providing an appropriate amount of time and fee to coordinate the ramifications of the changes.
8. Don't combine and condense the classic schematic and design development phases. This is often done to satisfy an owner's unrealistic timeline.
9. Don't provide a cost estimate without a proper supporting basis. Do not fail to revise this estimate when the design or

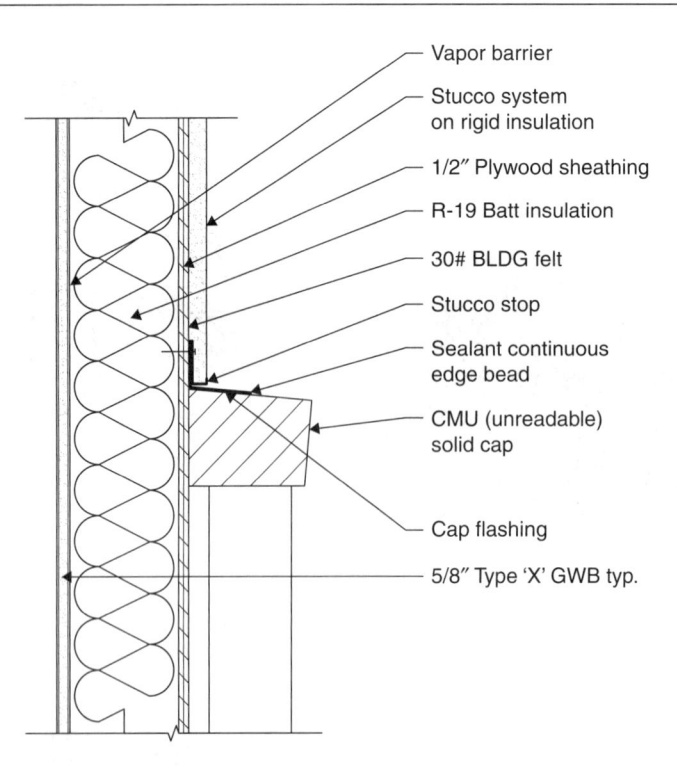

FIGURE 5.10 Detail 2 on the plans shows both the stucco stop/weep screed and the cap flashing over a sloped masonry cap; however, the weather resistant barrier is shown under the flashing and not shingled over flashings, thus allowing any trapped moisture to be directed into the substrate instead of out through the weeps. Additionally, there should have been a minimum of $1/2$-inch clearance between the weep and the sill flashing.

conditions change. Consider using the services of a professional cost estimator.

10. Don't rely on either governmental review or inspection processes, or the contractor, as your quality assurance program.

11. Don't ignore the turnaround time and thorough review of shop drawings or RFIs.

12. Don't let personality conflicts create problems on a job site.

13. Don't expect the contractor to interpret your intent.

SECTION SIX

ENGINEERING & CONSTRUCTION DEFECTS

or

Engineers Don't Make Mistakes

6.0 JUMP-STARTING SECTION 6

Engineers, you can't decide whether to love them or hate them. They talk a language all their own. To translate what they're saying, one needs to understand the importance of their role, duties, and responsibilities in the entire process. In reading this section, the following points will help one better understand what's expected of the engineering profession:

1. Prior to accepting any design assignment, the engineer must first ask the question: Am I capable and do I have sufficient knowledge and experience to take on this assignment?
2. The engineer needs to ask the owner and the architect all the relevant questions to understand the project *and its context* before commencing work.
3. The best design calculations in the world are rendered meaningless if they are improperly transferred to the drawings. Plan checking cannot be left to the most junior person in the office.
4. Reviewing shop drawings and submittals in a record short time and leaving that review to the most junior person in the office will lead to problems.
5. Site visits during construction need not address only errors. Good communications make good friends, good friends make a good team, and a good team can avoid litigation.

As shelters became more complex and the ability to construct greater things such as boats, roads, and bridges increased, the elementary physics evolved into the mechanics of what makes things happen, including the interaction of the various natural materials with one another. Engineering emerged through the application of science and mathematics, by which the properties of matter and sources of energy are made useful. Engineering evolved from the need to build to meet humanity's needs.

As time passed and new innovations developed, the need to broaden knowledge eventually brought us the three basic fields of engineering: civil, mechanical, and mining. From these three basic fields sprang numerous specialties, including agricultural, electrical, chemical, structural, aeronautical, nuclear, and forensic, to name a few. Engineering, in and of itself, is primarily the application of various criteria to known facts to implement an end result. Although this may sound simple, it is often quite complex, particularly if the engineer does not have sufficient knowledge of all the facts before commencing with the assignment. This section is devoted to what can and does happen when the engineer fails to address all of the issues that are needed to arrive at a final conclusion. The rest of this section reflects some of the more critical issues facing a practicing engineer developing the design for any specific assignment.

6.1 COMPANY PHILOSOPHY

A "mission statement" should reflect a firm's philosophy. However, it can be either ill conceived or structured as a marketing tool that sounds good but does not define the principles of quality practice. Even a good mission statement is too often ignored in the rush to "get the work out." The impact of a well-thought-out and implemented mission statement cannot be overemphasized. It sets the tone for the day-to-day operations of the firm, and these operations will ultimately determine if the firm is successful.

Business marketing decisions should not trump professional engineering obligations. For the owner of an engineering firm, the tone that is set by the business will affect the philosophy of the employed engineers—for their entire career. Establishing the philosophy of an engineering firm is a great responsibility that requires careful consideration and thoughtful action, not "greed, speed, and the need to please."

Engineering firms need to balance the requirements of good engineering practice with the successful operation of a business. The two requirements sometimes will clash. The firm must have sufficient work to keep everyone constructively busy, but not so much that a panic situation is created every time a project completion date approaches. Panic increases the probability of errors. If engineers are routinely so busy that the "panic" is allowed to become the norm, the chance of errors multiplies, because the mindset of the firm's employees changes to an attitude of "just get it out the door." There must be a balance between the desire to make a profit and the capabilities of the firm. Every firm is composed of finders, minders, binders, grinders, and flag wavers. The difficult part of the profession is keeping those forces in balance.

Engineers are constantly looking for a marketing edge that will separate them from the competition. One marketing phrase is that "we can do it faster" (speed). This is attractive to many clients. If their project can be designed in a shorter period of time, it can be generating income sooner. Unlike the promise of a "better" design, which is hard for many owners to visualize, accelerated cash flow is a real and measurable advantage. However, too fast a schedule compromises the engineering design process and may result in an overly conservative design or an under-design. The amount of time spent by the design team conceiving the system is often insufficient. A typical example is a 60,000-square-foot structural steel building constructed with 80 tons of excess steel, simply because the engineer in a rush initially chose a non-economical building system.

Most engineers have a natural desire to please their clients and their design teams. We all want to be appreciated for our work. Engineers can sometimes be asked, rather forcefully, to do things that go against good engineering practice. This request can come from the owner (There must be a cheaper way!), contractor (I've done lots of these projects and you are the first to require this!), or another member of the same design team (I need those beams shallower!). Reasonable requests should be evaluated with reference to good engineering principles, and unreasonable requests should be forcefully dismissed. Sometimes you have to know when to put your foot down. The "need to please" should never override sound engineering judgment.

6.2 INSUFFICIENT INITIAL PROJECT INFORMATION

Responding to the Request for Proposal (RFP)

A design firm will generally be contacted either for a proposal for technical services or to provide professional design services. Based upon our experience with a long list of failed engineering projects, we can say it is a known fact that many design professionals will enter into a technical design contract with less than the minimum required information. The reason for this can be attributed to any one or a combination of a number of factors:

- The design of the project may not be developed sufficiently for preparation of a reasonable proposal. For example, the structures to be placed on the site and their interaction with the site may not be well understood at the proposal phase.
- Various team members, including the owner and the architect, in an effort to keep consultants' costs low, may oversimplify the stated project requirements.
- The engineer or firm
  - May have failed to ask enough questions about the project.
  - May not have personnel with the ability to see the overall assignment and to determine the information needed to properly perform the engineering effort.
  - May not know all of the design team members and/or their responsibilities in the overall project. Even if the team is fully assembled, communication among the various team members at this early stage is likely to be minimal. This leads to omissions or misunderstandings that often plague the design.
  - May not even know what the project site or neighborhood looks like.

It is common for engineers to want to avoid giving the impression that they could be "uninformed" or viewed as asking "too many" and "too obvious" questions. Often, they are too anxious to get started and "solve problems." This is particularly true in the presence of a potential new client. Therefore, they do not ask enough questions to fully define and understand the project design concepts and restrictions. *Not asking these initial questions can, and often does, result in a construction defect or a failed project. The engineers who failed to ask necessary questions will look foolish when they testify in court and have to answer that, "NO," they "did not think to ask" these questions.* So **ASK THE QUESTIONS UP FRONT!!**

Preparing the Scope of Services

The scope of services sometimes does not match the scope of the project. This can occur due to an error in understanding the request for proposal or to the increasing market demand to lower professional service fees. Competition is sometimes so intense that some firms feel the need to reduce the scope of services in order to submit a competitive fee. When the scope of services does not match the effort required in the project, compromises are likely to occur that will negatively affect the project.

For example, in geotechnical services, there are at least four ways to cut the total fees that can negatively affect the project:

- Skimping on the number and depth of test borings
- Skimping on the amount of field sampling of soils or sampling of soils at intervals that will have no influence on the performance of the structure.
- Skimping on the number of laboratory tests that should be done.
- Assigning lower salaried/experienced staff to the project and not putting much, if any, senior-level staff and fees in the proposal for project review and oversight.

Typically for soils work, the main lapse in proper scope is that the boring depths are not deep enough to cover the foundation types, loads, and resultant stresses. We have seen instances where the soils engineer did not know that the first floor level would be 10 feet below the highest original ground surface. The result was that the limited sampling that occurred was in the portion of the soil that was cut away. No deep samples were obtained at the anticipated first-floor level and, therefore, adequate and appropriate soil design parameters were not developed. Another example occurred when the borings hit refusal on a thin rock layer over highly expansive clay, and, therefore, the highly expansive soil characteristics were not provided for in the design.

In structural engineering services, similar methods have been improperly used to cut the total fees for a project, for example:

- Skimping on the number of building sections and details drawn to illustrate the structure.
- Providing overly conservative, generic designs, which results in a structure that is expensive to construct.
- Designing for the worst condition and thoughtlessly using those components throughout the project, thus increasing the construction cost.
- Relying on standard details (drawn for other similar projects) to cover the typical conditions and not drawing details to cover atypical conditions.
- Over-relying on the architectural drawings to illustrate the structural components.
- Assigning the design of building components to the contractor. On various projects, we have seen the design of the structure for the roof, floor, walls, and foundations (including the layout of members) relegated to the contractor. This reduces the design and coordination time required of the engineer significantly but increases the potential for problems.
- Shortcutting the construction phase. Shop drawings are sometimes "reviewed" in record time, and the engineer who designed the project may never "see" the structure. Shop drawings are a last chance to effectively review the design prior to construction, yet this task is often left to the most inexperienced staff member.

All these compromises do not have an immediate effect on the designer or his or her firm. Often, similar shortcuts are being made by the other engineering disciplines on a project. Each of these compromises lowers the information available for everyone involved in the project and increases the chance of error, and thus the risk of lawsuits. Compromises in preparation of the construction documents can haunt the project through construction and often linger long after construction is complete. Compromises in the preparation of the construction documents negatively affect the engineer's reputation, may result in discipline by the board of registration, and often result in litigation or injury. With lawsuits and the reaction by its insurers, the engineering firm might eventually dissolve.

Establishing the Fee

There is pricing pressure in the industry. For the past 20 years or so, the name of the "professional services and design" has been to "MINIMIZE" up-front analysis of issues that will affect the design. The owner/builder/contractor/client wants to minimize the up-front design services costs because they have to pay for these costs out of their own pockets. They want to complete design and construction in the shortest time possible so that the project can begin to generate income. They also want to maximize the project design, including the appearance and functionality.

This creates a "no win" situation for these design professionals who "choose" to

- Minimize their fees.
- Minimize their total hours spent in design.
- Minimize their participation and communication with the other design professionals.
- Minimize or eliminate their services during the pre-construction and construction of the project.

For these minimum fees, the owner/builder/contractor/client expects the biggest, best, and most cost-effective and efficient project that will perform flawlessly for the full life of the project. This is an oxymoronic set of expectations since minimal design and construction do not equate to maximum benefits and long-term performance. One must ask themselves the following questions when negotiating fees with a prospective client who is more interested in squashing down dollars for a proper design in favor of one that is just enough to get a building permit.

- Does the client want the firm to work to industry design standards at less than prevailing fees? Is the client looking to cut design costs?
- If so, is the firm willing to take the risk of turning out a less than industry standard design in order to win the project?
- Is the firm willing to do less than a professional design in order to satisfy the client's desire to pay less?
- If so, then the firm should seriously consider the long-term consequences, costs, and liabilities associated with taking on this low-budget design project.

Current strategy on low-budget design projects is to utilize the least trained and lowest paid staff in order to bring the project in under the budgeted fee. The firm's principal intends to thoroughly review the project prior to completion of design by the junior firm members but what usually happens is that

- The junior engineers don't complete the project for review before it needs to go "out the door."
- Review time is limited and efforts are reduced to a quick once over!

Can you, as the engineer, or your firm, or the client afford this liability exposure?

The final question becomes: Are you, as the engineer, or your firm willing to take a low-fee project and run the risk of jeopardizing the future of the firm by incurring high insurance deductible payments and expert witness and attorney fees to defend a low-budget project when something goes wrong? And all of this exposure can be yours because your client wants it that way. Who is minding the store of design—the client or your design firm?

The low-budget project is a rat hole into which any prudent engineer or engineering firm doesn't want to send you or your staff's time and your hard-earned reputation!! **Don't do it!**

As engineers, we must stand our ground and say **"NO"** to those who want to rush the design process, minimize our scope of services, and hence reduce our fees.

The old adage "You get what you pay for" applies to the design profession. Minimizing fees generally ends up costing our profession and firms more than we want to admit.

A good example of both insufficient initial project information and the effect of minimized fees occurs when civil engineers do not perform a pre-design site inspection. When making the proposal, many times a firm does not take the time, does not wish to spend the cost, or uses survey drawings or the Internet to gain information on the site for its proposal in lieu of a pre-design site inspection. This can result in a lawsuit when the site is flooded by an obvious condition that a site inspection would have disclosed.

Another frequent example occurs when the survey, the utility drawings, or the Internet was in error or when the site and its surrounding conditions have changed. If there is no survey or geotechnical or environmental report, the civil engineer may be responsible for all the issues discussed in "Existing Site Conditions" in Section 6.3. The time and expense it takes to resolve these disputes is much more than the cost of any pre-design site inspection.

6.3 LACK OF ADEQUATE DESIGN INFORMATION

It is not uncommon for an engineer or engineering firm to take on assignments that are beyond their skills, experience, or capabilities. The following are some of the areas that are often overlooked by the engineering professional when accepting an assignment. As has been done throughout this book, the authors have selected examples from the fields of geotechnical, civil, and structural engineering, although examples could have been provided from other fields.

Schedule Requirements

It is important to establish a system to staff projects. The following questions should be asked for each project during the proposal stage:

- Can our firm meet the schedule with the current in-house workload?
- Can our firm meet the schedule with the current staffing level?
- Will the design contract require experience or expertise that the current staff does not have?
- Will we have to contract for services from an outside consultant?
- Can this consultant meet our schedule so that our firm can meet the client's schedule?
- What happens if our firm, or our sub-consultant, fails to meet the schedule? Will we be in default and will there be any monetary penalty?
- Should our firm contract for a limitation of liability clause to guard against schedule busts or other potential liabilities?

Existing Site Conditions

A thorough understanding of the site upon which the design is being placed is vital to ensure satisfactory performance for the life of the project:

- What is the surface character and topography of the project site?
  - Is the ground surface flat; slightly sloping; hillsides; cut with creeks, arroyos, canyons; and so on?
  - What does the ground cover consist of? Is it swamp, forest, desert, brush and thickets, and so on?
  - Is the site currently or has it been previously developed?
  - What kind of previous development is it? Stone or gravel quarry, agricultural, old landfill, heavy industrial site, raw materials handling or manufacturing plant? These kinds of sites may have a significant cost effect on the project due to significant site preparation requirements and/or environmental regulations.
  - Have there been prior structures on the site that may have only been partially removed, leaving existing foundations, septic systems, underground tanks, or pits?
- Are there environmental site complications or adjacent sites that could be a problem?
  - Next to a river and in a floodplain?
  - Upslope from a residential area?
  - Downslope from a public water impoundment?
  - Downslope from an environmentally compromised site?
  - Are there easements or covenants, conditions, and restrictions (CCRs) that need to be removed before construction can occur?
  - Are there restrictive air or mineral rights?
- What are the current site drainage conditions?
  - Will the new project drainage plan affect the current site drainage of property downslope and/or upslope?
  - How will the storm water from on-site development be handled, and where will it go?
  - Is there enough storage capacity on-site to handle the storm water?
  - Can the lot be adequately drained by designing to the minimum governmental design requirements?
- What is the geological setting of the site?
  - Earthquake prone?
  - Landslide prone?
  - Tsunami prone?
  - History of expansive or collapsing soils in the vicinity?
  - Sinkholes or other naturally occurring features that will require attention during design?

- What are the historic groundwater levels in the project vicinity?
- Are there any known springs, seeps, or other such features on or near the site?
  - If the site is cut into, will this expose springs, seeps, and so on?
  - If the site is filled, will the fill tend to trap seasonal water migration?
- Is there evidence of unstable slopes on the site?
- Is the building to sit near a ridge or escarpment or other natural or human-made structure that could affect wind forces on the structure?
- Are there below-grade features that will require attention during the design process so that they don't become surprises during the construction process?
- Are there any photos (aerial or oblique) available for review? This is especially helpful if the site is out of town or out of state. Failure to review these kinds of photos can lead to the discovery of really ugly surprises during construction. For example, coal tar tanks, dumps, and a buried 1956 Chevrolet have been found below the surface of various sites.
- Gather old topographic maps (city/town/county/state/satellite, USGS (US Geological Survey) quadrangle sheets) for review.
- Has a soil report been accomplished? If not, why not?
- What is the expected behavior of the structure(s) to be placed on the site? Does the structure's anticipated behavior interact well with the anticipated behavior of the site soils?
- If the proposed foundation system is not typical "spread footings," is the client aware of the cost ramifications? Has this possibility been considered when determining the fee to design the structure?
- Does the site have any access problems that could affect the arrival of large trusses, precast members, steel beams, and so on? Is there an adequate staging area for large items?
- How can a project design be completed if the engineer and/or the staff are not aware of the amount of surface elevation change across the site? What is the impact of these elevation differentials?
- Would you design a retaining wall on the assumption that
  - The passive lateral support of the footing base will be provided by a horizontal ground surface when in fact the site has 10- to 25-degree hillside slope and finished grading will result in 2 to 1 slopes down and away from the base of the wall?
  - No surcharge loads will be applied at the top back side of the wall and/or backfill soils?
- Could the engineer effectively design a project while knowing little or nothing about the site characteristics and/or soil/rock conditions?

6.4 LACK OF COMMUNICATION & COORDINATION WITH OTHERS

In the design of any project, it is essential that there be a project leader. If the engineering effort is performed under a sub-consultant agreement with a project architect, by contract the team captain is the architect. As such, it is the architect's responsibility to coordinate the design effort and provide the engineer with the overall design concept, scheduling requirements, and any outside information such as geotechnical reports, surveys, and so on; however, an engineer who does not receive the information and does not take assertive action to obtain it may become part of a lawsuit. Therefore, the engineer should address the assignment as if the engineering firm were in charge of the overall design. If the team captain (architect) is unresponsive, the engineer should initiate a letter requesting the required information for completing the assignment and for self-protection. The following issues need to be addressed, regardless of who is the team captain.

Design Concepts

- What are the approving agencies?
  - What are the agency requirements for submittals?
  - What are the reviewing agencies' time frames for review of each submittal?
  - Will the required review periods mesh with the client's time schedule?
  - If not, how does the firm plan to resolve the conflict?
- What will the project consist of? Usage, occupancy, and so on.
- What types of structures are anticipated? Wood or steel frame, concrete masonry units, concrete (cast-in-place, pre-cast, other)? Are these building materials in common use in the area where the project is to be constructed?
- How many levels for each structure?
- Are there basement or subbasement levels? Are the basements full or partial? If there are partial basements, is the layout economical, or could construction costs be saved by altering the layout?
- What are the anticipated minimum and maximum column and wall loads for each of the structures? Has this information been coordinated with the geotechnical engineer?
- Has the architect provided sufficient room, both vertically and horizontally, for the required structural and mechanical elements?
- Are there areas where significant structural and mechanical elements occur that could challenge ceiling height limitations?

Client's Expectations

- What is the client's expectation for performance of the structures? Movement and settlement are among the most common causes of litigation. Taking some special care regarding a client's expectations and the possibilities of such movement can avoid some horrific liabilities.
    - Do you, as the engineer, understand the difference among total, differential, and rotational/distortional movements of a structure? Have you conveyed that knowledge to the client and the other design professionals involved in the project? Does your design take these movements into consideration and provide effective joints in the structure to accommodate the movement?
- What is the anticipated magnitude of movement for each of the structures?
    - Does the soil report develop estimated total and differential settlements based upon the anticipated structural load conditions?
        - If no magnitude of settlement is reported, does your firm ask: Why not?
    - What is your firm's policy on allowable deflections? Do you even have a policy or guideline?
- Are the client's expectations (and budget) reconcilable with the expected level of performance?
- Is the client willing to accept the risk associated with using a less reliable foundation or site grading scheme to save money?
    - If the client is willing to accept this risk, does your firm put the client on **notice with a registered "Return Receipt Requested" written letter** that states that the client has chosen a less costly method and thereby also assumes the risk of consequences for a lower-performing system?
    - If your firm is not willing to send such a notice, then your firm should not take the project in the first place.
    - Should you design this system and sign and seal the construction documents required to build this system?
- Have you discussed the ramifications of these various structural movements and how they affect the architectural design?

6.5 DESIGN

Competence

Prior to accepting any design assignment, ask the question: Am I capable and do I have sufficient education and experience to take on this assignment? Too often an engineer looks at an assignment through rose-colored glasses, particularly if the client is an influential and significant one who would give the engineer regional or national notoriety.

A case in point is a large hotel/condo project where an international developer retained a nationally recognized structural engineering firm to design two 12-story structural cast in place concrete towers along the beach front in one of the U.S. island territories; however, the initial estimates to construct the facility were over budget, so the developer fired the recognized structural firm and retained a small local structural engineer to reduce the construction cost. Realizing the difficulty of the assignment, the local engineer retained an outside foreign firm to prepare the calculations and outside drafting help from an architect and the building contractor to prepare the documents. Needless to say, with "too many cooks," the end result was a disaster, particularly when the local engineer retained a relative, who had no engineering experience, to do the special structural inspection. The two structures were completed and partially occupied when an earthquake collapsed one structure, causing it to tip into the other. Both towers had to be demolished.

During the long litigation that followed, it was discovered that the local engineer of record did not know how to interpret the computerized, undocumented calculations, which had failed to include seismic forces; had failed to catch a number of errors in the design documents; and had failed to detect improperly placed reinforcing bars during the special structural inspection. The trial lasted for two years. The engineer got his notoriety. A lesson to be learned is this: Never take on an assignment that exceeds your capabilities, knowledge, and experience just to get your name in lights, because you may get your wish.

Assignment

The best-qualified engineer with major experience in this type of project should always be assigned to every project. Often this is not possible within the office because that person is already assigned to other projects. This presents an opportunity for another engineer to gain experience under the supervision of this qualified member. The qualified engineer must act as a mentor and answer questions, ask questions, and spot check or review the work periodically. In the best-run offices, the mentor checks with his or her student *daily*. Today this is difficult because many engineers upon graduation believe they can design anything. Thus, it requires much patience to mentor a new engineer on how to "do it right" the first time.

Good engineering design requires a "start-to-finish" commitment, because "a chain is only as strong as its weakest link." Some examples follow:

- In structural engineering, each component of the structure and attachments to the structure transfer all loads from their point of origin to the foundation. It is, therefore, essential to consider each and every component in the design for its appropriate load.
- If the civil engineer is concerned only about drainage issues within the confines of the project site, the design is subject to failure because the effects of impacting off-site conditions have not been considered.
- The architect may draw the plans and details based upon only the engineer's sketches, calculations, and information. It is the responsibility of the engineer to thoroughly review

and check the architect's drawings against the design calculations prior to sealing and releasing the plans for construction. Unfortunately, some engineers fail to "do it right" and often "pay the price" by failing to perform this critical review.

Quality Control

In years past, it was not uncommon for the engineer, after completing the design documents, to place them in the drawer for a couple of days before critiquing them with a fresh approach. The firms that want to "do it right" still check projects before they release them. The most efficient way to check a project is for the engineer to get a complete set of drawings, including those from all other disciplines, and mark every line on the engineering drawings yellow (reviewed and approved) or red (revise) while coordinating the engineer's work with the other disciplines' drawings and the calculations. All too often, the design goes "out the door" the very day that the drawings are, supposedly, "finished." To not "pay the price," have sufficient time and budget to "do it right" by performing a thorough and complete check of all the plans.

If you are an engineer, do not continue to associate with firms that are consistently "putting out fires" with their project deadlines. Even the best firms sometimes get caught short of time, but an ongoing pattern of missed deadlines may be an indication of deeper problems.

6.6 ENGINEERING ISSUES THAT PROMOTE CONSTRUCTION DEFECTS

Once the budget and schedule have been reviewed and the assignment has been accepted, the project must be produced. Due to time and budget constraints faced by the engineer today, the following problems have become too common.

Failure to Properly Determine Loads & Forces

Failure to properly determine the loads and forces affecting the structure (i.e., live, dead, wind, seismic, thermal) can cause a multitude of problems and can be expensive to repair. If the loads are incorrect, everything that follows is incorrect.

A classic example is the failure of a rotary kiln that was seriously damaged on initial start-up because the engineer failed to address the dynamic live loads generated by the blowers resulting in a complete failure of the structural frame under the blowers, delaying the production start of the processing plant. The engineering firm was charged not only with the cost of repair but also the loss of revenue to the plant resulting from the delay. Fortunately, this failure did not result in the loss of life; when the blowers were started, no one was standing on the maintenance deck next to them.

Another case of incomplete knowledge of the imposed loads involved a 35-foot-high, 90-foot-diameter steel tank that stored fly ash produced by the scrubbers of a coal-burning electric generation station. The normal load tables for contents of tanks did not include values for fly ash, so the engineer, whose fee was only $600, used values for grain, thinking that it was close enough. He guessed wrong, and the first time the tank was filled, it ruptured, spilling its contents over several acres. Luckily, again, no one was killed, because the failure occurred at 10:00 p.m. and the lone employee was in the restroom away from the tank when it failed. The forensic investigation discovered the facts of the case and consulted with a materials expert to determine the content loads for fly ash. The cleanup, lost revenue, new tank, and related equipment cost, reportedly, ran over $2 million. See Figure 6.1.

A common error occurs when thermal forces are not properly considered. Some engineers have a tendency to make each connection in a concrete or steel structure overly rigid, which does not allow enough relief for the expansion and contraction caused by thermal stresses We have seen countless buildings rip themselves apart because

a

b

FIGURE 6.1 These photos (a and b) show the storage tank before and after the collapse. Fortunately no one was in the area at the time of failure.

of a failure to accommodate thermal forces. Other errors in determination of loads and forces are unfortunately common:

- Failure to properly interpret the load and force requirements of the building code.
- Failure to provide a complete load path from the point of origin to the foundation.
- Failure to properly determine the tributary area supported by a member, including columns, beams, and foundations.
- Failure to account for the weight of the member being designed.
- Failure to consider all relevant load combinations or to properly determine the governing load combinations.
- Failure to determine and design for unbalanced live loads.
- Failure to understand the relationship at the soil/structure interface.
- Failure to consider stresses and movements produced by environmental effects.

It is extremely important to remember that often failures in structures happen because the engineer failed to focus on all of the elements affecting the design. It is important to remember that if the structure is not adequately designed in the beginning, it will not be adequate at the end. Building, bridge, dam, and other structural failures have killed people in the past and will do so again in the future. Do not become a part of a team of engineers that does not provide attention to and understand the overall design process.

Overreliance on Computers

Computers are very useful tools, but often the computer becomes the substitute for engineering judgment. Often one hears "the computer calculations confirmed that such and such is true and that's why we used it." Be cognizant of the fact that basing one's conclusion on the "judgment" of a computer is not without risk. Remember the old computer acronym "GIGO": "Garbage in—Garbage Out."

Calculations are often the bane of the engineering profession. Computers can spit out as much paper as you request. It is not uncommon in the evaluation of a structural failure to review binders full of finite element printout that was essentially worthless and then review succinct hand calculations that proved to be the basis of a good design.

Failure to Understand the Limitations of Computers

In our work to solve failures and litigation issues, we have found an overreliance on computers. Often the designers must take time to input so much information that interpretation of the results is not thoroughly performed. Once finished with all the input, they take the apparent answer and simply run with it. Most programs take too much input and give too many answers. Some engineers never heard of GIGO. It is still true that simple hand calculations are often faster and easier to find errors, and the answers are clear. The case involving the collapse of two 12-story towers on the island of Guam because the initial calculations addressed only wind load and failed to consider seismic factors is one example of the failure to check complicated calculations.

Another example is an investigation concerning a three-year-old public building in which the forensic investigator had to inform the city council that its building had problems. The forensic engineer knew the response would be: That's your opinion; we support *our* engineer." During their first meeting, the forensic engineer handed the city council members three sheets of computer printouts from their engineer's calculations and told them to read the 16th line on the right side of each page. The 16th line stated that the member calculation was "NO GOOD." The council asked if those "NO GOOD" members were in place in their building. The answer was "yes." It took $1.7 million to repair only the major errors on this project. We can only speculate that the design engineer, in a rush, did not thoroughly read the printout.

Another example occurred when the engineer used a popular steel design program that transports the answers directly to the CAD (computer aided design) drawings. This was not an issue in the litigation, so the investigation did not allow the forensic engineer time to investigate the cause. Regardless, the drawings had many different beam sizes to the same spans with the same loading conditions throughout the structure. The design engineer did not group the beams or consolidate the sizes. Again, one wonders if there was a review of the output or check of the drawings. This large number of beam sizes increased the cost of detailing, purchasing, fabricating, and erecting the different size beams. It also increased the waste material and the chance for errors. Too few or too many beam sizes on any large building are an indication of an inexperienced or careless engineer.

As another example, during a deposition on the collapse of a major building truss, the plaintiff's engineer testified that the truss failed at a certain joint because the "computer program said so"! After some searching, the defense's engineer was able to locate pictures of this specific joint in one solid piece with over 30 feet of unharmed truss lying on the ground. The truss had failed at another location and this joint fell to the ground intact. The defense investigation team was retained by the engineer of record who appropriately added mechanical units to the truss and had evidence that the owner later added additional loads in the wrong locations. Upon completion of the defense engineer's deposition and entering the picture into evidence, the case quickly settled.

In analyzing complex structures during forensic investigations using computer programs, it has been discovered that often one must make estimates on a large number of input issues and the answers can vary greatly depending upon those estimates. In some cases, it even depends on how the building was erected. This fact was discovered in litigation over a complex structure that the engineer of record allowed to be overloaded. As a part of the settlement, the owner was required to hire a nationally known engineering firm with years of experience in this type of structure to do the repairs. At the completion of their computer programming, a comparison of answers was made between those of the forensic investigator and the repair design engineer. The differences were enlightening. The finish repairs were based on engineering judgment and the computer outputs were only one of the factors considered.

Finally, a classic example of overreliance on computers during design was a multistory concrete office building in which the flat roof deflected 5 inches and ponded water, had numerous drywall stress cracks, and had major water intrusion throughout the building, all of which prompted an intense investigation. During the investigation of the post-tensioned concrete roof slabs, it was determined that they all had shear and tension cracks in critical locations. The investigating engineer performed simple hand calculations that confirmed that the cracks had formed in areas of overstress. This project was a developer design/build pre-stressed cast in place concrete structure. The building housed "Triple A" tenants on long-term leases who were threatening to break their lease due to the continual reappearance of cracks and the major water intrusion that had gone on for a long time. The developer was concerned about his reputation and required the engineer of record to cooperate with the investigation.

The design engineer had a two-way slab-beam computer program with columns up and down, ends fixed to determine the moments. This one design was then used for all the floors and roof. Then, due to architectural requirements, some concrete columns were changed to very small steel tubes and framed "fixed-end" beams into the side of other beams. All of these changes were made without rerunning the computer program to determine the revised design. The roof did not have columns going up as the original computer model assumed and the slabs were not cambered. The small steel columns were limber and did not provide the stiffness of the concrete columns used in the original computer program. It was obvious why the roof deflected so much and ponded water. The investigation disclosed a very large bow in the steel columns and many torsional shear cracks where the two beams framed into each other. It was also discovered that the dimensions of the slab-beam sections did not conform to PTI (Post-Tensioning Institute) Technical Notes Issue #3, October 1993, which is the standard for this type of construction. The original and only computer program not matching the final design and the geometry errors made the final design questionable. The design engineer insisted the computer program was right and there was nothing wrong with the final design.

To solve the impasse, a national design firm was hired to coordinate with the design engineer. They had many meetings over a long period of time, in which the design engineer developed a series of computer models to try and prove the design. Finally we were told they used a European finite element program that enabled the design engineer to analyze the entire structure and design the necessary repairs to the existing structure. The original investigating engineer was not involved in any of these sessions.

After four years of conflict and discussions came the realization that remediation of the structure was necessary. This effort required the reinforcing of a series of columns, filling the cracks with epoxy, and adding fiber carbon reinforcement to the concrete post-tensioned slabs at various locations. A very costly and interruptive remediation program was required that had to be accomplished at night and weekends in an occupied building. It is interesting that all of the reinforcement to the structure was in the same locations that the experienced engineer had identified with simple hand calculations.

Regardless of whether the design engineer was arrogant, stubborn, or just trusted the computer too much, the fact remains that the confrontational approach did not contribute to a timely resolution. The design engineer paid the price not only in dollars to cover the cost of repairs and damages incurred by the tenants, but also loss of credibility and a prestigious national client. Keeping an open mind and being proactive with others involved to better understand and resolve a problem, rather than being a part of the problem, go a long way in building lasting relationships.

Based on investigations of failed projects, it is evident that the designer must often input so much information and review so much output that interpretation of the results is often complex. Once finished with all the input, the inexperienced rushed engineer may take the apparent answer and simply run with it. Many programs take too much input and give too many answers that obscure the important issues. Regardless of the cause, overreliance on computers without experienced engineering oversight often leads to errors. It is still true that in many designs, simple hand calculations are often faster and easier to find an error, and the answers are clear.

Failure to Transfer the Results of the Design Calculations to the Drawings

The best design calculations in the world are meaningless if the results of those calculations are not properly coordinated with the structural drawings. It is not uncommon during forensic investigations to find recurrent errors that include the following:

- Transfer errors in steel, concrete, and timber beam sizes.
- Transfer errors in quantity and size of reinforcing steel in concrete beams, footings, and masonry walls.
- Frames designed with significant moment transfer at the columns, but details that failed to provide adequate strength in the connections to transfer the design moment.
- Transfer errors in footing size and location.
- Failure to draw footings to scale, such that overlapping footings exist that were not considered in the design calculations.

- Retaining walls designed without hydrostatic pressure, but no drainage system shown on the drawings. If the retaining wall is a basement wall, the drainage system also needs to be coordinated with other design professionals, including the architect and the civil engineer. We have seen drainage system outfalls shown by the civil engineer, but not installed in the field, behind the wall because both the architect and structural engineer were pointing at each other saying that it was the other's responsibility, thus none was shown on their drawings.

Lack of Judgment

It is essential that every young engineer strives to develop engineering judgment with increased design experience. Knowing "rules of thumb" and having a sense when things "feel right" are the most useful traits that one can develop in the engineering profession. This requires you to look, listen, process, and then question the results, *your* results.

It is the authors' opinion that there is a disconnect present in today's structural engineering profession between experienced engineers who developed their engineering judgment based upon allowable stress design (ASD) and younger engineers who have studied load and resistance factor design (LRFD). To avoid different design procedures within the same office that can lead to in-house conflicts, as well as conflicting opinions on how best to develop the design assignment, both the younger and the experienced engineers must adjust. Training is required to coordinate with the different design philosophies.

In reality, both designs have their respective use. This is recognized by the latest AISC (American Institute of Steel Construction) code that has both in one book. One must be careful when considering using LRFD in that serviceability often governs (deflection, fire proofing, and vibration). Using ASD may be more practical. Certainly on very small jobs, ASD is useful. Each firm must establish criteria so that the staff knows which design to use to create consistency.

Failure to Coordinate Documents

The lack of construction document coordination is becoming a major problem leading to lawsuits. The lack of quality control is one reason. See Section 6.5 above. Another is the inability of design professionals to stop making changes so as to leave time for coordination. Computers have made it so easy to change drawings that designers are often still making changes on the day of release. This does not leave time for coordination. Some designers believe that computers coordinate automatically. They do for only a limited part of the drawings, but computers are not capable of checking the details to ensure that they are compatible with the overall design. One must also realize that humans are operating the computers, and, therefore, the machines are subject to human error. A recent forensic investigation involved litigation wherein the structural drafter did not realize that the mechanical drawings use reflected ceiling plans. All the third-floor mechanical openings were shown on the second-floor structural drawings. Some architectural and engineering firms try dumping coordination of their designs on the contractors. The contractors are fighting back by writing many RFIs and then suing on the grounds of defective documents. One of the authors defended an owner on a project in which the contractor submitted 2,100 RFIs on a $38 million project because there was no coordination among the various disciplines.

Failure to Complete the Last 5 Percent of the Construction Documents

Over the course of time, the excitement of a new project wears off. When the deadline approaches, there is a tendency to finish the last 5 percent of the construction documents with less than a stellar effort. The engineer is often anxious to move on to new projects. It is important to resist this natural tendency and to *complete* the project to the best of one's abilities. This attention to detail separates good engineers from poor engineers. If there is a detail that needs to be drawn, draw it. If there is a calculation needed to check a beam size, do it. In school, getting 95 percent is considered a very good performance; in the real world, it's 5 percent wrong, and, in the worst cases, that 5 percent could kill someone or be the cause of a major lawsuit.

Specifications Must Be Consistent with the Drawings

Document precedence clauses are often fought over and are lengthy lists of 10 or 12 items. It is essential that the specifications be consistent with the drawings. Construction Industry Standards, unless otherwise noted in the General Conditions, are that the specifications take precedence over details and details over plans. It is essential that anything stated in the specifications be consistent with the designer's design, the calculations, as well as the details and plans. Too often, the specifications are a "catch all CYA" for any omission on the drawings and/or have become "standardized," which often makes them more confusing than helpful. The best specifications are precise, specific, and to the point, defining the scope of what is to be done; defining what materials should be used in the performance of the work; defining the standards that the work must conform to; and explaining any applicable approval process for acceptance of the work. In today's world, it is not uncommon for the sharp contractor to take an inappropriate set of specifications and turn them into a gold mine in construction claims because the specifications are not applicable to the work being performed. The design team should use the same precautions in reviewing the specifications as are used in reviewing the completed plans.

6.7 THE CONSTRUCTION PHASE

The Engineer During Construction

Unfortunately, many times the engineer is left out of the construction phase with the possible exception of the review and approval of the shop drawings. As the engineer of record, it is recommended that you make a concerted effort to be paid to remain completely involved throughout the construction process. Much of the litigation we see today could have been avoided if the design team had just maintained an active line of communication with the contractor. This line of communication takes many forms:

- The shop drawing process may be the last chance for the engineer of record to determine that a component the contractor intends to use in performing the work meets design. There is a prevalent misconception that by preparing a set of shop drawings the contractor assumes responsibility for that portion of the design, and if there is a failure it's entirely the contractor's fault. Nothing can be further from the truth. Shop drawings define how the fabrication shop intends to fabricate and how the contractor intends to construct components of the design prepared by the engineer so that they satisfy the design of the engineer. It is essential for the engineer to thoroughly review these shop drawings to ensure they do, in fact, meet the design. It is amazing how many problems could have been avoided had the engineer given the shop drawings more than a cursory look.
- Site visits should be made to observe the course of construction and determine whether or not the contractor has produced an assembly that conforms to what was designed by the engineer. Issues to take note of during these site observations would include the following:
  - Proper load transfer to the foundation.
  - Connection details including a concentrated check of critical and non-redundant connections.

Written communications of the site visit findings should be sent to appropriate parties stating the good as well as the bad. Too often, such communications address only what was wrong and not what was done correctly. Good communications make good friends, good friends make up a good team, and a good team can accomplish miracles, that is, avoid litigation. Avoid informal communications that have not been well thought out. Just because you can respond rapidly doesn't mean you should. Don't send fast e-mails unless you don't mind hearing them on the morning news. One of the easiest cases to defend was when one of the authors located an e-mail from the owner's office to the field saying, "Don't waste your time finding out who is responsible just xxxxx the contractor." We were defending the contractor who had followed the drawings. Before you hit that send key, reread that e-mail and make sure it is truly a professional document that can be read to a jury.

If your firm is going to undertake special inspection, be sure that your inspectors are properly qualified. It is essential that your firm thoroughly understand the ramifications and liability associated with special inspections. The following questions are the minimum that should be addressed when entering the field of special inspection:

- What are state and local requirements?
- Are special classes required from a recognized agency, necessitating the passing of an examination and being awarded proper certification for the inspections required?
- What are the local codes and technology that must be understood by the inspector and the methodology under which their knowledge of such codes and technology is kept current?
- Are they required to attend seminars and classes to keep them current on changes in their special fields of expertise?
- Do they thoroughly study and understand the project drawings before they go to the site and commence their inspection?
- Do they understand the contractual relationship between the parties and are they bound by law to conform to various clauses?
- Do they understand company policy and procedures? One must remember when conducting these special inspections that the employee conducting the inspection is on a job site away from the office, thus restricting any personal observance of their performance by a principal of the company.
- Are they basically honest and will not compromise the integrity of the company?
- Are they strong enough to stand their ground if the contractor or owner wants to compromise on an issue that is clearly shown in the contract documents?
- Are they experienced enough to address an issue that is not clear and may require an on-the-spot decision to avoid a costly delay?
- Have they undergone sufficient OSHA training to keep from being injured on the job?
- Have they been instructed on how to document and record their findings?

Post-Construction & the Engineer

There is a very good chance that the project that just completed construction had its share of problems—some caused by others, some perhaps caused by you. Learn from the mistakes, and share this knowledge with others so that they do not have to learn the hard way. It is a good practice to meet with the construction team near the end of a project to get input on problems experienced during construction and any recommendations that could be useful in future designs to minimize problems. Some of their suggestions may not be feasible because they could affect the structural integrity. However, communications between parties will often lead to a better understanding of how a structure works and how it is assembled.

Some of the most spectacular failures in structural engineering have certainly prevented others from making the same mistake, simply because they were well publicized. Create the same type of program within your firm to share information on the little problems that are revealed during the course of construction and each subsequent project will benefit.

6.9 DO'S IN THE PRACTICE OF ENGINEERING

1. Do obtain, as early as possible, full program information prior to completion of the schematic design phase.
2. Do document all of the input provided during the preliminary design phase.
3. Do start design only when you understand the interrelationship of the different disciplines.
4. Do create construction details considering the whole context of the project, not just the one detail.
5. Do recognize that computer screens limit the amount of detail that can be seen or considered at any one time.
6. Do ensure that conditions with no redundancy are given a high priority in the review of calculations and detailing. These conditions include cantilevered retaining walls and beams, tension rods, shear in concrete beams, and the like.
7. Do coordinate the calculations with the drawings via a systematic approach to ensure that the requirements of the calculations are accurately represented on the drawings.
8. Do coordinate the efforts of the design team so as to ensure a well-designed project. This includes the project geotechnical and other specialty engineers as the project develops.
9. Do allow an appropriate internal and external coordination review before the project is issued for construction.
10. Do take the time to stay on top of the plan approval process.
11. Do make the effort to properly and timely review shop drawings.
12. Do coordinate with the contractor on the importance and timing of any RFI submitted to the office.
13. Do use experienced people for contract administration instead of the most junior person in the office.
14. Do consider all relevant disciplines when reviewing proposed change orders.

6.10 DON'TS IN THE PRACTICE OF ENGINEERING

1. Don't be isolated from the rest of the design team.
2. Don't ignore the client's input.
3. Don't hire the cheapest consultants that you can find. They are often the most expensive in the end.
4. Don't allow the client or yourself to change the project after the preliminary design drawings without providing an appropriate amount of time to coordinate the ramifications of the change.
5. Don't allow a second or third party to communicate engineering issues with the primary client.
6. Don't let outside salespeople control product selection or dictate specifications.
7. Don't let the code minimums control the design.
8. Don't use canned documents without adjusting them to project-specific conditions.
9. Don't provide a cost estimate without a proper supporting basis. Do not fail to revise this estimate when the design or conditions change. Consider the services of a professional cost estimator.
10. Don't rely on the governmental review process to serve as your quality assurance.

SECTION SEVEN

CONSTRUCTION ISSUES

or

Defects & Litigation—Why Me!

7.0 JUMP-STARTING SECTION 7

Anybody can be a contractor. Nothing to it! After all, what does it take to dig a ditch, pound a nail, or paint a wall? A few days in a licensing school will qualify you in most states to hang out your shingle as a contractor. To help the reader fully understand the importance of this section, the following points need to be kept in mind to better understand what's expected of a contractor:

1. The importance of competent, knowledgeable, open-minded, and inquisitive field supervision cannot be overemphasized if a contractor is to survive.
2. Continued education on new products, methodologies, codes, rules, regulations, languages, and people-handling skills is a must in all fields of construction.
3. Communication, documentation, and pictorial records are not an after-thought; they are everyday needs. They should be both copious and timely.
4. Sequence planning and time scheduling are not just words; they are the lubricant that makes the entire effort move like a well-oiled machine.
5. As a contractor, you are responsible for the care, custody, and control of your work area. If you are the general contractor, you are responsible for the care, custody, and control of the entire work area. Not paying attention will make your lawyer and your insurance company sad.
6. Take the bull by the horns and develop a mindset that you and your personnel will excel in your specific field of contracting; don't settle for second best.
7. Insist that your clients, subcontractors, and material suppliers provide you with a performance analysis of your effort, management, and communication skills as you progress through a project.

The construction industry is unique in many respects, particularly when compared to other businesses, be they manufacturing, mining, public utilities, medical, and so on. Most other businesses are developed and managed around specific goals, utilizing a knowledgeable group of permanent employees, who work together to develop well-defined policies in order to produce end products that will bring in repeat business. The fact that these employees have been working together for years allows the creation of sound management and control systems, allows for the employees to be continually educated, and allows performance methods to be continually improved and, most important, the circulation of new ideas that improve the overall quality of the businesses to be circulated.

Construction, on the other hand, often pulls together employees with different capabilities to accomplish a given project; then when the project is completed, they all go their separate ways without any interaction that will lead to improving techniques and methodology. Failing to take advantage of lessons learned can have a detrimental effect on both a contractor's bottom line and reputation.

This section deals with a series of construction issues that manifested throughout the industry like a plague and have devastated the credibility of many and led to the financial destruction of companies. These issues cover a spectrum, from contractual responsibilities, through constructing the work, to final acceptance and warranty. It would be impossible to address each and every issue that has resulted in construction professionals being at the short end of the stick; however, after reading this section, we hope you will have a better perspective on the importance of approaching every job with an open mind, as well as a jaundiced eye.

7.1 UNEDUCATED, UNAWARE, OR INCAPABLE FIELD SUPERVISION

Over time, the process of supervising the work in the field has changed dramatically. In centuries past the "master builder" was the architect who designed as well as supervised work on some of the most colossal projects the world has ever known. As the years passed, certain skilled trades with leading roles in the construction of the work took more control of their portion of the work. Most prominent among these were the carpenter, the stonemason, and the general laborer. This happened because most of these colossal projects were constructed of wood and stone, which were moved into place by the laborers under direction of the carpenter and stonemason, bringing about the advent of field supervision. Even today, these two trades remain prominent in third-world countries.

When the architect's role as the general overseer of the work diminished, coordination was taken over by contractors, who would assemble the needed trades and would supervise the work. Initially, it was fairly easy to coordinate the work of a handful of trades. With the development of new and more sophisticated building materials, the trades grew from carpenters, stone masons, and general laborers, to include specialists in many other fields: plasterers, teamsters, operating engineers, glazers, tile setters, ironworkers, millwrights, boilermakers, steamfitters, plumbers, electricians, sheet metal workers, air-conditioning specialists, roofers, insulators, acoustical specialists, and alarm specialists, just to name a few. Soon the industry developed into a series of general contractors, who through competitive bids competed for work. Upon being awarded contracts, these general contractors continued to supervise the overall work but employed primarily carpenters and laborers, while subcontracting the other work out to trade contractors, who competitively bid for it. Generally, the field supervisors employed by the general contractor rose through the ranks of the carpentry trade and had "hands-on" experience working with other trades in the field. This methodology for constructing projects remains even today.

As the construction industry grew and became more sophisticated, the governing laws, regulations, building codes, and other outside influences became more a part of the day-to-day construction process. These factors, with tighter construction budgets and time schedules, complicated not only the construction side of the building process but also the design side. Drawings and specifications were not as precise and much was left for interpretation by the general contractor and/or the trade contractor. Build it "per code" or "pursuant to manufacturers' printed instructions" became "buzz words" that were part of the specifications and/or general notes found in the design documents

The industry-wide deferrals to governmental authorities increased the burden on contractors' field supervision to such an extent that many of the larger general contractors began to employ specialists to address the ever-mounting responsibilities, for example, project engineers, safety engineers, scheduling engineers, material control engineers, expediters, assistant superintendents, trade superintendents, quality assurance engineers, and so on. This approach may work for those general contractors who have volume and/or very large projects that can warrant such luxuries. However, for the average general contractor, the new burden of governmental requirements rests with the person directing the work to be done, be it the contractor's field superintendent or the contractor. In view of these developments, it is essential that the contractor and/or the designated field supervisor become more sophisticated and knowledgeable in how to approach each project.

All too frequently, particularly in the custom home industry, general contractors are, in reality, "brokers." In essence, they have brokered out all of the construction effort to trade contractors needed to complete the defined scope of work. This approach often leads to a lack of personal involvement in the daily work activities taking place on the project, putting the trade contractors in a position of running the work themselves. Even worse, they are often responsible for checking the quality of their own work.

In being detached from the daily job activities, the contractors lose the intimacy they previously had with what is actually going on. This lack of personal involvement is producing poor work as its result.

The following reflect a series of issues that can lead a contractor down the litigation trail when field supervision is either not prepared for the project or nonexistent.

Insufficient Knowledge of the Work

It is essential that the contractor and/or the designated field supervisor have the opportunity to thoroughly review the contract agreement, all plans, and specifications that make up the contract documents, including the estimate and/or budget and all subcontract agreements, before proceeding with the work. Every contractor should make it mandatory that any question dealing with any of the contract documents be presented to the client, architect, engineer, and/or representative in charge with a formal Request for Information (RFI) and/or clarification early in the project. Each RFI must have the date the request was made and the date when a response is expected back. Each RFI must be recorded in a RFI Log Book and include a tracking number.

If the contractor and/or the designated field supervisor has insufficient knowledge of the project going into the work and mistakes are made that result in actual and/or alleged construction defects, or delays affecting the satisfactory completion of the work, the contractor will be held liable, particularly when the information was available in the contract documents or noted as a recognized industry standard.

This general knowledge can be attained in a number of ways by making a thorough review of the plans and specifications, with particular attention given to the following:

- Understanding the requirements and responsibilities contained in the specifications.
- Reading the general or structural notes on the plans.
- Reviewing the plan details and cross sections defining how the work should be accomplished.
- Pay close attention to geotechnical or other special reports prepared to address specific project-related issues.
- Becoming familiar with building code reports defining how materials and/or products should be applied.
- Reviewing ASTM or ANSI, material and product specifications, and standards.
- Reviewing manufacturers' printed instructions on how to install and apply their respective products.
- Reviewing shop drawings and erection plans prepared by the subcontractor to define what that work will look like and how it should be assembled.

- Checking manufacturers' information sheets to verify that the product is recommended/guaranteed for the intended purpose.
- Asking questions of the trade contractors who are performing the work.

Time and cost are undoubtedly the two most important components of any contract. As these requirements are fundamental to the contract's "bargain," it is important to note that the simple phrase found in so many contracts "All time limits stated in the Contract Documents are of the essence of the Contract" is binding. Despite the simplicity of this statement the contractor or the designated field supervisor must understand this statement as a warning, not to be taken lightly, for without this statement, the time periods given as needed to complete the work may merely be guidelines and not contract requirements. However, when this phrase is incorporated or implied in the contract, all time is "of the essence." This statement magnifies the importance of a properly prepared construction schedule.

Inexperienced Supervision

In today's economy, inexperienced supervisors have been a problem for many contractors. In the past, the supervisors came up through the trades, first as an apprentice helper or apprentice, then as a journeyman, eventually rising to foreman, general foreman, and superintendent. Because of their extensive personal experience, these supervisors were usually highly qualified.

Today, we find that many of the supervisors have little experience regarding the work that they are supervising. It is not uncommon to find young supervisors, fresh out of college, who are talented in many ways, but lack the experience in the trade that they are supervising. They, therefore, are estranged from the opinion of the worker in the field as to how the work should be performed. By not being able to discuss the work with the tradesperson assigned to perform it, the inexperienced supervisor is at risk of either accepting the work as performed or not being able to provide instructions as to how the work must be performed. This scenario often leads to accepting the work as performed and facing the possibility of litigation if the work at a later time is found to be defective and/or improperly performed. It is, therefore, essential to the success and survival of the contractor to ensure that the field supervisor is totally familiar with the skills and techniques necessary to perform the work in accordance with the contract documents. This does not mean that an in-depth knowledge of all tradespeople's skills is necessary, but a good general knowledge of how the work should be performed can help avoid future problems.

Little If Any Knowledge of Quality Standards

Every field supervisor should have a thorough knowledge of the quality standards that relate to the work. Architects will normally incorporate in the specifications a section containing a listing of "applicable standards," whether or not these standards apply to the project being constructed. Little is done to identify those portions of any standard that may apply, leaving it up to the contractor to decipher its relevance. The suspicious reader usually sees this listing of standards as "CYA" for a time when the job is in serious trouble and someone produces a copy of a given standard to quote when assessing blame.

Frequently, the architect listing the standards doesn't even have copies of the standard and has never seen nor read the whole standard; yet the contractor must ostensibly discover and know what the standard requires.

To stress the importance of better continuity, the following is an example of leaving things to chance. A document from a manufacturer of floor tile had been provided to the project architect listing 19 separate specifications relating to the setting of floor tile in hazardous work areas. The architect was then very reasonably instructed to edit the list as required for the project conditions and include the correct set of specifications into the contract specifications. Unfortunately, the architect simply passed the entire list to the contractor, who then provided it to the flooring subcontractor without any thought, believing that the flooring subcontractor would be able to interpret the specification. When something goes wrong, as it did in this case, the finger pointing can get very nasty.

To make it worse, the buck is passed to the last person in the chain, the actual tile setter; the one who is probably the least able to understand the intricacies of the 19 specifications and is also perhaps the least qualified to decide which specifications ought to be applied in this specific instance.

The responsibility for knowing industry standards is universal, and there is no question that such standards are important; however, they have become too indiscriminately used. These all-inclusive lists of standards, most of which are not applicable to the work, are overly used in construction documents today. The fact that most specifications are loaded with standards doesn't mean that the contractor and/or the field superintendent and foremen know the meaning of many of them. Normally, the contractor, the field superintendent, and foremen have an intimate knowledge about the standards of the trade they started in, but little else. This is not to imply that standards of the trade will or should take precedence over applicable manufacturers or industry standards set down by various associations.

One such venerable association is the American National Standards Institute (ANSI), which coordinates and administers the federated voluntary standardization system in the United States, which is composed of over 1200 national trade, technical, professional, labor, and consumer organizations; governmental agencies; and individual companies. Another prominent standards organization is the American Society for Testing and Materials (ASTM), which is the world's largest source of voluntary consensus standards for materials, products, and systems and services, and it currently has a listing in excess of 10,000 standards serving virtually every industry including construction. The International Conference of Building Officials (ICBO) also has a library of reports on materials and products used in the construction industry. There are as well various materials, trade, and professional organizations, including the following:

- American Institute of Architects (AIA)
- American Concrete Institute (ACI)
- American Institute of Steel Construction (AISC)
- National Fire Protection Association (NFPA)
- National Roofing Contractors Association (NRCA)
- Sheet Metal and Air Conditioning Contractors' National Association (SMACNA)

The important thing for the designer is to know and list those that *really apply* to the project being constructed. In reviewing the construction details on the plans, it is always good policy to verify the detail with the standard. If there is a noticeable difference between the standard and the detail on the plan, the contractor should request clarification from the architect. For any given product, the supplying subcontractor should really know the applicable standards that affect the work being performed and comply with those standards unless directed otherwise in the contract documents. The subcontractor should, in addition to verifying that the standard is in compliance with the contract documents, advise the contractor of any deviation, prior to commencing the work.

Lacks Understanding of Value Engineering

While "value engineering" has become a familiar and important element in project development, it carries with it the need for careful study and analysis of the proposed changes. Value engineering was initially intended to provide *equal value* for less cost. Value engineering is a process whereby total cost, including maintenance and operation, is evaluated on different systems. One must look at not only the initial cost of the completed structure but also the operational cost over the life of the structure. A simple but important example would be the installation of fluorescent T8 light fixtures versus inexpensive fixtures that use twice as much electricity for the same amount of light. In the flow of design, there is a natural pause at the design development phase that is an ideal spot to address how value engineering can best be applied to the project.

In today's world, value engineering is misused to cheapen the finished product in order to reach a lower "bottom line" by providing products and/or work that is not of equal value. It is not unusual for a proposed value engineering change to affect other portions of the design and/or work, resulting in a total added cost not initially contemplated. Sometimes the term is misused to substitute a product that is not of equal value and fails to meet the requirements of the original specifications. A decision by the architect and the team of sub-consultants regarding the acceptability of the proposed products or processes must be rendered and documented before the work is performed. When in doubt, request performance test results and inquire about the product's performance from those who have used it. Verify that the product applicator is certified and properly trained in the use of the product, particularly if such certification is a condition of the manufacturer to secure a proper warranty. If necessary, have the manufacturer's representative verify the application techniques and certify in writing that the work was done correctly.

Although savings may be a consideration when the contractor proposes a new product or method, it should not always be the overriding determination. Costs and time of completion are important considerations; however, the most important consideration is the ability of the new method or product to perform as intended. Failure to meet the performance requirements mandated by the contract documents or other authority can result in a costly experience for all parties involved.

Inability to Coordinate Work Activities

Coordination among the sub-trades is one of the biggest problems in construction today. Coordination of the work may also be referred to as "superintendence" of the work. Irrespective of the terminology, the inability to properly coordinate the work is caused by the lack of experienced contractor supervision, followed closely by the inability of the supervision to understand the dynamics of proper scheduling. The contractors, by educating the field supervisors, as well as providing the resources for them to learn how to schedule properly, will be repaid tenfold. Understanding the importance of how to properly sequence the various work activities will make coordination and/or superintendence of the work much easier.

Remember, coordination does not mean that one directs or controls how the work must be done by a subcontractor. Coordination is seeing that the predecessor activity in the construction timeline is completed in a timely manner, accurately and completely, so that the successor activities can begin as scheduled. In a typical construction project, numerous activities may be working in parallel at any one time. By interconnecting these activities, the contractor can easily determine the consequences of downstream activities, should one of the early activities be delayed or not completed on time. Failure to recognize the effect of the predecessor activities to the successor activities can have a deleterious effect on other portions of the construction schedule. The point being, once the first domino in a schedule starts to fall, it can have a major effect on the entire schedule.

The ability to coordinate the work is expected of every contractor and will be found in virtually every standard contract. The inability to coordinate the work can, and often does, lead to costly delays, which lead to acceleration, which can be a breeding ground for construction defects and litigation. We find the following are the primary factors influencing the inability to coordinate work activities:

- *Lack of "hands-on" field experience.* This often happens when a young person with little practical experience is put in charge of a complicated project involving many specialty trades.
- *Lack of sufficient general knowledge of the work.* This often occurs when a superintendent with plenty of

commercial experience is placed in charge of directing the operations of a processing plant or other major industrial project.

- *Fails to take an active interest in watching how the work is performed.* This occurs when the superintendent spends the bulk of the working hours in the trailer and fails to take an active interest in the field activities.
- *Lack of scheduling experience.* This normally points to a superintendent who allows the subcontractors to take control or lets the project run itself.

Lack of Scheduling Experience

A lack of scheduling experience or experience with scheduling software on the part of the contractor or the designated field supervisor can be devastating, particularly on the more complex projects or those projects that have tight completion milestones with the usual "Time Is of the Essence" clause. Every contractor should consider sequence planning and time scheduling experience to be one of the most important field supervision responsibilities. Failure to do planning and control scheduling has often led to serious consequences, from costly litigation to bankruptcy.

Unrealistic & Improperly Prepared & Monitored Schedules

Schedules have long been the nemesis of many a contractor. Having a clear knowledge of the work to be performed and maintaining timely and accurate documentation of the issues affecting the performance of the work are integral to any given project. In actuality, the project schedule has many times been referred to as the "heart of the project," as it defines how and in what sequence the work will be performed to ensure that the project will be completed in time to meet a defined end date.

By developing a good project schedule, the contractor and/or the designated field supervisor can build the project "on paper" before a shovelful of dirt is moved. When schedules are properly prepared and maintained, they become the "heartbeat" of the project. When the contract documents call for the contractor to prepare and maintain a job progress schedule (e.g., under the AIA Document A201, General Conditions of the Construction Contract), the contractor becomes contractually obligated to provide a complete project schedule that reflects when key segments of the work will commence and will be completed. Once the schedule is accepted by the client, the contractor becomes bound to the time elements and work activities shown on the schedule, unless modified through an appropriate change order. In doing the sequencing of the work activities, one sees when labor, material, and equipment are required to be present to complete the various work segments. The schedule will also show what other work would be affected, should a key work activity not be completed on time.

In preparing a schedule, the contractor must develop not only the timing and sequencing but also how the project will be constructed. In so doing, it is important to take the following performance needs into consideration:

- When to conduct a specific safety talk that relates to an upcoming activity
- When to have shop drawings approved to meet fabrication and installation activities
- When and how many specific workers, by trade, are needed to complete a major work activity to avoid delays in successor activities
- When RFI responses are needed in order to avoid delays in the work, as well as what activities will be affected if the information is not received on time
- When, what, and how much labor, material, and equipment are needed to complete any specific activity
- When and what effects will result from technical specifications (e.g., curing time, backfilling requirements, sequence of structural loading)
- When special inspections or meetings are required to avoid delays of successor activities
- When subcontractors are needed on site to avoid delays of successor activities

When proper scheduling information is available, the contractor or the designated field supervisor can avoid surprises and better plan, manage, and adjust the flow of work. It is imperative that any schedule developed be properly monitored and frequently updated so as to reflect the current condition of the work. The schedule indicates when the project is healthy and running smoothly, as well as when and where it's in trouble and what will happen if those troubles are not corrected.

Obviously, the size and complexity of the project will dictate the sophistication of the schedule; whether it is a simple bar chart or multi-activity person-hour loaded "critical path schedule." The key point is, without one, the contractor and/or the designated field supervisor lack the primary tool needed to perform their assigned responsibilities of completing the project in accordance with the contract documents and local codes, on time, and within budget.

In addition, it becomes nearly impossible to recreate an accurate account of what happened on a project after the work has been completed. This is particularly important, should claims be made by the contractor against others or if the contractor has to defend against claims. If you are to respond to a legal action or dispute regarding the construction process, a detailed and accurate critical path schedule will be essential. See *New Pueblo Construction vs. State of Arizona,* 144 Ariz. 395, 696, P2nd, 85 (1985).

Failure to Complete or Properly Update Schedules

A failure to update the schedule can have a stifling effect on the project, as well as the end of preserving any possibility for time extensions or of properly presenting a claim for additional dollars due to unanticipated delays. Many contractors rely upon a standard bar-chart schedule that reflects the key items of work but never develop relationships between the key work activities. They provide the document to meet a requirement under the contract but never intend that it be put to use as a management tool.

Schedules that are incomplete or are not maintained become a useless piece of paper and a major heartburn. They become a critical piece of evidence in litigation against contractors, not because the contractors are necessarily wrong, they just don't have the necessary evidence to prove they're right. Damages can become very large, particularly if the project failed in one or more of the following ways:

- Failed to complete on time, thus delaying tenant or leasehold improvements that the client had negotiated with others
- Failed to have utilities on line; requiring temporary power be provided
- Failed to have areas ready for fixtures and/or stocking to meet published opening dates

Such delays can lead to damages being assessed against the contractor: See *Alpine Industries, Inc. vs. Gohl,* 637 P.2d 998 (Wash.Ct App, 1982), which upheld the jury award of the client's lost profits for the contractor's delayed completion of a manufacturing plant and *Ralph D. Nelson Co., Inc. vs. Bell,* 671 P.2d 85 (Okla. Ct App, 1983), in which the owner recovered lost rents from the contractor due to the delayed completion of an office building.

When properly prepared and maintained, the project schedule can provide the necessary detail to support delays that have affected the construction progress and resulted in additional dollars expended to accelerate the work, as well as the additional days required to complete the work or support the field office overhead and general conditions resulting from the delay. Representative examples of those issues that can be tracked with a properly prepared and maintained schedule include the following:

- *Tardy Shop Drawing Reviews:* The schedule should reflect turnaround time on shop drawings for all key activities on the critical path, such as reinforcing steel and structural steel. Many contracts have used 20 days as a reasonable turnaround time on shop drawings.
- *Delay Due to Tests and Inspections:* Many contracts will call for special inspections before the next phase of the work can commence, such as verification of the rein forcing steel placement before the concrete can be placed. Indicating the various placement dates on the schedule and providing a copy to the inspectors puts them on notice as to when their inspections will be required.
- *Weather Conditions (Acts of God):* Each weather day can be recorded on the schedule, pinpointing the activities affected, such as a heavy thunderstorm that washed out a series of foundation forms and reinforcing steel that were almost completed and ready for concrete.

The time it takes to remediate the problems can be properly documented to reflect the successor activities affected by the damage and provide evidence to the owner supporting the actual number of added days the contract should be extended.

One of the most important features of a schedule has to do with how the contractor intends to make up for the delays affecting the project. Such a schedule is more commonly known as a "recovery schedule." In essence, this schedule properly documents the activities delayed and their effect on the critical path, then it implements a recovery plan that incorporates how the contractor can bring the schedule back in line. This may require a number of steps to be taken, such as the following:

- Increasing the labor force
- Overtime or shift work
- Moving crews from noncritical activities
- Outsourcing some of the activities

Irrespective of how the contractor intends to recover and bring the project in on time, it is essential to create a series of "what if" scenarios when preparing a recovery schedule. By doing this, one can be better prepared to quickly react if a problem occurs, such as when the activities from which crews were pulled start to fall behind and become critical. By not considering the "what if" scenarios before submitting a recovery schedule to the client, you are exposing yourself to serious consequences. This is particularly critical when faced with a tight labor market, lack of skilled tradespeople, and material shortages.

In any respect, it is essential that the contractor properly documents the delay and indicates how the schedule was affected. It is always good business practice to provide an updated schedule that spells out a recovery plan with the delay documentation to the client as quickly as possible after the delay has occurred and submit it with a well-thought-out recovery schedule.

7.2 THE FALLOUT FROM POOR FIELD SUPERVISION

Poor field supervision creates a series of other maladies that lead the unknowing contractor down the litigation trail. The following are representative aftereffects that are due to unknowledgeable, incompetent, and/or incapable field supervision.

Cannot or Will Not Follow Plans & Specifications

Inability to follow the plans occurs much more often than one believes and usually results from the contractor's having succeeded in ignoring the contract documents on numerous projects, thus forming a belief of immunity from any consequences. Illustrations are many; the following are a few typical ones that have come back to haunt those contractors who ignored the plans and specifications:

- Using a lower grade of reinforcing steel
- Mild steel bolts instead of high-strength bolts
- Box nails instead of common or ring-shank
- No. 15 instead of No. 30 felt
- Noncertified materials rather than the ASTM products

All these happen to avoid paying a few more dollars for the correct product and are all too common a practice in the industry today. Cutting corners on labor is also prevalent in today's economy. It is not uncommon to visit a job site and see a concrete contractor adding water to the mix in order to "shoot" the watered-down concrete across the area rather than having to spend the few extra minutes to buggy it over and properly place it. Not only does this technique compromise the strength of the concrete, in reality it also increases the labor expense. By adding the additional water, the curing and finishing time is lengthened and finishing labor is normally a great deal more expensive than the general labor used to place the concrete.

Unfortunately, some of the blame should be placed on the architect or engineering professional for not making the plans and/or specifications clear. These problems often develop on low-budget projects when the design professionals and contractor are manipulated into performing their services for a low or marginal profit. When this scenario develops, the plans and the resultant work will reflect it. Because of this lack of information, the contractor begins to make decisions to expedite completion of the work that should have been specified by the design professional.

"Shoot 'er a little more water, Joe — you'll never make this corner!"

(Courtesy of Western Technologies, Inc.)

Allows Others to Control or Approve the Work

In today's world, in which many contractors become inundated with paperwork, get engrossed in computer-age gimmicks, employ less-than-qualified field supervision or support personnel, are confronted with unusually difficult client relationships, or act primarily as a broker, it is tempting to partially forget the work. The contractor may leave it in the hands of the owner, subcontractors, and governmental inspection agencies to control the work. Scenarios that partially or totally shift control of the work to the others on the project create a ticking time bomb.

By shifting control, the contractor is abandoning the primary responsibilities, which can lead to serious consequences. The contractor must realize that assuming the role as prime contractor on any given project means agreeing to the terms and conditions of the contract entered into, as well as any federal, state, and municipal construction laws. Examples of these statutes and principles follow:

- It is the duty of the prime contractor to provide a safe place for all workers, irrespective of whose employees they may be.
- The care, custody, and control of the site rest with the prime contractor. Hiring a subcontractor to perform the work does not relieve the contractor from responsibility to protect the subcontractors and the general public from injury.
- Just because a contract becomes difficult, expensive, or inconvenient or turns out to be for an owner who is totally unreasonable, that does not provide an excuse for poor performance on the part of the prime contractor, allow an interruption in the performance of the work, or push the responsibility for the work on to others.
- Letting the owner or others assume control of the work does not create a waiver of their rights to collect damages against the prime contractor for failing to properly manage the work.
- As prime contractor, one is responsible to the owner for the work of the subcontractors, unless they are in direct privity with the owner; therefore, any remediation of construction defects that resulted from their work will be charged to the contractor.

Cheats on Inspections & Testing Requirements

The building code is a safety code, not a performance specification, and does not address many of the critical areas for which inspection and testing are required during the course of construction. Unless inspection and testing are required under the contract documents, these critical areas are left to the discretion of the contractor.

On the majority of projects, geotechnical investigation is done before the contractor is retained, as the information contained in the geotechnical report is needed in the structural and civil engineering calculations to develop the design of the project. The geotechnical report is normally referenced in the design documents but may not be included as a part of these documents when presented to the contractor for bidding and/or construction of the project. Additionally, it is common practice for the client to pay for geotechnical testing and inspection services while requiring the contractor to set the date and time for such services. This contractor responsibility leads us to one of the most important areas in which cheating on inspection and testing requirements can be found: site preparation.

Site preparation includes making the subgrade ready for the building foundation as well as subsequent backfilling. Site preparation is often the most important part of the construction process, as all structures are only as good as the underlying soils they are built upon.

Covers Up Mistakes

Unfortunately, it is not uncommon for a contractor to cover up mistakes, and to brush them off in a "make do" scenario. Many times, this issue is brought about when a project starts to fall behind schedule and everyone is in a panic to complete the work to avoid facing liquidated damages or other consequences. This type of approach to the work will often result in more problems. Examples of such instances are many:

- A contractor constructing a food-processing plant failed to both install required fire blocking and to apply a fire-resistant coating over the sprayed-on polyurethane insulation. The contractor then covered the walls with fire-rated gypsum board pre plan. Unfortunately, when pipefitters were installing process piping, sparks from the welding operation entered the wall cavity and ignited the polyurethane. Although the plant was equipped with fire sprinklers, the water could not reach the fire as it was protected by the gypsum board. Without the fire blocking in place, the fire almost instantaneously flashed through the entire substrate into the ceiling area consuming the total building. The result was complete collapse of the structure.

- A framing subcontractor installed steel brackets upside down on a three-story wood-frame structure. The brackets were to support a steel walkway and stairs. The exterior finish work was completed before the contractor realized that the brackets were installed incorrectly. Faced with a tight schedule, the contractor requested that the miscellaneous iron subcontractor change the connections in the field rather than shop fabricate a

WHY DIDN'T THAT $10 SOILS REPORT SAY ANYTHING ABOUT A SINKHOLE??

modified walkway. The fire sprinkler contractor had not completed his work; therefore, there was no primary fire suppressant system in service. The cutting and welding effort used to field modify the steel framing ignited the wood and the entire structure burned to the ground.

- A stucco contractor failed to properly flash around windows, reverse lapped the waterproof membrane in various locations, failed to repair tears in the waterproof membrane, and when stapling the mesh into place failed to hit a framing stud. After creating these flaws, he proceeded to cover them up with stucco. No one noticed these mistakes until the first heavy rain. Water intrusion into the substrate resulted in microbial growth and severely damaged the interior paint around the windowsills.

The most incomprehensible feature of these events is that the general contractor had to be aware of the problems (or to have done little on-site field supervision), thus becoming a party legally at fault in the cover-up. Instead of stopping the work and demanding that the subcontractor remove and replace the defective work, and then verifying that the work was done in accordance with the contract documents, the contractor (knowing or not) allowed it to continue. Most important, it led to expensive litigation that dragged on for years, ending in costly settlements, and resulted in the cancellation of the contractor's general liability insurance, making it difficult to secure new insurance at a reasonable rate.

It is essential that all parties always be alert and on guard, particularly when a project is being "fast tracked," completion bonuses are offered, or a normal schedule is compressed. These conditions are breeding grounds for mistakes.

It is good practice for the contractor to make use of scheduling to determine what activities are planned for the next week or month. By doing so, one can focus on the effort to be performed and review the details, specifications, and/or any standards that may be applicable for the upcoming work. The work should then be discussed in weekly meetings with the respective subcontractors so they become aware that the work will be inspected to ensure that it will conform to the contract documents.

Plays Architect & Engineer Rather Than Seeking Professional Assistance

When the plans for the project are incomplete, and the specifications do not reflect the details, it is not uncommon for the contractor to proceed unassisted, rather than seek professional help. This problem also surfaces when a project falls behind schedule and the contractor does not want to take the time to consult with the appropriate professional and unilaterally produces a solution. By taking on this design responsibility, the contractor also assumes the liability for the design and allows the architect and/or engineer to avoid responsibility should litigation occur.

"Hadda put them pipes someplace, Sonny... figured you wouldn't mind."

(Courtesy of Western Technologies, Inc.)

7.3 OTHER FACTORS THAT AFFECT JOB HARMONY

It is quite easy to recognize a harmonious project. The job site, as well as the on-site field office, will be clean and free of debris. Safety signs will be conspicuous, along with job bulletins. Cooperation and coordination among the trades will be excellent, and the work will normally be ahead of schedule. You can talk to anyone related to the project and he or she will express pleasure in being able to work on the project.

On the other hand, you can visit a project and know almost instantly that there are numerous problems and tensions among the workers, as well as among the higher echelons. Many of these problems can be traced back to unknowledgeable field supervision and a smattering of the following items.

Heavy-Handed Tactics with Subcontractors & Suppliers

It is not uncommon on dysfunctional jobs to find a large and/or influential general contractor using size and authority to threaten, intimidate, and/or otherwise coerce smaller subcontractors and suppliers into providing services and materials not included in their scope of work. It is not uncommon for such a contractor to threaten a material supplier or smaller subcontractor with termination, withholding of current or future payment, or the bringing in of other suppliers or subcontractors to handle portions of the contracted effort. These contractors use various clauses within their custom-designed purchase order or subcontract agreement that allow such action. Examples of such clauses can be found in Section 2, Contractually Speaking. Unfortunately, many a supplier and subcontractor have been led down this trail and into major financial distress. First the supplier or subcontractor jumps through hoops to satisfy the demands and threats but then finds out that "the problem" belongs to others, such as the general contractor's failure to schedule the work, or the lack of sufficient information in the contract documents.

Failure to Make Use of Pre-Construction Meetings

Pre-construction meetings can be one of the most effective tools in bringing about a smooth and properly run project. This meeting should have all in attendance, including, but not limited to, the following: owner, architect, engineers, special inspectors, prime contractor, subcontractors, and major material suppliers. The general outline of the meeting should cover, at a minimum, the following issues:

- A brief description of the project, its projected start date, mandatory completion date, the standard workweek, and recognized holidays should lead off the session. In conjunction with this leadoff, issue a copy of the latest project schedule to all in attendance.
- Introduction of all the parties who will be involved in the construction effort and what their respective responsibilities are.
- A review of standard operating procedures, including pointing out those sections in the contract documents that cover these procedures presented as a separate handout, which at a minimum addresses the following:
  1. Time and attendance requirements of weekly meetings (i.e. progress meetings, safety meetings)
  2. Shop drawing procedures and turnaround times
  3. RFI procedures and turnaround times
  4. Change order procedures and turnaround times
  5. Inspection procedures and notices
  6. Closeout procedures and punch lists
  7. Warranty work and procedures
  8. Dispute resolution procedure
- Presentation of the latest set of contract documents to confirm that all parties are working with the latest information.
- Open discussion on any concerns or issues that need to be resolved before the work commences.

Everyone present should be encouraged to participate in the open discussion and bring any concern to the forefront. Failing to conduct and/or make use of a pre-construction meeting can and often does lead to serious misunderstandings between parties at a later date. Memorializing the results of the pre-construction meeting in writing ensures that everyone is aware of the project scope, current status, schedule, and its rules and regulations. The importance of this meeting should be transmitted to all concerned prior to starting construction.

Failure to Include Design Professionals in Weekly Progress Meetings

It is not uncommon to leave the design professionals out of the weekly progress meeting. Many times, the reasoning behind this may stem from the owner's reluctance to pay for their time, or the contractor may be working under a construction management agreement whereby the contractor controls the design and plans to relay any issues to the design professionals that require input. Irrespective of what the reasoning might be, it does not make good sense to leave a key member of the construction team out of weekly progress meetings. The earlier a design professional is aware of any issue that relates to the design, the quicker it can be resolved. In fact, many times the resolution will take place during the meeting, after reviewing the issue with the trade involved and receiving ideas on possible solutions.

An additional benefit from having the design professionals at the meeting is that it allows them to walk the job and provide their comments on the progress and quality of the work being performed. In so doing, the design professionals have provided more sets of eyes and can gain confidence that the work is being performed in accordance with the contract documents. This tactic will also encourage the development of a bond of respect between the contractor and the design professional rather than create an adversarial relationship, which happens all too frequently when the design professional is left out of the loop until an emergency arises and then is blamed for a poor design or a delay in resolving the problem.

Failure to Understand the Value of Inspection & Testing Services

When contractors fail to conduct the testing and/or inspections either called for in the contract documents or needed to confirm the strength and safety of equipment or materials that become an integral part of the project, they expose themselves to serious liability should a failure occur.

These inspections and testing services cover a multitude of areas, and if not done correctly, could have disastrous aftereffects, including severe structure damage or failure resulting in injury or death. The old adage "an ounce of prevention is worth a pound of cure" still holds true today. The following are examples of issues that could or should require professional inspection and testing services:

- When the contractor discovers that there was no geotechnical testing performed on the site, make it a standard practice to require a soil test in order to determine what the structure will be resting on before construction starts.

"And we can save 700 lira by not taking soil tests."

(Courtesy of Western Technologies, Inc.)

- Before placing foundation concrete, inspect the cuts to be sure the concrete will be resting on solid undisturbed native material.
- Take density tests on fills at maximum lifts of 1 foot or less to ensure against settlement and/or sliding.
- Take slump and concrete tests to ensure that the concrete mix designs meet the requirements of the specifications.
- Inspect scaffolding and shoring to be sure it is safe before allowing tradespeople to commence work.
- When manufactured joists are delivered and dumped on the site, inspect the panel points to ensure that the pressure plates or welds are secure before they are erected.
- Inspect materials such as reinforcing steel, bolts, mesh, and so on to ensure that they meet the specifications.

Leaving such things to chance presents a disaster in the making and results in devastating consequences for the contractor.

7.4 LACK OF OR INSUFFICIENT DOCUMENTATION

Undoubtedly, one of the most preached about requirements in the construction industry, yet the least adhered to until "after the horse is out of the barn," is proper documentation. Even when there is a considerable amount of documentation, it may not address the proper issues. It is, therefore, essential that the contractor be aware of what and how to document all of the issues that affect performance of the work.

Every contractor and subcontractor should develop a standard procedure for properly documenting the work effort. The documentation should extend from the time a contract is awarded through final acceptance of the work. We stress this point, as it is not uncommon for litigation to take place long after the project has been completed and memories fade fast. Be aware that the sued prime contractor's attorney will bring all of the subcontractors into the litigation as third parties at fault. This is often done to create more pockets to reach into when attempting to settle the case.

Who can remember what happened three, four, or more years ago, particularly when the supervisor who ran the work is no longer available? If there is no reliable record, then the ability to properly defend one's actions is greatly weakened. On the other hand, if an accurate set of documents has been maintained throughout the work proving that the work was properly performed in accordance with the contract documents, a strong defensive position exists.

Representative documentation that supports a contractor's position would include the following:

- *The Daily Progress Report or Daily Log,* if properly maintained, has proven to be one of the most effective ways of defining what happened, when it happened, and how it happened.
- *As-Built Drawings,* if properly maintained, have been used in numerous cases to show exactly how the work was completed. Be explicit regarding any significant change and provide accurate dimensions and elevations.
- *Job Progress Photos,* are essential to verify that the work was done in accordance with the contract documents. Additionally, a photograph is worth a thousand words when it comes to substantiating what happened in a structural failure, accident, fire, or other malady that may occur during the course of construction. A photo of one phase of the work may capture something in the background that reveals the cause of a problem later. Progress photos can be a lifesaver when a dispute develops. With the advent of advanced technology in videography, it's relatively inexpensive to set up a centralized roving camera that records the daily work effort as well as acts as a security devise after work hours.
- *Correspondence Files,* should be set up chronologically for each party that had a responsibility in completing the project, be they the owner, architect, engineer, subcontractor, or supplier to the governmental authority having jurisdiction. Documents will include e-mails, letters, "speed" memos, change order requests, directives, stop notices, and so on.
- *Project Schedule File,* including all updates, impacts, minutes of weekly scheduling meetings, test results, RFI log, shop drawing log, and any other documentation that may affect the progress of the work.

Incomplete or Lack of Daily Reports

Daily progress reports and/or a "daily log" should be maintained by each contractor and subcontractor pertinent to the respective efforts. Such reports must be tied into the project schedule and/or the time requirements stipulated in the agreement between parties. It should document on a daily basis what happened, who worked and where, specifically noting names of the people involved in any discussion concerning the work, any delays in the work that may affect the project, along with what caused the delay and who was responsible. The daily report should provide supporting evidence of any serious problem and/or job site accident with photographs, drawings, sketches, memoranda, letters, or other such data that will memorialize the problem or accident, in case of future claims or litigation. Be sure to incorporate any in-house problems such as strikes or late material deliveries that might affect the progress of the work and state which scheduled activities may be affected.

No Weekly Progress Meetings

Failure to conduct a weekly progress meeting with all concerned parties can leave the project like a ship without a rudder. The weekly progress meeting must allow an open exchange of information among all of the parties and bring pending problems to the surface. For best results, distribute an agenda of what will be covered a couple of days before the meeting. Passing the agenda out at the time of the meeting does not provide sufficient time for the parties to react or be prepared to discuss the issues that need to be covered.

A weekly progress meeting agenda might include the following:

- Priority issues should be the first order of business, categorized by trade, description, mandatory resolution date, and what consequences will occur if not resolved by the resolution date.

- Next is old business. Categorize each issue by trade and activity description, indicating the date each issue was first recognized and who was responsible for resolving the issue, along with the original resolution date, followed by the proposed resolution.

- Next is new business. Categorize issues that have come about since the last meeting and need resolution. Again, categorize each issue by trade and activity description, indicating the date the issue was first presented along with who is responsible for resolving the issue, and the date by which the issue must be resolved before it affects successor activities.

- Safety topics to be discussed at the next safety meeting that relate to forthcoming activities.

Once the weekly progress meeting has been completed, the contractor should ensure that the minutes of the meeting are properly prepared and distributed to all concerned by the start of the next business day to allow those in attendance to review and act on the issues presented.

Requests for Information (RFIs)

RFIs are written communications with the client, architect, and/or the engineer and are an integral part of the contractor's responsibility. They are a means of defining job-related problems that affect the work, and they require a response from the party to whom the RFI was addressed. RFIs may be caused by incomplete information in the contract documents, failure to return shop drawings on time, failure to have integral work of others completed correctly and on time to accept other work, another contractor's working and/or blocking access to a work area, late inspection of the work, or damage to the work by others. These are all issues to be documented.

Documentation & Construction Claims

Construction claims come in many forms and are an integral part of survival to all who do business in the construction industry. As claims are such an integral part of doing business, the contractor needs to have a working knowledge of how to address a claims issue. Unfortunately, most contracts will include clauses that attempt to discourage or totally eliminate claims. One of the most common clauses found in contracts is the "No Damages for Delay Clause." Over the years, this clause has been argued in courts throughout the United States, and unfortunately for subcontractors, many courts have developed what is known as the "Severin Doctrine" (*Severin vs. United States,* 99 Ct. Cl. 435 [1943]). In general, the doctrine limits the circumstances under which a contractor can support or pass through a subcontractor claim to the owner unless the contractor has suffered actual damages. It is, therefore, essential that the contractor and subcontractor thoroughly understand the ramifications of how to properly substantiate any type of delay claim. Fortunately, there are exceptions to the Severin Doctrine that many courts have upheld, including the following:

- *Uncontemplated Delays:* These are delays that were not contemplated by the parties when they entered into the contract. Compensation has been awarded in such instances when the owner failed to make the site available in time for the contractor to commence work and/or the owner provided defective plans and specifications. *Hawley vs. Orange County Flood Control District,* 27 Cal. Rptr. 478, 481-84 (Ct. App.1963).

- *Protracted or Unreasonable Delay:* This is an owner-caused delay that could include an unusually long delay in starting work on a project or stopping work for an unusually long period of time for the benefit of the owner. *McGuire & Hester vs. City & County of San Francisco,* 247 P. 2d 934, 936-38 (Cal. App. 1952).

- *Owner Interference:* The owner owes the contractor an implied duty not to hinder the contractor's performance. This statement is found in many litigations involving construction. If an owner breaches this duty, it may be argued that the contractor is then released from any existing no damage for delay clause. Such a breach could be implied when an owner willfully and for his or her own benefit employs another contractor or has his or her employees commence work in an area in which the contractor was to perform his or her contractual obligation, therefore failing to make the site available. *Coatesville Contractors & Engineers vs. Borough of Ridley Park,* 506 A. 2d 862, 867-68 (Pa. 1986).

- *Fraud, Misrepresentation, or Other Wrongful Conduct:*   It is generally understood that a "Covenant of Good Faith and Fair Dealing" is implied in every contract. *Wagenseller vs. Scottsdale Memorial Hosp.,* 147 Ariz. 370, 383,710 P. 2d 1025 (1985). By the owner's failure to disclose material facts, such as existing and/or a changing site condition that would have an effect on the contractor's quantities of work, the no damages for delay could be voided. Such misrepresentation could set aside the contract allowing the contractor to recover in quantum meruit. *Murdock-Bryant Construction, Inc. vs. Pearson,* 146, Ariz. 57, 60-63, 703 P.d 1026 (App. 1984).

- *Delay Outside the Scope of the Clause:*   As the no damage clause is uniformly interpreted by most courts, it may not become an issue if it does not cover delays, which usually will favor the contractor. In general, a delay outside the scope of the clause covers issues that include delays on the part of other contractors, subcontractors, and material suppliers but not specifically the contractor performing the work, such as a steel supplier being unable to provide the materials on time. *U.S. Industries, Inc. vs. Blake Const. Co., Inc.,* 671 F. 2.d 539, 543 (D.C. Cir 1982).

These are some representative examples of the complex issues surrounding just the "No Damages for Delay Clause" of a contract and are presented to inform the contractor of the absolute necessity to properly document and provide timely notice of any delay, irrespective of any contract clauses. Numerous publications are available to contractors covering the documentation and presentation of claims, some of which have been presented in the appendix of this book.

7.4 QUALITY ASSURANCE/ QUALITY CONTROL

"Quality assurance" and "quality control" were buzzwords that started in the manufacturing industry and worked their way into the construction industry during the late 1940s, 1950s, and early 1960s when the federal government pumped millions of dollars into the construction industry for rehabilitation after World War II, the interstate highway system, thousands of new homes, and the rush into space. This influx of funds created, and expanded, new agencies to monitor how these funds were to be controlled. Agencies such as the Federal Housing Authority (FHA), Housing and Urban Development (HUD), Occupational Safety and Health Administration (OSHA), Federal Highway Administration (FHA), and National Aeronautics and Space Administration (NASA) would develop new (and revise old) manufacturing and construction standards, which had to be met in order to continue to receive funding. Many of these standards were well developed and became the basis for improving or developing state statutes and municipal building codes in existence today.

Additional and more sophisticated standards developed as the construction of missile sites used air-frame contractors who retained "quality control engineers" who established programs on how these missile sites were to be constructed. Construction work that had previously been the responsibility of the Corp of Engineers and the Bureau of Yards and Docks, with experienced staff architects, engineers, and contractors, and with trained and educated personnel, was now transferred to personnel who were trained in the building of airplanes in accordance with governmental regulations.

Standard nuts and bolts used in construction for years now had to be certified that they were manufactured in accordance with a federal specification number. Without this certification, the work would not be accepted by the responsible quality control inspector. Although many of these inspections were unnecessary, costly, and often challenged, and they eventually disappeared, they still left an indelible mark.

The principle left by these governmental programs was that quality was truly a key ingredient of any construction project. We now find that more and more local municipal governments have taken up the push for more quality in construction. "Special inspections" by knowledgeable professionals have now become a requirement for work that exceeded the capability or expertise of the local building official.

As more new and sophisticated materials continue to arrive on the market, the latest information is required to ensure their proper use, prompting those involved to take a greater interest in quality assurance. This interest has involved engineering and trade associations, building officials' organizations, local municipalities, and others and has resulted in new laws being passed, and codes and standards being upgraded, leading to an ever-expanding need for quality testing programs on new construction work. Most tests in the construction industry are conducted in accordance with the standards of the American Society of Testing Materials (ASTM). On governmental projects, the specifications may require that testing be conducted in accordance with either federal or state standards. Most state highway departments and many counties have their own standards.

Testing is a recommended and typically required means of assuring that the materials incorporated into a construction project meet the specifications and will perform the intended purpose. On many projects, the testing is separated into two categories: quality control and quality assurance.

Quality control involves testing by the material supplier to maintain uniformity of the product. For example,

Portland cement manufacturers test the raw materials upon entering the plant, test the material in several stages of production, and test the finished product. Such tests are used to control and adjust production to ensure consistency and quality of the product shipped from the plant.

Quality assurance is provided and paid for by the owner/buyer of the product. This may include testing the material delivered to the project for compliance with the specifications and testing after it is incorporated into the project for compliance with the construction requirements. Generally payment for the contract work requires compliance with the quality assurance testing.

The difference between quality control and quality assurance may not be clearly understood, and all too often the line between them becomes fuzzy and may result in disputes. Differences in results between tests performed by the supplier and those performed by or for the owner/buyer often are the basis for dispute with both sides claiming their results are accurate. On some governmental projects, the specification requires formal submittal of the quality control and quality assurance methods and procedures. This requirement normally defines the differences and responsibilities of the parties and stipulates that the quality assurance tests govern acceptance of the work.

The scope of the quality assurance program may be reduced by the owner/buyer as a cost-saving measure, which often results in a dispute between an owner and contractor. A good example occurs when an owner or contractor, who has assumed or been given the role of determining the quality assurance requirements for a project, limits the number of compaction tests during placement of an engineered fill. (After all, why test every lift as the fill is made? Let's save a little time and money!) It is *necessary* to conduct a representative number of tests in each lift to ensure that the compaction is in conformance with the specifications.

Not only must the proper number of tests be made during the quality control and quality assurance phases, but also the results must be properly interpreted. Unfortunately, many people in the construction industry, including architects, engineers, specification writers, contractors, and, in some cases, suppliers, are not knowledgeable about the properties of specific materials or how to interpret the test results. As previously noted, most tests in the construction industry are conducted in accordance with ASTM standards. The American Association of State Highway and Transportation Officials (AASHTO) has standard test procedures that are referenced on highway and other projects. Many of these standard test procedures are the same or similar to ASTM test methods. Rigorous conformance to the test procedures in project requirements is essential. Investigations made in the course of dispute resolution often find earlier testing was faulty or interpretation of test results was erroneous.

Quality Control Testing

Quality control is the effort put forth by the manufacturer to achieve an end product that conforms to a given standard or specification. Quality control, also referred to as "product control," involves more than just testing. Properly trained personnel, adequate equipment, and suitable raw materials along with an appropriate manufacturing procedure are all essential parts of the control of a product's quality. Quality control testing is only one step that determines if a product conforms to a standard.

Quality control testing also provides evidence of the end product's properties. Many manufactured products incorporated into construction projects are accepted on the basis of the manufacturer's quality control test results or warranties of the products' conformance to certain standards based upon these test results. Mechanical, electrical, plumbing, roofing, structural steel, masonry units, and similar products are often accepted and installed without any further verification of their conformance to project requirements.

Some other products such as Portland cement, liquid asphalt, hydrated or quick lime, and mineral aggregates that are incorporated into the project are tested on some projects and accepted without further testing on others. The ability to satisfactorily serve as a component in the manufacture of products that are incorporated into the project is often measured by testing the end product. The quality of the end product will normally be tested by the supplier and often further tested by the owner's representative. The ability to evaluate the contribution of the component parts to the quality of the end product is a valuable asset.

Quality Assurance Testing

Quality assurance testing is performed on behalf of the owner to verify that project construction conforms to the project specifications. It is often directly related to acceptance and payment for the work. It is essential that the project specifications clearly state product requirements and the manner in which compliance will be measured. If the referenced standard test methods are to be modified, the modifications must be clearly stated in the contract documents. The frequency or amount of testing should be given in the contract documents. Alternately, a sampling and testing plan should be provided by the owner's representative prior to the start of the project.

Sample selection for testing of product acceptance should be carefully performed. Sample selection confined to only the best, worst, or representative portion of the product indicates that the choice is biased. Random selection of samples is the preferred method to obtain an unbiased measure of product quality. Random sampling is not the same as haphazard sampling. Random sampling should be performed such that all portions of a production lot have the same probability of being selected as the sample location. The method of selection of random samples and the resulting sample selection should be documented in the project records.

If there is reason to question the acceptability of a portion of a production lot, that material may be sampled and tested; however, that test result should not be used as a measure of overall quality of the product. Oftentimes, visual inspection of work in progress is a better indication of conformance to project requirements than performance of a test. However, documented test results are generally necessary to provide evidence that the work meets specification requirements. Trench backfill is a good example of the need to observe the performance of the work while in progress but should never overshadow or replace the need for compaction testing.

Products that are manufactured at or near the project site, such as Portland cement concrete, hot mixed asphalt pavement, base courses, soil embankment and foundations, stucco, mortar, and similar products, typically generate the most construction testing problems. Such problems are often the result of the lack of sophistication of control and testing conditions when compared to products manufactured under better, that is, factory controlled conditions. Adherence to proper procedures for obtaining samples and preparing, collecting, and handling test specimens is paramount. The use of proper test equipment and rigorously following specified test methods are essential.

Nuclear gauges for measuring in situ earthwork densities are often a source of error. In addition to the need for the nuclear gauge to be properly calibrated, it generally is not the specified test method. However, due to its ease of use, it is often relied upon for quality control tests and, in some cases, final acceptance testing. When used for these purposes, the nuclear gauge test results must be correlated to the specified standard test by performing companion tests. The difference in test results becomes a correction factor used with the nuclear gauge. In some instances, the nuclear gauge reading may be closer to the actual measure of the product's property; however, correction of the test results to correspond with the specified test is required. Reestablishment of the correction factor must be performed on a regular basis. In addition, the correlation must be repeated whenever a new material or a new project is encountered.

As the construction industry continues to develop new products and methodologies, the need for minimum acceptable standards grows. Testing and inspection updates of these new products are needed to ensure continued quality and performance. These constant changes in products and standards prompt the need for construction supervisory personnel to be continually educated in the proper use and application of those materials and products. Without this knowledge, there can be no assurance as to the quality of the finished product, be it a cast-in-place concrete beam, a waste and vent plumbing system, an exterior insulation finish system, a built-up roofing system, a surround sound home movie installation, or the completed hi-tech smart buildings being constructed today.

It is, therefore, essential that each contractor fully understand the attributes, responsibilities, and boundaries of any particular trade and what is expected under governing law, contract specifications, and applicable industry standards. Without this understanding, there can be no assurance of quality and without quality there will always be speculation and discontent.

7.5 FEDERAL, STATE & MUNICIPAL LAW

Federal, state, and municipal law may vary in many respects when it comes to issues involving the contractor and the construction project itself; therefore, it is essential for the contractor to be aware of which law applies and which has precedence over the other. Examples of these follow:

- When working on federal lands or in the confines of a federal institution or military base, you are first and foremost working under federal law. Any other laws are incidental unless specifically made applicable in the contract documents provided in support of the work. Those who have not had the experience in working under federal guidelines should thoroughly familiarize themselves with FAR (Federal Accounting Regulations) and, if working on a military installation, ASPR (Armed Services Procurement Regulations), in addition to any federal regulations applicable to the work being performed.

- Federal law governing construction crosses state boundaries has limited effect on state law, unless the project is federally funded or a state of emergency exists. State law, on the other hand, has the option to adopt and/or modify the federal law to comply with state statutes. State law has superiority over municipal law. Some states have adopted their own state construction codes, such as California, Florida, Michigan, New Jersey, and New York. These codes take precedence over local municipal codes and ordinances.

- Municipal law normally addresses construction-related issues by adopting those federal and state regulations that are applicable to the specific needs of the municipality and may use them in total or as a supplement to state building codes and ordinances. Most municipalities will adopt such standards as found in the International Building Code (IBC), Building Officials Code of Administration (BOCA), National Fire Code (NFC), Uniform Mechanical Code (UMC), Uniform Plumbing Code (NPC), the National Electrical Code (NEC), or others and then revise them through city ordinances to fit the needs of the municipality.

It is essential to the success of the contractor's business to become familiar with that section of the law that best applies to contracting. It should be noted, however, that the interpretation of any applicable code, law, and/or ordinance is left to the local building official having jurisdiction. This can, at times, conflict with newly developed technology and/or methodology being used in the industry, but never yet incorporated into the existing law, code, and/or ordinance; or the building official may have little knowledge of the issue and render an inaccurate interpretation, which may cause harm to the contractor and public in general.

To challenge the building official's interpretation of any existing section of the law, code, and/or ordinance or to claim that there is no current law, code, and/or ordinance that addresses the issues in dispute, a contractor must have carefully researched the issues to be challenged. This is best accomplished by developing a matrix of the various segments of the issue in order to better determine which portions, if any, support your position, partially support your position, and/or don't support your position. The contractor bears the burden of proving the position taken and must be prepared to provide credible supporting evidence of this position.

7.6 LACK OF SKILLED TRADESPEOPLE

The apprenticeship training programs, which were provided by the various trade unions, for all intent and purposes have disappeared in most areas of the country with the demise of the unions and the growth of open-shop contractors. Some contractor organizations are attempting to develop trade schools to teach the needed skills for construction trades. However, many of the tradespeople are arriving on the job with little to no knowledge of the work they are to perform and many have little knowledge of the English language. This unfortunately presents problems in communications regarding not only what must be done, but also how it must be done. This lack of training, skill, and communication has unfortunately led to a growing number of construction defects and job-site accidents, resulting in higher operating costs for the contractor and increased construction litigation. It is essential that any contractors when confronted with a workforce that has little knowledge of the English language ensure that they develop a means of communication with the workers.

Improving communications may include hiring supervision and/or leaders who are bilingual and have the ability to communicate with the workers, posting safety and job notices in both English and Spanish, and conducting on-site training sessions or working with local agencies or trade association in the development of skill training. In any event, contractors should ensure that their field supervision has sufficient knowledge of the workers' language to stop them from committing an unsafe act or performing defective work. Pocket books covering English to Spanish translation are available through trade associations, hardware stores, etc. that include a number of construction terms to help the field supervisor better communicate with a worker who does not speak English.

This issue of communication with an unskilled worker should not be taken lightly as it remains the responsibility of the contractor to maintain care, custody, and control of the work site; therefore, any injury or construction defect will fall back on how this responsibility was managed.

7.7 KEY POINTS TO REMEMBER

In this section, we have provided the reader with a series of factors that, when not taken into consideration during the daily course of business, can lead to costly consequences. The following are key points to remember as do's and don'ts.

Important Do's

1. Maintain competent and qualified supervision by requiring and providing continual management, quality control procedures, and technical education in their respective fields.

2. Hold monthly supervisor meetings to discuss issues that affect quality and performance in the field or office, and, most important, listen to the concerns and needs of those people you have entrusted to perform the work.

3. Insist that every field supervisor maintain a daily log that covers the work performed that day, the number of trades on the job, visitors, inspections, tests, weather conditions, defects, accidents, failures, and delays.

4. Insist that every field supervisor have a tool kit including a camera, notebook, and pencil to allow immediate documentation of issues that affect the proper performance of the assigned work, including but not limited to defects, accidents, failures, and delays.

5. Use the project schedule when conducting weekly progress and safety meetings; this will help ensure that the topics discussed relate to the work to be performed and are essential to meeting the critical path milestones.

6. Keep accurate meeting minutes, prioritizing those issues needing attention, the parties responsible for resolving and/or acting upon the issues, and the date and time the issues need to be resolved to avoid a schedule delay.

7. Maintain an active liaison with the architect and engineers responsible for the design by involving them in your progress meetings and through a well-documented Request for Information program.

8. Keep the client informed of your progress and any delay that may affect the time of completion, including those actions being taken to address such delay.

9. Maintain an accurate set of As-Built plans that clearly show any deviation in the contract set, recording the date such change was made, who authorized it, and cross references to any Request for Information or change order issued.

10. Insist that your field supervisor practice quality control by walking the project at least twice a day to observe the quality of the work and record the progress.

Important Don'ts

1. Don't enter into an agreement that you have not thoroughly read and understand.
2. Don't start construction until you have seen the soils report and know what you are building on.
3. Don't start construction until you are totally familiar with the site conditions, including property lines, easements, environmental requirements, drainage, utility locations, emergency ingress/egress, and so on.
4. Don't start construction until you have reviewed the plans and specifications and are assured that they reflect accurately what is to be constructed, leaving no outstanding issues that could disrupt, delay, or otherwise affect the work.
5. Don't ever take on the responsibility that rightfully belongs to the architect or engineers.
6. Don't ever cover up mistakes.
7. Don't proceed with placing your work on the top of others' work when you know or believe that such work is or may be defective.
8. Don't dump slash, trash, and debris in engineered fill zones or drainage channels.
9. Don't water settle fill material in lieu of mechanical compaction.
10. Don't add water to concrete to avoid having to transport it.
11. Don't give out impossible completion dates to be impressive.
12. Don't allow others to run your work.
13. Don't allow work to be performed that you know is unsafe or incorrect.
14. Don't believe that because the city or other governmental agency approved your work that it was correct.

SECTION EIGHT

CARE, CUSTODY, & CONTROL

or

Who's Watching the Store

8.0 JUMP-STARTING SECTION 8

This section has no "Jump Start" introduction like the other sections because it presents 22 vignettes, each of which demonstrates the drastic consequences of an architect or engineer failing to provide the necessary engineering, required drawings, and specifications, as well as the contractor's failure to maintain *care, custody, and control* of the work.

Every architect and engineer should have the care, custody, and control of the design of a project from its conception through completion by providing clear instructions to the contractor throughout the course of the work. These instructions are in the form of construction plans, specifications, shop drawing reviews, Requests for Information (RFI), supplemental instructions, and quality assurance observations. Unfortunately, many owners do not understand the value of qualified professionals during the construction phase of the project.

Every prime contractor who takes on the construction of a project has a responsibility for the care, custody, and control of everything that transpires within the boundaries of the project site. In other words, the contractor has a duty to properly manage, coordinate, inspect, oversee, and control the entire work taking place on-site. In so doing, the contractor has, in essence, committed to providing a safe working environment for all who enter the work site.

The care, custody, and control of the project pass to the owner once the work, as defined in the contract documents, has been successfully completed and accepted by the owner. This includes the responsibility to maintain and properly care for the various components, as may be required in operating instructions provided at the time of acceptance, and appropriate recognized standards of care, including the establishment of a safe environment for all who enter within the boundaries of the completed project.

Many architects and engineers have attempted to transfer numerous construction details (and their liability) to the contractor on the unjustified basis that the contractor provides the "means and method" of how the project will be constructed. Many contractors in turn attempt to transfer this liability and responsibility to their subcontractors. This does not work since the subcontractor is not in privity with the owner or other subcontractors, nor does the subcontractor have the authority to coordinate the work of others. While the prime contractor may, by contract, require his or her subcontractors to indemnify him or her from their actions, the contractor is still open to liability.

Likewise, the owner often attempts to transfer blame back on the contractor for problems that are directly related to the owner's failure to properly maintain the project in accordance with the contract documents, operation manuals, or recognized standards of care.

This section presents a series of actual events that have occurred because the architect, engineer, prime contractor, or owner failed to take the care, custody, and control responsibility seriously, the results of which can be quite costly, not only in the payment of fines that may be imposed by governmental authorities, but also in defending him- or herself through litigation by the damaged party, along with increased insurance premiums and, most important, the destruction of property, injury, or death. It is interesting that many of the events covered in other sections of this book could easily been made a part of this section.

Many of the following events fall under recognized guidelines or standards, but one must look deeper to understand why the event took place and what could have been done to prevent it. Obviously, the foremost point in presenting these events is that had the architect, engineer, contractor, or owner done it right, they wouldn't have had to pay the price. During the course of looking through the following actual events, what parts of Sections 2, 3, 5, 6, and 7 reflect the primary issues that led to the disastrous results associated with the *care, custody, and control* on any one of these projects?

8.1 GONE IN 40 MINUTES!

During the construction of a large processing plant, a pipefitter was welding fittings on a spool piece that was to be inserted through a hole in the exterior wall. The hole in the wall had already been made. The pipefitter was working on the interior of the building, next to the exterior wall, and had properly placed a freestanding welding shield to protect the nearby workers from the arc. The structure was a prefabricated metal building complete with an active fire sprinkler system. The interior side of the exterior wall was constructed of 2 x 6 wood framing fastened to the steel girts and covered with a $5/8$-inch gypsum board up to the ceiling. The interior gypsum wall was covered with a $3/4$-inch plywood wainscot.

The ceiling was a suspended gypsum board with glued-on acoustical tiles. The attic space was fully fire sprinkled. The exterior wall substrate had sprayed-on polyurethane insulation applied to the interior face of the metal siding, which also covered the underside of the roof decking.

Sparks from the welding operation entered the wall cavity and ignited the wood framing and the spray foam polyurethane insulation (SFPI). Within 40 minutes, the entire structure was on the ground. What happened?

SFPI is a highly combustible material under adverse conditions. Once it's burning, it can generate temperatures to 10,000 degrees Fahrenheit, making it extremely dangerous. The design details called for a protective film on the surface of the spray foam. However, the contractor, for financial reasons, convinced the local fire marshall that the spray foam was encapsulated in the wall, and the building was sprinkled, leaving little risk. In reality, the contractor's reliance on wall encapsulation and sprinklers resulted in catastrophic consequences. The second major consequence occurred when the contractor did not provide proper fire blocking within the wall but relied solely on the steel girts to act as a fire-stop. The lack of a proper fire-stop created a chimney effect within the wall cavity, allowing the fire to go unchecked until it flashed across the underside of the roof. Since the fire was within the wall cavity, the fire sprinklers didn't activate until it reached the attic space. By then, the fire was so hot and intense the sprinklers could do little to stop the fire contained in the wall cavity. The excessive heat soon collapsed the bents, bringing down the roof and creating a multi-million-dollar tragedy.

This is a complex situation and much can be learned from a review of the events. Consider the following facts:

- The original plans and specifications had been prepared showing a protective fire coating over the polyurethane insulation and had been approved by the local building officials and the fire marshall.
- The fire marshall made a field decision to approve the change in construction procedures as presented by the contractor, which left out the protective coating.
- The work was constantly being inspected by the contractor and local building inspectors.
- The contractor and subcontractor gave safety talks and thought they were doing the right thing.

Then, which of the following reasons caused this fire to occur, injuring three workers and destroying an entire building?

- The contractor was trying to save money for the client.
- The fire marshall relaxed the plans and specifications by allowing the contractor to install the polyurethane insulation without a fireproof coating.
- Both the local building official and the fire marshall lacked sufficient knowledge of prefabricated steel buildings; otherwise, they would have realized that additional detailing was required to ensure that the fire blocking would, in fact, prevent a fire from moving up the interior wall cavity.
- The architect failed to provide sufficient detailing that reflected the variable relationship between the tapered bent sections and the uniform size girt that was intended to provide fire blocking on the exterior wall. Had this been considered, the architect would have increased the width of the girt to match the expanding wall cavity.
- The subcontractor should have had a fire-watch while he was performing the welding of the pipe.
- The subcontractor should not have cut the hole in the exterior wall until after he had completed the spool piece.
- The contractor, the local building inspector, and the fire marshall should have recognized that the fire blocking was inadequate and did not meet the requirements of local or national fire codes.

The answer, of course, is all of the above! Needless to say, there was sufficient blame to go around from the architect, to the contractor, subcontractor, and local building officials. As the local building officials are protected from liability, the burden fell primarily onto the contractor and subcontractor.

Photographs taken during the forensic investigation covering the cause of this fire can be found in Section 9.

8.2 DORMITORY FIRE DELAYS OCCUPANCY!

During the course of construction of four new three-story dormitory buildings at a university campus, a fire broke out virtually destroying one of the buildings and damaging two others.

The buildings were wood frame with a one-coat stucco system over oriented stand board (OSB). The floor system was lightweight gypsum concrete over OSB decking. A hallway ran down through the center of each floor with student living quarters on each side of the hallway. The hallway was to have been constructed to meet the requirements of a two-hour fire-rated assembly. The buildings were all fire sprinkled. Walkways between buildings and stair towers were steel frame with metal decking and poured in place concrete. The steel walkway framing was attached to the wood frame structure by a series of steel brackets bolted to the wood framing. The steel brackets were installed by the framing contractor during the course of construction. The building was framed and the one-coat stucco system applied, and much of the interior work was completed before the walkway framing was to commence.

When the steel erector came on the project, it was discovered that many of the brackets had been installed incorrectly. The project also had dimensional errors between buildings, with the result that the prefabricated walkway steel sections could not be properly connected. Because the new semester was to begin shortly, it would cause a major delay to open the walls to remove, remediate, and reinstall the brackets to receive the walkway framing. The general contractor directed the steel subcontractor to modify the brackets and steel framing in the field. This decision required that cutting and welding on these brackets be done in place.

During the course of the bracket modifications, the fire sprinkler system was coincidentally shut down for repairs. The cutting and welding of the brackets was not stopped during the time the fire sprinkler system was out of service. This work created sufficient heat to slowly ignite the wood framing. The remediation crew had moved to work on other brackets before the fire was discovered. The fire moved quickly up the exterior wood framing and proceeded to spread through the wood floor joists and into the attic framing. By the time the fire department reached the site, one dormitory was totally engulfed and the fire was beginning to attack the other buildings.

Applicable OSHA standards for conducting such work can be found under OSHA General Requirements § 1910.252 Welding, Cutting & Brazing, in which the basic precautions for fire prevention in welding or cutting work include the following:

1910.252(a)(2)(iii)(A) *"Fire watchers shall be required whenever welding or cutting is performed in locations where other than a minor fire might develop, or any of the following conditions exist:"*

1910.252(a)(2)(iii)(A)(1) *"Appreciable combustible material, in building construction or contents, closer than 35 feet to the point of operation."*

1910.252(a)(2)(iii)(A)(3) *"Wall or floor openings within 35-foot radius expose combustible material in adjacent areas including concealed spaces in walls or floors."*

1910.252(a)(2)(iii)(A)(4) *"Combustible materials are adjacent to the opposite of metal partitions, walls, ceilings, or roof and are likely to be ignited by conduction of radiation."*

1926.24 Fire Protection & Prevention *"The employer shall be responsible for the development and maintenance of an effective fire protection and prevention program at the job site throughout all phases of construction, repair, alteration, or demolition work. The employer shall insure the availability of the fire and suppression equipment required by Subpart F of this part."*

1926.352(f) Fire Prevention *"When welding, cutting or heating is performed on walls, floors and ceilings, since direct penetration of sparks or heat transfer may introduce a fire hazard to an adjacent area the same precautions shall be taken on the opposite side as are taken on the side in which the welding is being performed."*

Upon investigating this fire, a number of interesting issues were uncovered:

- The general contractor did not have a "Hot Program," as required by OSHA, to address the cutting and welding needed to correct improperly installed brackets and extreme deviation in the face-to-face dimensions between buildings.

- The general contractor also allowed the fire sprinkler contractor to turn off and work on the sprinkler system while the welding was being done.

- The subcontractor did not field measure between the buildings prior to fabrication or delivery of the miscellaneous iron. Shop drawings were prepared from the design documents and approved by the general contractor with no indication that field verification was needed before fabrication.

- Brackets were installed by the framing contractor who worked from the design documents and not the erection drawings prepared by the miscellaneous iron subcontractor.

- The general contractor used aggressive tactics with the miscellaneous iron subcontractor to have him expedite the changes to meet the scheduled completion date.

- The turned-down building footings were found not to be square and parallel with one another resulting in the face-to-face exterior wall deviations.

- As the school was being constructed on state property, the local building officials had no authority concerning compliance with the building codes. The on-site inspection was conducted by the school's maintenance personnel, who had minimal construction experience.

After reviewing Figures 8.1 and 8.2, it should be evident that there are a number of deficiencies in both the contractor's and subcontractor's work. The state was protected from any liability, and, therefore, the burden fell to the contractor and subcontractor.

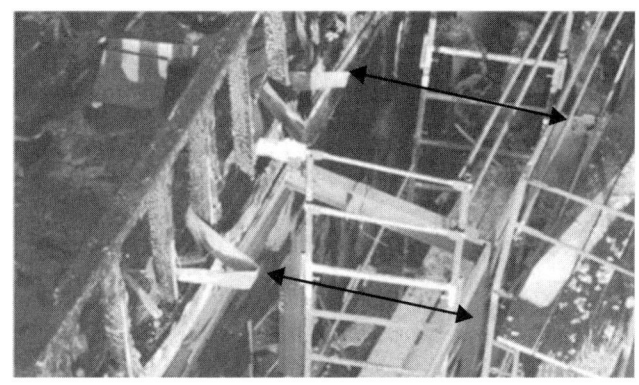

FIGURE 8.1 Variation of the width between the exterior face-to-face building dimensions affected the length of the cross beams, photo a necessitating continual in-field modification as noted in photos b, c, and d and Figure 8.2.

Care, Custody, & Control 101

b

c

d

e

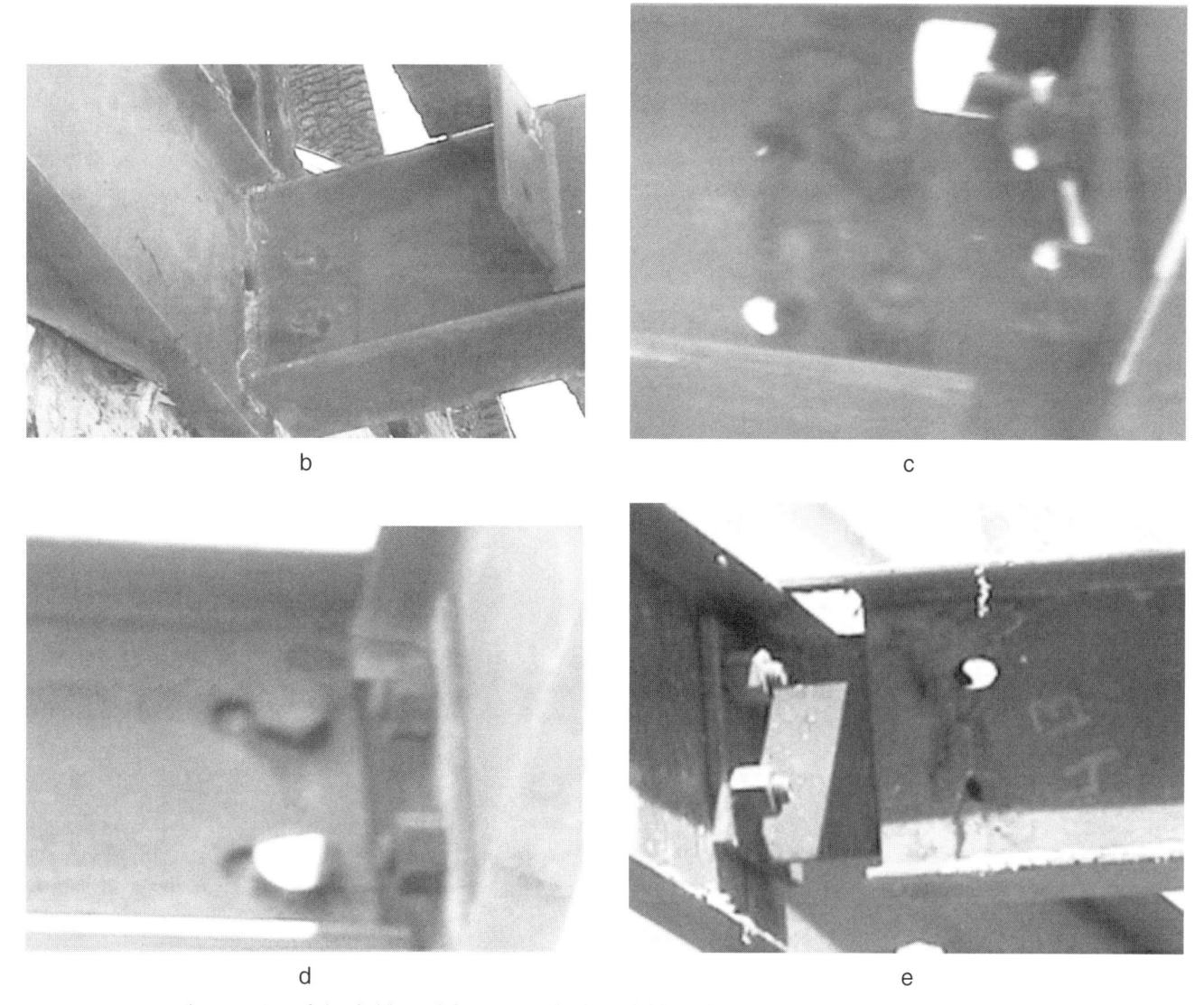

FIGURE 8.1 Note the severity of the field modifications, which could have been made in the shop after field measurement of the various width differentials. Not only were the modification very crude they also exposed the wood framing and exterior finish to extreme heat from the cutting and welding of the steel members which eventually cased ignition of the flammable materials and a major fire.

a

b

FIGURE 8.2 Photos a and b reflect typical field changes resulting from incorrectly installed brackets by others. Special note should be taken as to how close all of the steel framing is to the wood asking the question why would anyone attempt to make these changes without a fire watch and approved plan outlining all of the precautions that needed to taken before even commencing with the work.

8.3 FIRE LEVELS SITE, FIRE ENGINE, & SURROUNDING HOMES!

During the final framing stage of a 120-unit 10-building condominium complex, a fire broke out when a plumber was installing copper water piping on the second floor of one of the units. Within a matter of minutes, the fire was raging throughout the site, moving from one building to the next. The first fire truck to arrive on site was soon overcome by the intensity of the fire before the crew had a chance to roll out and connect the hoses and had to be abandoned. Homes in the surrounding neighborhoods were starting to be affected by the intense heat, requiring units from two other cities to respond to contain the fire. The neighboring school had to be evacuated, as did a number of homes. When the smoke cleared, all 10 buildings, in various stages of completion, were gone, as was the fire truck, various construction equipment, personal vehicles of the worker and three neighboring homes. See Figures 8.3 and 8.4.

Upon investigating this fire, a number of interesting issues rose to the surface that should be addressed, if nothing more than a reminder to others who may be considering constructing wood frame structures close to one another. These issues included the following:

- The general contractor never held any safety meetings that addressed fire safety, protection, and/or control, yet the entire complex was of wood frame construction.

- The general contractor had no fire-watch policy for contractors who would be using torches in their work, that is, the plumber, roofer, and miscellaneous iron and mechanical subcontractors.

- The plumbing subcontractor had no fire watch when performing his work, nor was a fire extinguisher, water hose,

a

b

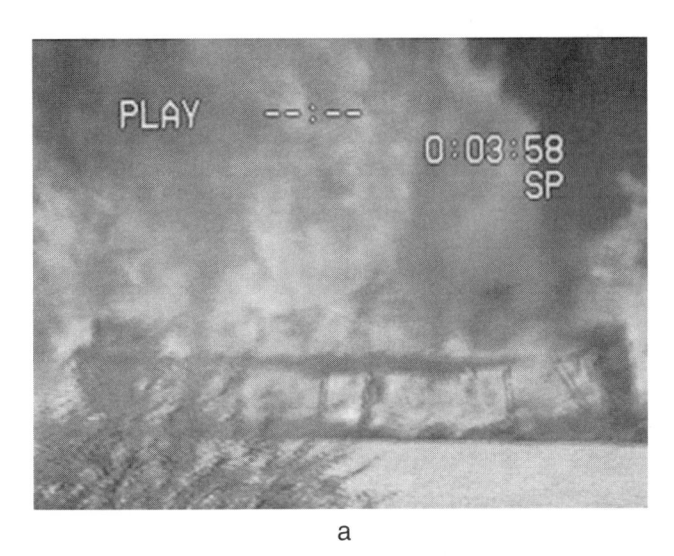

FIGURE 8.3 Photos a and b taken only about four minutes apart show multiple structures completely engulfed in fire. These photos give the reader an idea of just how fast and intense a simple mistake by one worker can lead to a total disaster affecting the lives of many.

a

b

FIGURE 8.4 Photos a and b depicts additional advancement of the fire within minutes of initial ignition. Note that virtually all of the structures were in the framing stage and that there was minimal separation between them, making it impossible for any type of early containment of the fire.

or other means of fighting a fire in the immediate vicinity.
- The plumber who caused the fire did not have a helper, and when the fire got away from him, he attempted to hide the torch in one of the chemical toilets as he ran from the site.
- The general contractor had all 10 buildings in the framing stage. Had the buildings been in different stages of completion, there would have been less danger of fire spread.
- There was no all-weather access road around the site for emergency vehicles to maneuver to attack the fire properly.
- Because the contractor didn't stage construction, there were open plumbing trenches that prevented adequate access for firefighters and created a hazard for the workers trying to evacuate the area.
- Combustible materials were stored close to the structures.
- As a result of this calamity, the jurisdiction now requires that a fire hydrant system and fire access lanes be installed prior to a stick of wood being brought on construction sites.

After investigating the fire and collecting sufficient evidence to confirm that the fire was started by the plumber, and pointing out the negligence on the part of the general contractor, significant fines were imposed by the State Industrial Commission. The general contractor also was involved with the litigation that followed from home owners who suffered the loss of their homes and property, but the local municipality was responsible for the destruction of the landscaping on the median and damage to the local street.. Remember, you as the general contractor have the care, custody, and control of the project during the course of construction.

8.4 CONCRETE ISN'T WATERPROOF?

Deferred maintenance is becoming endemic in our nation. Figures 8.5 and 8.6 are of a cast-in-place concrete underground condominium garage where it was more important to protect the cars than solve the leaking problem. The parking garage extended beyond the tower above and the waterproofing covering the garage had failed. This was in a desert region, so they stopped watering the landscaped areas, believing that would solve the problem. It did not, so

a

b

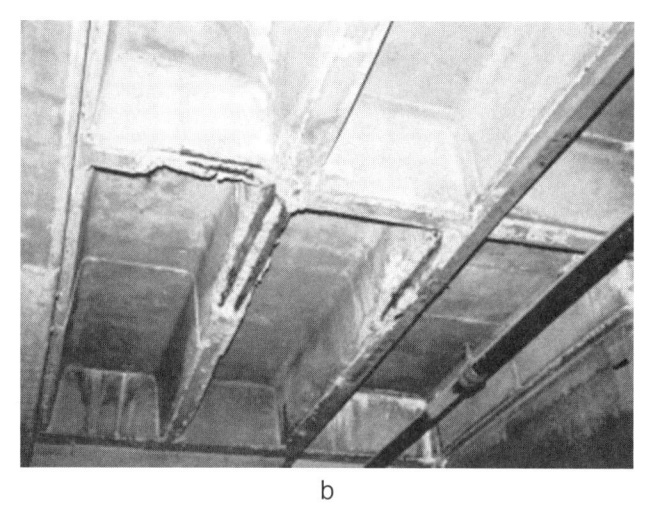

FIGURE 8.5 Photo a depicts a beam with rusted stirrups that had to be shored. As well as sheet metal used to collect water. Rust in the joist system as shown in photo b was common.

a

b

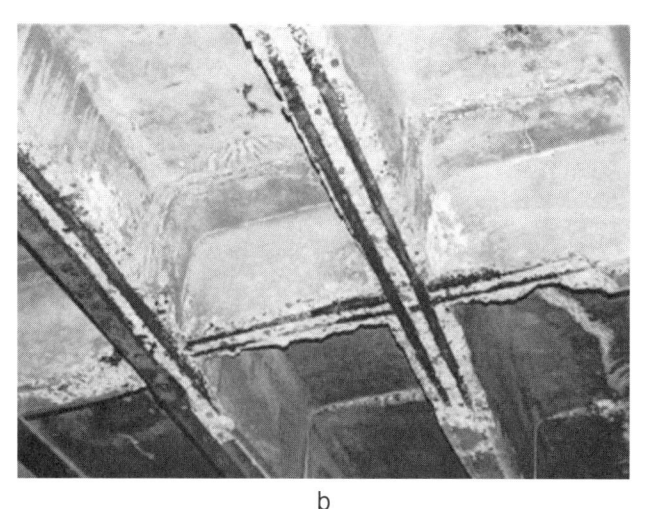

FIGURE 8.6 Photos a and b show typical rust in the joist and other damage to the concrete.

the owners installed an elaborate collecting system of sheet metal, troughs, and drains to take the water to the sewer system. Because the sheet metal was nearly flat, it kept the moisture in the system for an extensive amount of time. The vapor helped maintain the trapped moisture within the beams that increased the rusting of the reinforcing steel. In one beam, several stirrups were rusted completely through and it had to be shored for safety. To date, the condo members are deciding if they will remodel the first floor area as part of the concrete rehabilitation and how to finance both.

8.5 WATERPROOFING & DRAINAGE! WHO NEEDS IT?

This is a case of a concrete sports stadium that was power washed with water to remove the debris after games. The problem was they had structural steel members embedded in the concrete and a concourse that was level without any waterproofing. The concrete seating area also did not have waterproofing. The power washing forced the water deep into the concrete, and it took a long time for the water to come out. Meanwhile, the steel structural members were rusting severely. A vehicle broke the concrete at an expansion joint. When a forensic engineer was retained to investigate the concrete at this joint, it was discovered that the embedded steel members were corroded.

The repairs are simple to understand, but very complex to execute and the consequences are very expensive. The repairs consisted of exposing the steel by removing the overlying concrete, removing the rust, sandblasting the exposed steel, welding on steel cover plates to replace the missing steel, protecting the steel with epoxy paint, replacing the removed concrete, installing more drains, and coating the concrete concourse with a walkable waterproof membrane. Removing the concrete, shoring the members, and replacing the concrete were expensive. To date, the owner has spent $14 million repairing the stadium and has another $20 million to go. See Figures 8.7, 8.8, and 8.9.

a

b

FIGURE 8.7 The material below the pencil in photo a is rust and that above is the remaining steel of the bottom flange of this main girder. Photo b is the top flange of an embedded steel beam. Notice the indentation to the left of the pencil. In some locations, over half the top flange is gone due to rust. To the right of the pencil is a chunk of the rust.

a

b

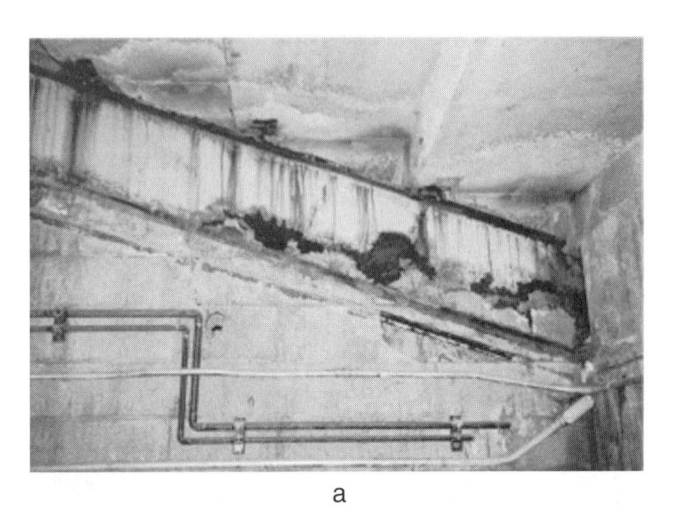

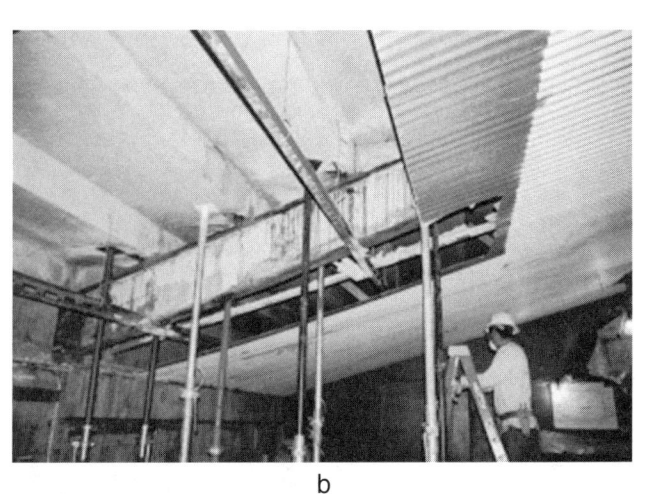

FIGURE 8.8 Photos a and b show several more rusted beams. Note again where they used sheet metal to divert the water and hide the rust.

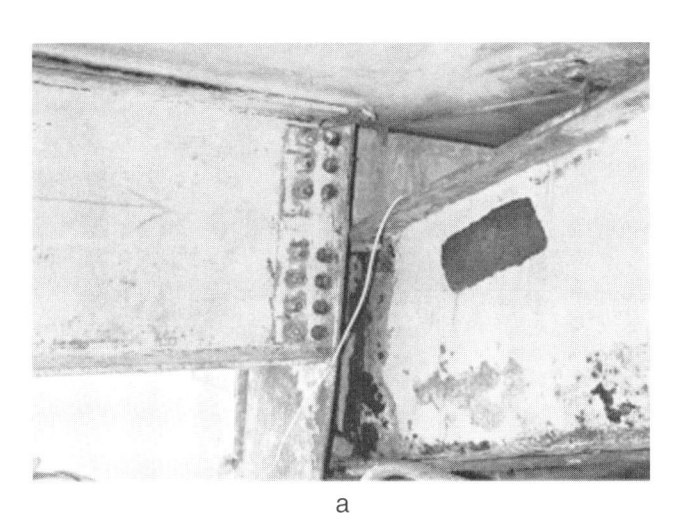

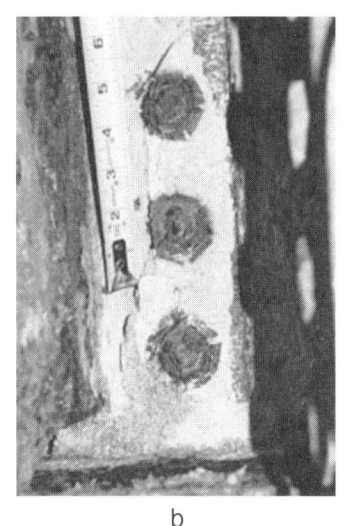

FIGURE 8.9 Photo a shows how the sloping beam on the left diverted the water to the bottom, allowing water to collect where the beam is connected to a column. Photo b shows the results of this water. The nuts at the connection are completely rusted away.

8.6 A BOW STRING TRUSS THAT LOST ITS BOW

Wood bow string trusses were used in the western United States from the 1950's until about 1962. One of the authors has repaired over 100 buildings with failed wood bow string trusses. The main failures have been in tension in the bottom chords, which split. The top chords (when properly braced), the web members, and the bolted connections have seldom failed. When the bottom chords split, the common method of repair is to install steel rods full length on each side connected to a steel anchorage at the bottom heel joint on both ends. Added wood members at the split resist local loading. Because the steel rods are difficult to install, thus expensive, an inexperienced engineer used cables that are easier to install. Figures 8.10 and 8.11 show how he failed to understand that cables stretch with time.

8.7 WHO NEEDS AN ENGINEER FOR SHORING?

The following project reflects when a contractor fails to determine the load factors that will be applied before he rents the shoring. The title says it all. Luckily no one was killed and workers suffered only minor injuries. The concrete subcontractor did not hire an engineer to design the shoring. The subcontractor sized the tower shoring properly and installed it correctly because the loads on this part of the system were reasonably easy to calculate. The design of beams to carry the decking is also standard. The challenge is to design the beams to transfer loads from the deck area and other beams to the towers in this complex system. One of the beams was overloaded during the concrete placement, causing the entire forming system to collapse. He would have saved money by doing it right and hiring an

FIGURE 8.10 Photo a shows the general collapse. Photo b shows the cable on the bottom chord of the collapsed truss. It is visible on the far right side adjacent to the bottom chord next to the steel anchorage.

106 SECTION EIGHT

FIGURE 8.11 Photo a shows the other end of the failed truss with cable. Photo b is interesting in that the adjacent truss that did not collapse was repaired correctly with a steel rod.

FIGURE 8.12 General overview of the collapse.

engineer to design the complete system. Sad to say this happens all too often. See Figures 8.12 and 8.13.

8.8 FALLING LIKE DOMINOS!

Wood trusses are common throughout the construction industry and come in various sizes and shapes. It is not uncommon to see two framers walking the top of an exterior wall setting trusses in place and toe nailing them to the top plate, then walking back to get another truss to set in place with little thought of what may happen if one of the trusses racked and tipped over. In some installations, a 2 x 4 is placed on the top cord of two trusses and held in place by one framing nail in each cord. No thought is given to the fact that when one truss starts to rack and fall, it creates a chain reaction with increased force that would affect all of the other trusses that had been set in place. This situation happens all too often. Good truss manufacturers ship instructions on how to properly brace the trusses during erection.

8.9 SUPERMAN SYNDROME!

During the construction of a series of upper-scale condominiums, the general contractor was pressing forward to meet the marketing plan of the owners. In general, the project site was well maintained, relatively free of construction debris, and with all of the required signage to meet both state and federal safety standards. The contractor was conducting weekly progress and safety meetings and for all intents and purposes one would believe that

FIGURE 8.13 Photo a shows the debris with the concrete. Photo b gives you a perspective of the height of the collapse. Shoring towers can become very unstable with height.

there was no way that anyone would experience a major injury, let alone a death.

When faced with unrealistic time demands and poor management decisions, something has to give. It is not uncommon for the construction team to disregard safety precautions in order to save time and finish the project. Unfortunately, the consequences can be expensive, or even fatal.

This case involved the installation of 10-inch schedule 40 steel-chilled water piping in a cast-in-place concrete utility tunnel. The plans called for the chilled water recirculation pipe to be hung from trapeze hangers on one side of the tunnel. The Uni-Strut hangers that would be supporting the pipe were secured to the ceiling of the tunnel by two threaded rods that were anchored into the concrete with a standard expansion shield, as called for on the design documents. These anchors work well when statically loaded vertically. The anchors will become loose under repeated horizontal movement.

The proper equipment for installation of the pipe would not be available until the following week. Rather than wait for the equipment to raise the 20-foot-long, 10-inch pipe sections into position, two workmen decided to muscle the pipe up onto fully assembled trapeze hangers. Muscling the pipe into place created cyclical horizontal movement, which loosened the anchors. While installing the second length, the anchors on the first hanger failed, followed by progressive failure of three adjacent anchors. This failure dropped both pipes onto the two men, killing one man instantly and sending the other to the hospital, where he later died of internal injuries.

Upon interviewing others, it was discovered that a hydraulic lift had been requested but would not be available until the following week. Because of the pressure put on the mechanical contractor to get the system on line, safety procedures were ignored. The consequence was two fatalities.

8.10 WHO NEEDS A BODY HARNESS?

A telephone company wanted to remove its old relay tower. The work was bid out to a series of rigging contractors, each of whom was given the opportunity to visit the site and investigate the condition of the old tower before submitting a bid. The successful bidder arrived on site with a two-man crew and began dismantling the tower. The crane service, also an experienced rigging organization, arrived with two men and set up the crane in accordance with the requirements dictated by the rigging contractor. The tower was a triangular tube type consisting of five 20-foot prefabricated sections bolted together and had a freestanding height of 100 feet. After years of service, the bolts connecting each section had deteriorated and some would prove difficult to remove. Due to the size of the tower, only one man could handle the disassembly. The workman assigned to disassemble the tower used a safety harness to tie himself off to the tower as he disassembled and removed the relay cones from the tower.

The operation went well until it came time to disassemble the tower. The workman proceeded to disassemble the bolts on the upper tower section before taking the time to attach the choker from the tower to the hook on the crane's line. After loosening some of the bolts, it became evident that the rust had indeed corroded the threads to a point that would require a cutting torch to remove them. After advising the crane operator of the need for a cutting torch, the workman unhooked his lifeline from the tower and started to climb to the top of the tower to attach the choker to the crane line. Unfortunately, before he could reach the choker, the entire top section of the tower broke loose and both tower and workman crashed to the ground. The injuries sustained were severe, resulting in the paralysis of the workman who filed litigation against the phone company for negligence. Needless to say, the only people who came out ahead were the attorneys.

8.11 DON'T FENCE ME IN!

It is not uncommon to see open trenches on construction sites, and yet, virtually every general contractor is aware of the dangers of open trenches. State and federal safety standards all have code sections that address the requirements of excavating, trenching, and shoring, including specific safety precautions not only for the workers but the general public as well. There are standard safety videos and handouts that are used to emphasize the safety requirements that must be taken when trenching is required. Yet it is not uncommon to visit projects and find trenches unprotected for days on end. How many times have serious injuries or death claimed construction workers due to the negligence of the subcontractor or the general contractor in not insisting that applicable safety precautions be taken?

It is imperative that the general contractor understands all of the conditions surrounding any type of trenching operation associated with work being performed on the project site. This includes, but is not limited to the following:

1. Type trench and usage, that is,
   a. Plumbing trench (depth, width, size of line, and type of service)
      i. Water (New, Existing, Potable, Storm, Pressure, Active, Inactive)
      ii. Sewer (New, Existing, Sanitary, Storm, Pressure, Active, Inactive)
      iii. Gas (New, Existing, High or Low Pressure, Active, Inactive)
   b. Electrical trench (depth, width, size of line and type of service)
      i. Power (New, Existing, Active, Inactive, High or Low Voltage)
      ii. Communication (New, Existing, Active, Inactive)
2. Location and site conditions, that is,
   a. Soil Conditions (rock, gravel, loam, alluvial, clay, sand)
   b. Topography (flat, hilly, mountainous, rivers, lakes, streams)
   c. Surroundings (roads, structures, utilities, railroads, schools, playgrounds)

It is one thing to subject workers, who have the opportunity to decline working in an unsafe environment, to the dangers of the open trench, but it is totally unconscionable for a contractor to subject the general public to such dangers. Children particularly love to explore and play in trenches. The photos in Figures 8.14 and 8.15 were taken as part of the evidence collected during the investigation of two trench-related cases, each of which involved young people, one a young boy and the other a teenage girl. The young boy died when the trench collapsed on top of him. The young

a

b

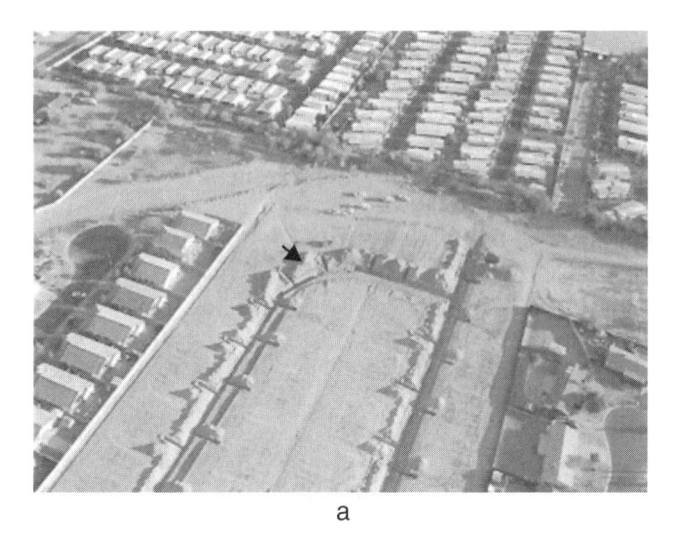

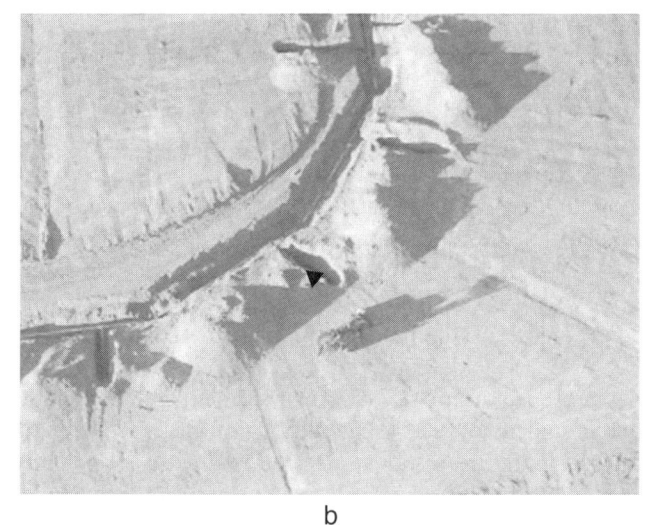

FIGURE 8.14 Notice in photo a that the only fence surrounding the construction site are the block walls of the surrounding homes and a chain link fence along the trailer park boundary line. Take special note in photo b of the amount of unprotected open trench, the excavated soil angle of repose, and the sluffing off of the sides of the embankment. Red arrows points to location of trench collapse.

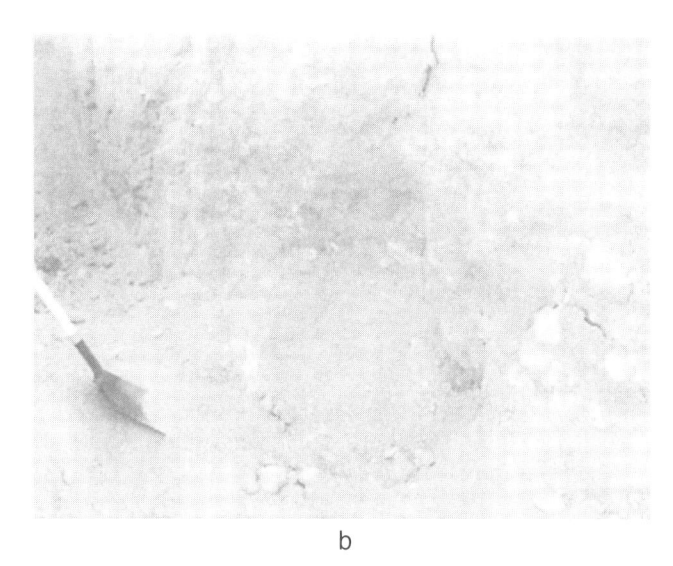

FIGURE 8.15 Photos a and b show the area of the trench that collapsed, killing a nine-year-old youngster playing in the trench with other children. The depth of trench was over 5 feet. An estimated 900 pounds of soil fell on the youngster. Notice that none of the embankment was terraced as required by code for the classification of the soil found in the area. In other words, the trench as constructed was not safe for workers let alone children.

girl was running around a construction site with her boyfriend and fell into an open trench, breaking her neck and causing serious spinal injury. If you are the general contractor, keep in mind when reviewing these photos that you have the care, custody, and control of the project site.

In the case where the young boy died, the contractor/developer had not only failed to fence the entire site, his plumbing subcontractor had been cited earlier by the State Industrial Commission because the 700-foot-long sewer line trench, which varied in depth from 4 feet to 12 feet was not terraced or protected, and the contractor/developer was cited for the lack of dust control. Employees who had been working on the sewer line had been moved to another site due to severe dust conditions associated with the excavation effort.

During the ensuing investigation, it was discovered that an earlier geotechnical investigation of the on-site soils found them to be identified as variable silty sand with a trace of gravel, with a plasticity index that varied greatly across the site. Because of the collapse potential of the soil, the geotechnical engineer advised the contractor/developer that caution should be exercised by closely monitoring the soil type as excavation work progressed. The geotechnical report was never provided to the excavation contractor or taken seriously by the contractor/developer. In fact, the contractor/developer admitted that neither he nor any of his employees had ever received any type of Competent Person Training in the state-mandated job-site safety program required for licensed contractors. It was also admitted that the only reason that a soils investigation was performed was because it was a requirement of the local municipality and the architect. The superintendent stated that he used his judgment to decide if the soil was stable but could not point to areas along the trench line that would require terracing as required by state standards and outlined in the geotechnical report.

Further investigation uncovered that the workers on-site had allowed the children to play in the dirt piles with instructions to stay out of the trench. Hard to believe, but true.

The details of the case in which the young girl had been severely injured were much the same, although no earlier citations had been issued. Again, neither the contractor nor the subcontractor had anyone with Competent Person Training. The response to why the trench was not fenced or protected was simply: "We saw no need for it; after all, the site was miles away from an active community, and none of the homes were anywhere near completion or ready for a sales walk through. Who would have thought that kids would be running around out here in the middle of the night?"

8.12 WHERE DID ALL THE SUPERVISION GO?

The previous Section 7, Construction Issues, addressed a number of key issues that are affecting the construction industry today. The following involves a design-build project for a multi-million-dollar custom home that literally had a 4-foot by 15-foot long section of wall fall away two years after the home had been completed and occupied. It is a classic example of what transpires when the prime contractor, who also designed the home, fails to have full-time supervision on-site, allowing the subcontractors to run the work. It also reflects the incompetence of local building officials in not understanding the proper methodology and limitations of building with adobe block, as well as being less than observant of what was transpiring on-site during the course of construction. See Figures 8.16 to 8.22.

110 **SECTION EIGHT**

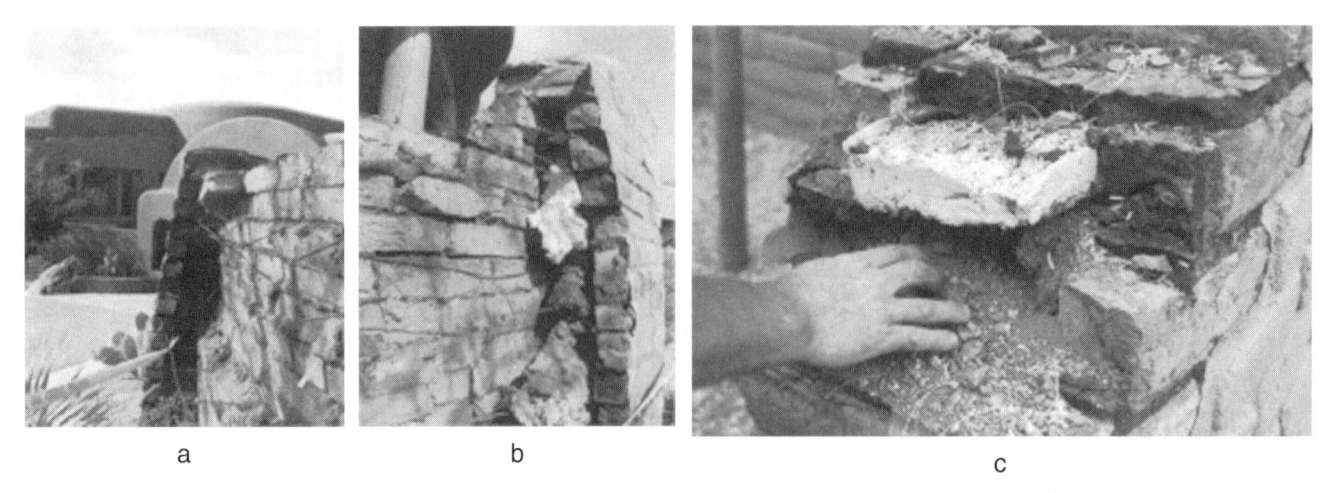

a b c

FIGURE 8.16 This wall did not comply with the contract drawings or recognized adobe construction practice. The battered wall was not shown on any of the architect's details and was incorrectly constructed. The in-fill consisted of dirt, grout, Styrofoam, and polystyrene as illustrated in photos a and b. The reinforcing was totally useless. At various points, plants, as shown in photo c had started to root inside the wall from water infiltration.

a b

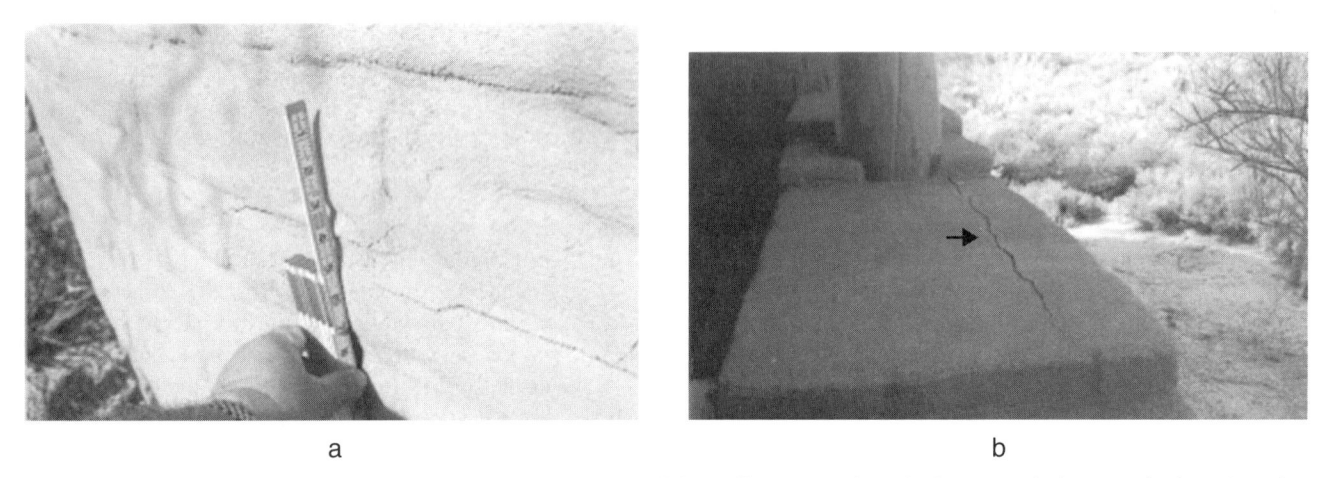

FIGURE 8.17 Major lateral stress is starting to show on various adobe walls surrounding the home, including cracks found on the top and sides of adobe walls throughout the residence, allowing water to enter the void, consolidating the fill mass and exerting lateral pressure against the adobe block. Photos a and b illustrate a few of the cracks found throughout adobe block work.

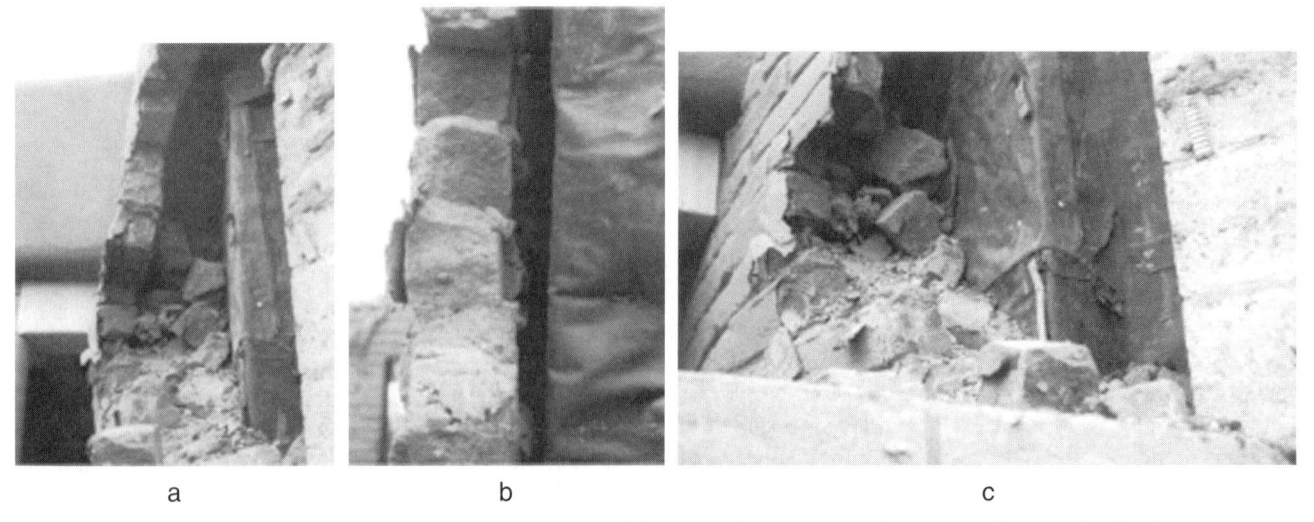

a b c

FIGURE 8.18 Typical examples of the adobe veneer on wood frame walls. Notice in photos a and b that the block ties were never used and the 1-inch air space has increased considerably by the consolidation of the debris dumped behind the wall. None of the work complied with any detail shown on the plans. Photo c clearly shows how continued water intrusion through cracks in the adobe block helped consolidate the debris and aided in pushing out the adobe veneer.

Care, Custody, & Control

a   b

FIGURE 8.19 The battered wall at the entry courtyard in photo a had no foundation under the concrete masonry units or the adobe veneer. This wall also does not conform to any plan, code, or standard adobe construction practice, all of which were referenced in the contract documents. Cells containing vertical reinforcing steel in a 6-foot-high retaining wall shown in photo c were filled with dirt instead of grout.

a   b

FIGURE 8.20 The detail in the contract documents for this retaining wall showed adobe/CMU with veneer ties. Detail called for "*8 solid grouted CMU (All Cells)*" This lack of adherence to the design documents was found throughout the work. Notice in photos a and b no block ties are present and debris in the cell of CMU. Neither of these constructed assemblies conforms to any design requirements, code, or standard.

a   b

FIGURE 8.21 Photo a shows a very unusual assembly of a steel beam, plywood, and gypsum board soffit holding up a soldier course of adobe block. The in-fill between the framed exterior wall and the battered adobe veneer was crammed full of broken adobe block and other debris. Photo b shows the same area shown in photo a before the wall was opened up. The adobe block had already started to move in a number of places and could easily have resulted in a major failure and injury to the home owner. Many of these blocks could easily be moved by hand.

FIGURE 8.22 No waterproof membrane was found in the area noted by the arrow in photo a. The picture on the right the masonry work shown in photo b is a fairly representative example of the workmanship found on the CMU work performed by the masonry contractor.

After reviewing these photos, it is evident that there are a number of deficiencies in the contractor's, subcontractor's, and municipality's organizations. Based on these findings, what parts of Sections 2, 3, 5, 6, and 7 reflect the primary issues that led to the disastrous results associated with the care, custody, and control on this project?

8.13 IT'S ONLY A SEWER LINE TRENCH! WHAT'S THE BEEF?

It has become virtually standard operating procedure to excavate an interior sewer line trench in tract home construction, bed the line with dry soil, then backfill the entire underground sewer line with dry soil, wet the soil a little after the placement of the ABC (aggregate base course), and dampen the pad with a termite pre-treat just before the placement of the concrete floor slab. See Figure 8.23.

This lack of properly preparing the soils and bedding followed by improper compaction has led to water intrusion into the subgrade under the floors of many homes, resulting in settlement or heaving of the slab on-grade. Additionally, as the water enters the trench and the bedding material, in many cases it also increases the chances of affecting the flow of effluent through the sewer lines. The proper bedding and compaction of soils around sewer lines cannot be overemphasized.

8.14 COMPACTION, WHO NEEDS COMPACTION?

The following is a classic example of what happens when there is little attention given to how one should place and compact the fill portion on a cut/fill lot. In this case, the home was constructed in such a manner that approximately 50 percent of the structure was built on cut and 50 percent was built on fill. The home owner reported major cracking in the walls and tile floors in the master bathroom. A manometer survey of the slab on-grade revealed that the home had in excess of a 6-inch differential in the floor surface from one end of the home to the other.

In reviewing the contract documents, there was virtually no information regarding the appropriate way to prepare cut/fill lots prior to commencing construction of the home. A soils report made recommendations on the type of soil to be used as fill, verifying that the native material was suitable provided that all large cobbles in excess of 6 inches

FIGURE 8.23 Photo a and b reflect the typical way utility trenches are backfilled with no regard as to the type of soil or need to properly compact the fill material.

were to be removed before placement. Additionally, the fill area was to be constructed in lifts not to exceed 12 inches in depth and recommended that field density tests be conducted on each lift for every 2000 square feet of fill area.

Investigation of the circumstances surrounding the conditions found that in most instances, no terracing or subgrade preparation was done prior to the fill operation. In fact, interviews with tradespeople confirmed that slash from clearing the site was at times deposited in the fill areas. A review of the density tests revealed that no tests were taken of the subgrade before the fill operation commenced, and only a limited number of tests were taken at some of the sites with the majority of tests being conducted in the upper 2 feet of the 10- to 15-foot-deep fill area.

When confronted, the contractor advised that the developer had a representative visit the site daily, and the geotechnical engineer was under contract to the developer. Therefore, the contractor claimed he had no control over the testing requirements. The geotechnical engineer's excuse was that the contractor determined when and where density tests would be taken.

This type of finger pointing is prevalent in most litigation and, in this instance, required the remediation of virtually all of the hillside homes in this development.

After lengthy litigation, the architect, engineer, and contractor were all contributors to a multi-million-dollar settlement. In reviewing Figures 8.24 to 8.28 in this vignette, take a moment to list the actions that could and should have been taken by the owner, architect, engineer, and contractor that would have prevented this disaster from ever taking place.

8.15 CARE, CUSTODY, BUT NO CONTROL!

Split responsibility on a construction project can result in disastrous consequences. A school district thought it could save money by building a new school in phases. It hired a civil engineer to develop the site plan, including the grading and drainage. That plan also included building pads and a special compacted pad to support a large chiller tank. A compacted aggregate base, installed on the pad and confined laterally by a steel ring, would support the tank. A contract was let and the grading, drainage, and pads were completed. See Figure 8.29.

An architect was hired to design the buildings, including a mechanical building located behind the chiller tank. The architect hired a firm with structural, mechanical, and electrical divisions in one company. Although the chiller tank

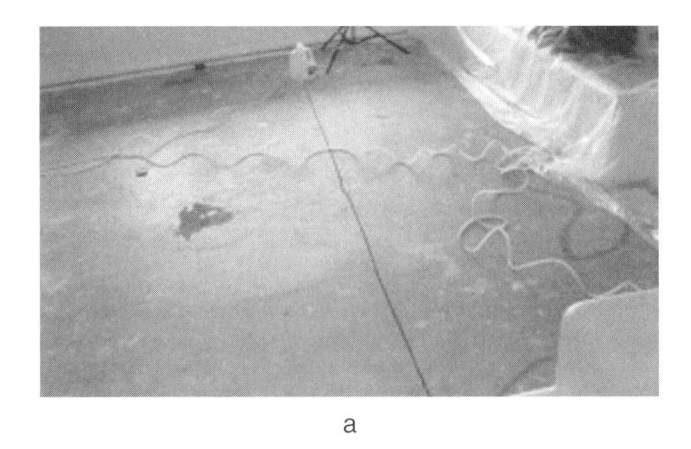

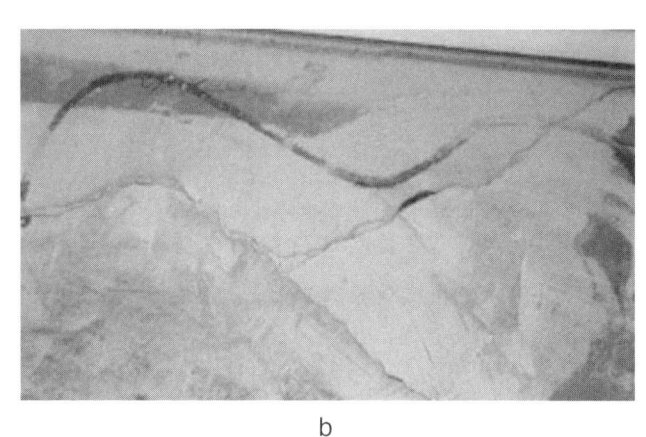

FIGURE 8.24 Floor slopes from control joint to the south as noted in photo a. Settlement cracks are found around the return air ducts. The contractor's extensive use of floor filler shown in photo b was a meager attempt to fill cracks.

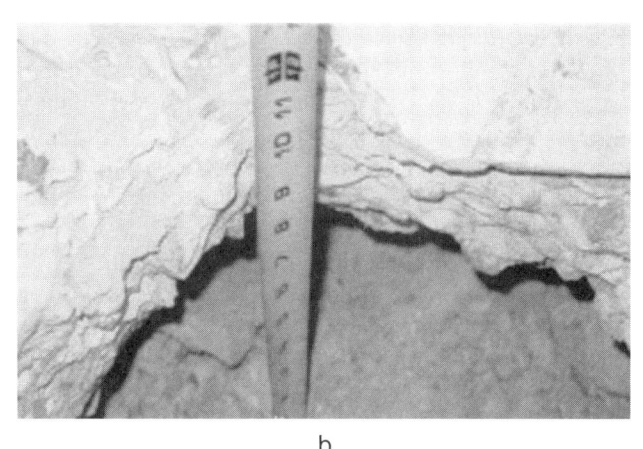

FIGURE 8.25 Under-floor voids ranged from 2 inches to 16 inches in depth as illustrated in photos a and b.

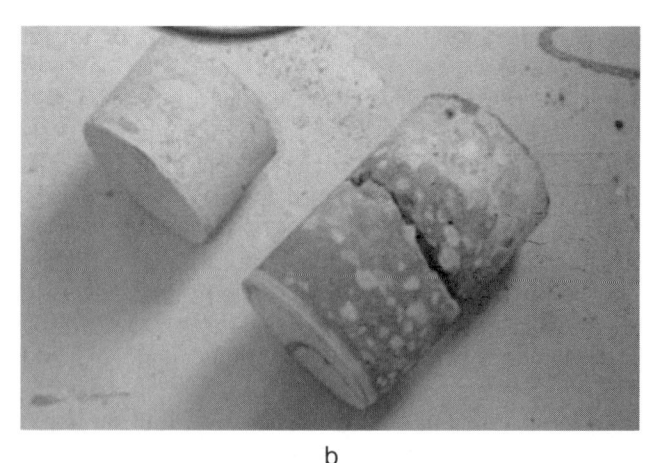

FIGURE 8.26 Cores were taken in numerous areas to measure the depth of the concrete slab as well as the depth of the void under the slab. The cores varied from 3 inches to 7 inches, plus a floor topping, indicating that the contractor was well aware of the poor fill condition and tried to hide it. Photos a and b show a location where on of the cores were taken and the findings when the cores were pulled.

FIGURE 8.27 As can be seen in Photo a, the under floor void caused by the poorly compacted fill is not limited to small areas; it covers an extensive area and varies greatly in depth. Density tests and soil probes were used to check the soil density. In one corner of the master bedroom, the soil probe could be inserted its entire 6-foot length with one hand, as shown in photo b.

was shown on the site plan, it was not shown on the building plans. Although the mechanical engineer was aware of the tank location, it was not shown on the electrical plan. It is interesting that the mechanical engineer and electrical engineer were in the same office but apparently did not coordinate their work and the architect failed to cross check the completed plans before releasing them for construction.

A contract was let for construction of the buildings, including the installation of the chiller tank. During the course of construction, the electrical contractor dug a large trench between the school building and mechanical building beyond the chiller tank pad. The 3-foot-wide and about 4-foot-deep trench cut through near the middle of the tank pad. The specifications required the trench fill be compacted to a density of 95 percent. Investigation confirmed that this requirement was not strictly enforced for trenches outside of the buildings. The trench collapsed during the first significant rain, resulting in undermining and destroying the integrity of the specially constructed pad under the tank. As expected, the finger pointing went the complete circle, resulting in costly litigation, not to mention the cost of removing the tank, reconstructing the pad, and reinstalling the chiller.

8.16 COMPACTION AROUND UNDERGROUND STORAGE TANKS TOO!

The following demonstrates the fallacies and misconceptions of how to properly compact soils around underground culverts, pipelines, and storage vessels, and the failure of the architect, engineer, and contractor in carrying out their respective duties. A local city ordinance required that all storm water runoff be contained on-site. The developer wanted to make maximum use of the land for a condominium project. As the lot was of insufficient size to allow the construction of both the condominiums and retention basins to store the storm water runoff, the developer decided that underground storm water storage was the best alternative. It was decided that the use of a series of 8 foot in diameter

Care, Custody, & Control 115

FIGURE 8.28 Floor fill in excess of 1 inch on top of the finish floor after the walls were framed and before the sheet rock was installed on the walls resulted in severe stress cracks as the floor continued to settle. These findings again proved that the settlement condition was known by the contractor during construction.

FIGURE 8.29 Panel 1: Collapsed trench extending under masonry wall and under three chiller tanks beyond.

corrugated galvanized iron pipes with bulkheads with a bleed-off system into a series of drywells was the way to go. The underground storage tanks were installed and the condominiums built. After five years, the residents were faced with cracked walls, floors, and ceilings, along with roof leaks, driveway settlement, and a series of other problems. Figures 8.30 to 8.34 are reflective of only a few of the defects that were attributable to the lack of proper compaction on the part of

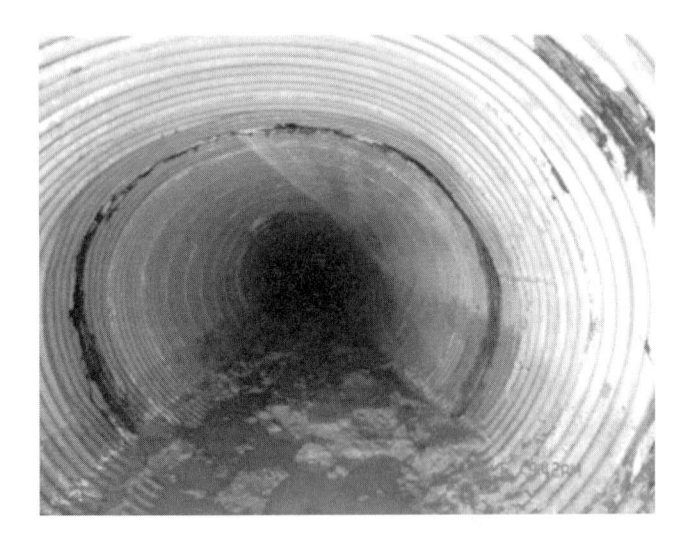

FIGURE 8.30 Lack of ensuring that the soils were properly compacted to the engineered required densities under and around the underground storm water storage tank began to have an adverse effect on the tanks as the overburden loads began to be applied. These loads began to cause the tanks to deflect into the poorly compacted fill. In some cases, this deflection exceeded 10 percent, resulting in the upper-level grades dropping 4 to 8 inches.

a   b

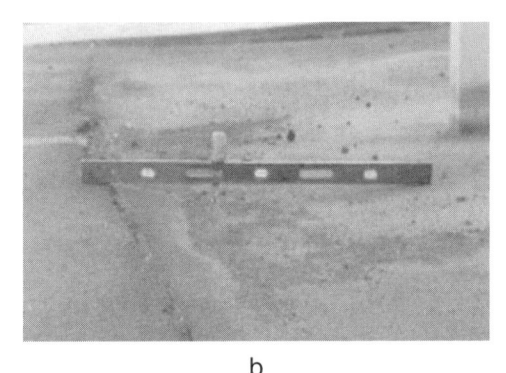

FIGURE 8.31 Stair and sidewalk settlement of 8 inches plus occurred in some areas as noted in photo a. Parking areas became duck ponds even after a miner rainstorm, as settlement around the underground storage tanks totally changed the drainage patterns within the parking areas as shown in photo b.

a   b

FIGURE 8.32 Stair towers adjacent to the underground storage tanks started to separate and pull away from the housing units as the poorly compacted fill soils began to consolidate around the tanks. As the stair towers broke away from the housing units, (photo a) the roofs and ceilings started to split open, (photo b) exposing the substrate and interiors to rain damage.

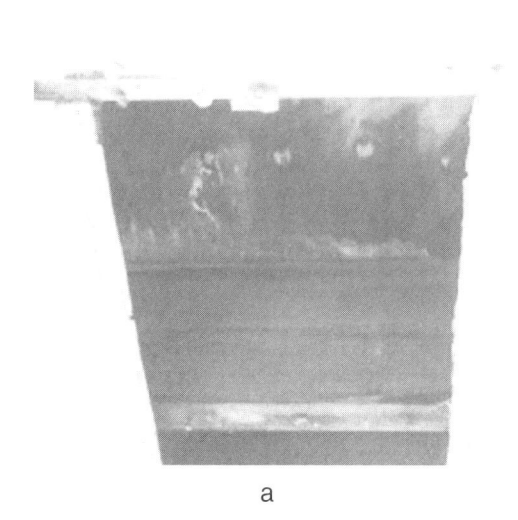

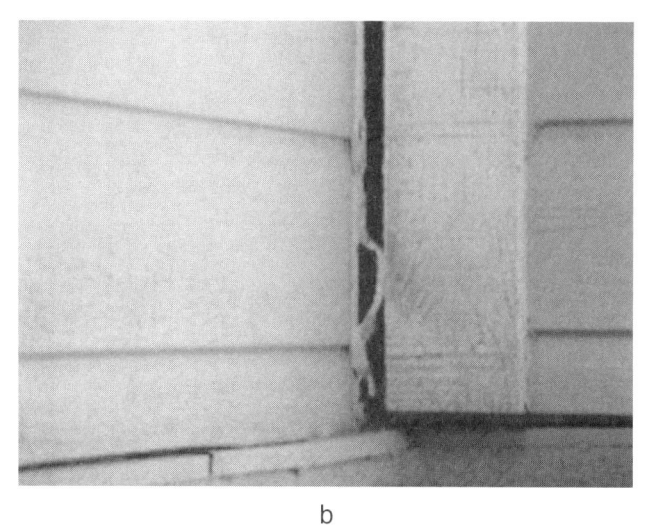

FIGURE 8.33 Exposing the underside of the roof decks revealed a clear picture of the mold, mildew, and dry rot that was affecting much of the structures, as documented in photo a. The closer a structure was to the underground storage tanks, the greater the settlement damage was as noted in photo b.

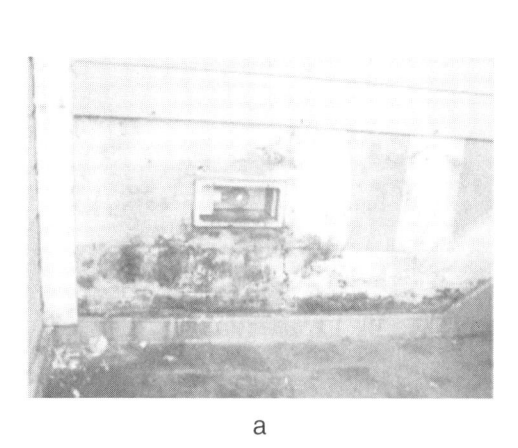

FIGURE 8.34 The greater the settlement next to the tanks, as noted in photo b the greater the damage to the structure became, and the greater the amount of microbial growth and deterioration of the substrate that was found, clearly shown in photo a.

the contractor responsible for installing the underground storm water storage tanks.

The evidence in this case pointed to all parties—the architect, engineer, and underground storage tank contractor. The contract documents showed no details of the direct relationship of the underground storage tanks to the building foundations. Although the engineering drawings required compaction tests, the compaction tests were only taken within the top 2 feet of fill. No density tests or samplings were taken at or below the spring line of the underground storage tanks. No inspection was conducted to ensure that the work was performed in accordance with the engineered plans, and the city took no responsibility.

8.17 IT'S ONLY A SMALL LEAK!

When constructing sewerage treatment plants, one of the requirements is to fill the tanks with water to determine if they are watertight before you backfill the soil around them. The primary purpose is to check the quality of the concrete for leaks. Of course, the walls of the tank have to be strong enough to take the water pressure exerted against the sides of the wall. When the prime contractor filled the tanks on this design-build project, the 168-foot-long, 22-foot-high walls bowed out. The top of the tank bowed out 16 inches. Leaks were present all over the face of the tank. The only element that saved the tank from complete collapse was the nonstructural concrete walking platform built into the top of the tank. Figures 8.35 and 8.36 show just how serious the situation was soon after the tanks were filled with water.

Three conditions must be taken into consideration when designing a sewerage tank: (1) when the tank is filled with water and under test as noted above, (2) when the tank is in normal operation and soils in place surrounding the tank, and (3) when the tank is empty and undergoing maintenance and is subjected to the lateral forces of the surrounding soils.

Investigation into the circumstances surrounding the cause of this failure uncovered the following: The design-build

118 **SECTION EIGHT**

a

b

FIGURE 8.35 Photos a and b show the wall leaning out at the top. Note the cast-in-place nonstructural walkway on the top that kept the wall from collapsing. It spanned as a horizontal beam from the end walls that are perpendicular to this wall. The main failure crack was on the inside of the tank at the bottom where it met the footing.

a

b

FIGURE 8.36 Photos a and b show the tension cracks in the concrete wall that allowed the water to spill from the tanks. The end walls also restrained the long wall, and these are the cracks at that restraint. They were not major structural cracks and were sealed with epoxy during reconstruction.

contract was let to an out-of-state firm that was required to follow all local laws. The contractor retained a local engineer who had little to no experience in designing sewerage tanks and used a computer program for the cantilevered retaining wall to design the tank walls. Subsequent investigation into the design revealed that the wall was designed and detailed properly, but the interface between walls and footings was incorrect. The computer program was for a different footing configuration, leading the engineer into designing the interface incorrectly, which resulted in creating the point of failure.

Because it was a design-build contract, the contractor had to correct the problem, as well as pay the liquidated damages as assessed under the terms of the contract.

To address the issues, the contractor retained an experienced forensic engineer who was able to determine the cause and develop an engineering protocol that saved the tanks without complete demolition and reconstruction.

8.18 OUT OF SIGHT, OUT OF MIND!

While a family was on vacation, a crack occurred in the CPVC fire sprinkler piping in the ceiling of the middle south bedroom and resulted in water damage to the ceiling, floors, and furnishings. It was estimated the approximately 10,000 to 20,000 gallons of water were released during this

a    b

FIGURE 8.37 The shorted out flow switch photo a and the "out of sight, out of mind" fire alarm controls, photo b.

occurrence. No alarm was sounded at the time. The home owner utilized the services of a plumbing contractor rather than a fire sprinkler contractor to make the repair to the CPVC line. The repair was accomplished by utilizing a PVC friction type union that compresses around the pipe rather than a glued or screwed positive connection. The interior of the home was remediated and no further investigation was conducted.

Approximately one year later, the family was again on vacation when the PVC friction union failed, resulting in approximately 87,000 gallons of water being discharged into the home. Again, the fire alarm did not sound. Water coming out of the home was noticed by a neighbor, who was able to turn the water off. This time water from the fire sprinkler system inundated the entire home.

Upon investigation, it was discovered that the flow switch controlling the fire alarm system had been hidden in the garage behind cabinetry, and whoever had wired the flow switch had allowed too long of an exposed conductor tail that shorted out on the flow switch activation spring. This dead short tripped the breaker feeding the current to the flow switch.

In checking out the circuitry from the breaker panel, it was discovered that the flow switch had not been wired as a dedicated circuit to the breaker panel. Sometime during the course of construction, this circuit was abandoned to allow service to flow to other lighting and receptacles. The flow switch that activates the alarm is out of service and the short cannot be detected via a tripped breaker. The control valve that releases the water into the lines is now open to any unusual occurrence within the fire sprinkler system. See Figure 8.37.

What went wrong? Why did a family have to go through two devastating floods before finding out why?

- Did the general contractor have a responsibility to ensure that the fire sprinkler system was installed in compliance with code and the local municipality authority having jurisdiction?

- Did the electrical subcontractor have a responsibility for hooking up the electrical control system associated with the fire sprinkler system and have a duty to ensure the life safety system worked, as required by code and the local municipality authority having jurisdiction?

- Did the local fire department have a responsibility for inspecting the fire sprinkler system and testing the fire alarm system, including a duty to the home owner to ensure that the life safety system worked before signing off on the certificate of occupancy?

- Did the plumbing contractor have a responsibility to the home owner when repairing the fire sprinkler system to ensure that the life safety system worked and a duty to verify that the system was in compliance with the current code and the local municipality authority having jurisdiction?

The answer is yes to all of the questions; however, the plumbing subcontractor and his insurance carrier took the biggest hit, followed by the electrical contractor and the general contractor. The fire department got off scot-free.

8.19 BUILDING IN A BATHTUB DOES NOT MAKE ONE CLEAN!

Expansive clays can be a disastrous and devastating obstacle in the construction of virtually all types of structures when the designers do not comprehend the destructive forces of Mother Nature. The following case revolves around a new community college structure that had to be torn down due to the negligence of the architects, engineers, and contractors involved in its design and construction.

The campus in which the new facility was constructed was located in the northeastern part of Arizona, where it has been a known fact that the area contains expansive clay soils, known as Chinle Clays. The college was a typical masonry structure over a concrete footing and stem wall, conventional concrete slab on-grade with long span trusses, metal deck, and built-up roof.

The interior had utility trenches for water and sewer to avoid placing these mechanical systems underground where any leak would cause a reaction with the clay soils. Bifold partitions were used to expand interior space as needed. Many hours of planning had gone into the facility to ensure maximum use. It was a far cry from the manufactured modular units that the college had been living with for a number of years. As reflected in Figures 8.38 to 8.44, the structure was constructed at the base of a long, gently sloping terrain. The geotechnical engineer recommended that the site be over-excavated by 3 feet and re-compacted with select fill to a nominal density of 95 percent. The contractor complied, ensuring that the site was over-excavated, not only in depth but also in width to a point that some of the over-excavation was 7 to 9 feet out from the outside footing (not wall) edge. The select fill was an approved aggregate base course material. The lecture auditorium at the northwest corner of the structure sloped downward to a depth approximately 7 feet below the finish floor of the balance of the structure. As with the rest of the structure, this area was also over-excavated.

Soon after the fall opening of the new facility, one of the students stepped off the entry walk and twisted her ankle as

FIGURE 8.38 Photo a shows the entryway to the facility is the only section of roof that has a gutter; then note where the water from the gutter is discharged. Note the 3-foot roof overhang in photo b. Take special note on how the surrounding topography all slopes toward the structure.

FIGURE 8.39 Notice in photo a the amount of erosion in the surrounding soils as the sheet flow from the surrounding slopes was all directed toward the structure. Drainage swales were minimum at best. Notice in photo b the long horizontal crack in the masonry parapet.

FIGURE 8.40 This view defines how the area immediately surrounding the facility had virtually no protection from storm water runoff, irrigation water, or a broken waterline outside the building envelope.

a

b

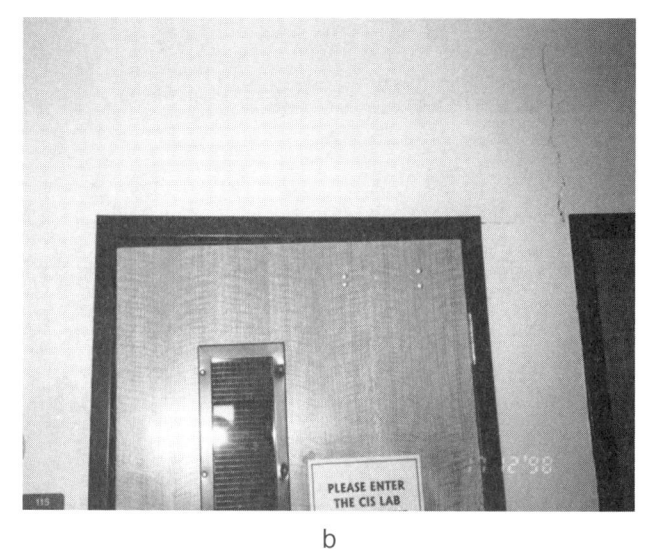

FIGURE 8.41 Interior racking and cracking of both interior load bearing and dividing walls, as noted in photo a, were found throughout the interior of the building. Virtually every door frame was racked as illustrated in photo b.

she went nearly up to her knee in some mud at the front entry; the school administrators blamed it on a broken irrigation line. Within a week, water was entering the lower portion of the auditorium through the electrical floor outlets, after which the doors started to stick, walls started to crack, and all of the sudden some of the exterior masonry exploded, floors began to heave, and some cords in the steel trusses snapped after which the school was evacuated.

It didn't take long for the fingers to be pointing in all directions—landscaper, contractor, architect, engineer, subcontractors, and so on. The attorneys were having a field day, as the complaints and cross complaints were being filed at the county courthouse. What went wrong? How could an entire new building be shut down? Who was at fault?

The photos are only a smattering of the hundreds taken throughout the course of the investigation. In reviewing these photos, keep in mind that you are constructing a facility on soils with highly expansive clays, that when subjected to moisture will swell considerably.

Four years of costly litigation followed the initial complaint filing, involving numerous parties. The storm and landscape water intrusion into the expansive clay soils, due to the negligence of a number of parties, was the primary cause of the resultant damage to the structure in catastrophic proportions, which resulted in the various contractors' general liability insurance carriers having to step up to the plate and cover the majority of the loss. Nearly all parties involved decided to settle out of court. The general contractor and his insurance carrier, however, in an attempt to recapture some of

122 SECTION EIGHT

FIGURE 8.42 Complete separation between load bearing and dividing walls, as shown in photo a, was the rule rather than the exception, resulting in the electrical EMT conduit coming apart or experiencing sharp bends. Photos b,c and d reflect how the door frames and ceiling was affected resulting in extensive damage to the entire roof assembly.

their settlement dollars, decided to pursue litigation against the landscape contractor, claiming that the leak in the irrigation line caused the problem. Detailed forensic facts uncovered during an in-depth investigation confirmed that the responsibility of the entire soils-related issue rested squarely on the shoulders of the geotechnical engineer, architect, and general contractor. The landscape contractor was awarded his attorney and expert fees.

8.20 A SLIDE PLANE BY DESIGN!

We have often read about and seen the news media covering landslides and other related disasters; however, little is said about the fact that many of these types of disasters are not caused by Mother Nature but by the negligence of humans. The following reflects how such negligence caused more than 200 home owners in the Greater Los Angeles area to become involved in a class action litigation that resulted in a hefty multi-million-dollar award.

The site topography was rolling hills found along the California coast, more commonly known as the Coastal Range. The problem started with the initial geotechnical investigation, when the developer cut the scope of services initially planned by the geotechnical engineer, which resulted in insufficient soil borings to determine the type and condition of the underlying geological configuration. The problem became exacerbated when the civil engineer developed the overall site plan, which included significant

Calendar Day 55 = 3/26/1996 to
Calendar Day 62 = 4/2/1996
Rain this period = .08 or less than 1/8"

Recommended over-excavation of 3' of clay soils from below bottom of foundations and to a lateral distance of 3 feet each side of the footing. Therefore the trenches range from 7'4" to more than 9 feet.

Orginal Ground Surface
Sandy Clay, Medium Plasticity

Sandy Clay, High Plasticity

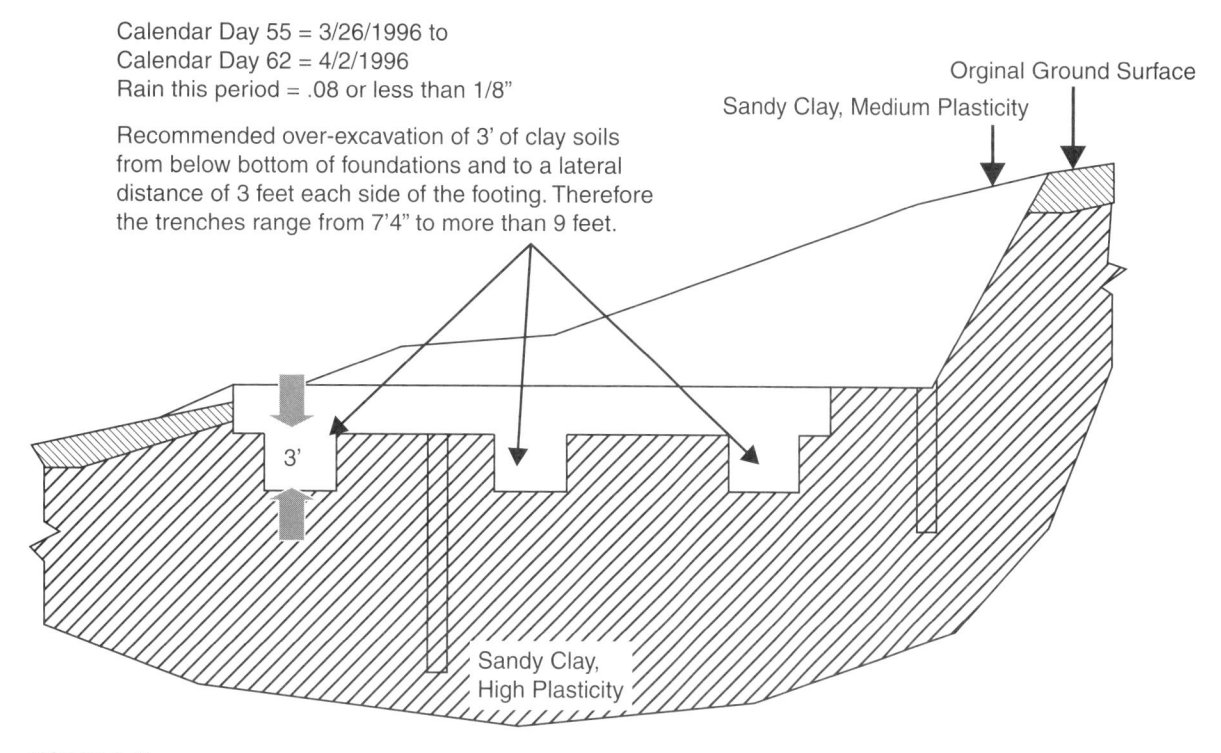

FIGURE 8.43

Calendar Day 167 = 7/17/1996 to
Calendar Day 257 = 10/15/1996
Rain this period = 2.74 or 2¾ " of rain.

Trapped moisture in footing excavation channels and granular footing and floor fill zones continues to move downward to contact with the clay soils. These clays have a high affinity for water and will continue to draw the water in as the clay swells lifting the structure.

Orginal Ground Surface
Sandy Clay, Medium Plasticity

Sandy Clay, High Plasticity

FIGURE 8.44

cuts in the surrounding hillsides for access roads and elegant large lots that overlooked the city lights below. In so doing, much of the stable soils were removed, opening the door for irrigation and storm water runoff to enter between the soft alluvial soils and the hard clays and bedrock, creating a slide plane. To top things off, the landscape architect decided to dress the hillsides with vegetation that required significant watering, the ideal trigger to set a landslide in motion, particularly after a 50-year storm descended on the area.

One cannot express too strongly the importance of conducting a thorough geotechnical site investigation prior to starting the architectural and engineering development phase of the work. Remember the structure is only as good as its foundation and the initial foundation is the soil it sets on.

8.21 YOU MEAN THERE IS A DIFFERENCE IN CONTRACTORS?

The following reflects how home owners, by not taking the time to determine the qualifications of those parties selected to build their dream home, can be trapped into serious consequences and unending stress. The cost of not doing it right can many times involve more than money as reflected in this single-family residence built on a hillside lot.

The two-story wood frame structure has a walk-in at the first floor in the front and a basement walk-out at the rear of the house. A concrete patio deck is located adjacent to the rear of the house. Five freestanding columns support the first floor suspended patio deck and the roof trusses. These columns are adjacent to the outside edge of the lower concrete patio slab.

The building pad was developed by cutting into a hillside that slopes from the front to the rear of the lot. This resulted in a fill zone that underlies almost the entire lower-level concrete patio slab and, most important, the five columns supporting portions of the house. The resultant fill slope was estimated to be 1.3 to 1 (H:V). Information developed from talking with a neighbor who took photographs during construction revealed that the original ground slope underlying the fill had not been properly prepared. Slash from clearing and grubbing operations was buried in the fill zone. This burial was then hidden as the fill was pushed from the excavated portion of the building pad. Drainage around the sides of the house was very poorly developed, particularly at a column that had substantial movement. It was visually determined that uncontrolled storm water runoff had caused severe erosion, in the form of gullies that discharged runoff toward the column and the edge of the fill zone.

No soil investigation report was secured by the builder prior to commencing construction, and no on-site testing was conducted during the fill operation. There were no site grading and drainage plans prepared by the designer or required by local building officials. All work was required to be accomplished according to the local building code. The footing excavations had been inspected and approved by the local jurisdiction.

When called to the site, the forensic geotechnical engineer discovered that the columns were starting to move down and outward away from the concrete patio slab edge. Additionally, a concrete patio slab at basement level was starting to pitch down and away from the house. Upon discovering this condition, a meeting was held with the geotechnical engineer, builder, and owner to review the findings. It was the engineer's opinion that the footings had not been seated into the native surface slopes but were most likely founded partially or wholly within the pushed fill. The builder agreed that something should be done and he would do it. However, after several phone calls and another meeting, in which remediation methods were presented, the builder chose to skip town and move to Hawaii, where he is apparently plying his trade. See Figure 8.45.

In view of the circumstances and seriousness of the problem, the owner chose to fund the remediation.

The remediation plan consisted of the following items:

- Excavation of the toe portion of the fill zone and placement of a series of reverse batter concrete slabs (6 inches thick) to form a buttress.

- Excavation of a portion of the fill slope surface soils to form benches upon which the excavated soil was replaced and re-compacted. The entire fill mass was not removed due to economic and staging factors. There was no good level area upon which all of the fill could be placed. The owner decided that if the house, columns, and patio slabs were stabilized, she would live with the possibility of some slope surface movements.

- Excavation of helical pier pits at each of the five column faces. Upon excavation of these pits, it became obvious that the upslope portion of the column footing pads had been placed on or just into the native surface soils. The down-slope sides of the footings were in the fills.

- It also became apparent that there was no quality control of the column footing sizes.

- Vertical helical piers were then inserted to lift and stabilize the vertical column movement and batter helical piers were inserted to minimize the potential for lateral column movement.

- The area of fill underlying the existing concrete patio slab was then pressure grouted to aid in tying the fill to the underlying native ground surface and to consolidate and minimize future settlement of the fill soils. Significant amounts of grout were needed to accomplish the fill stabilization.

- As a result of the pier installation and pressure grouting, the original constructed house was re-leveled and moved laterally to pre-movement conditions.

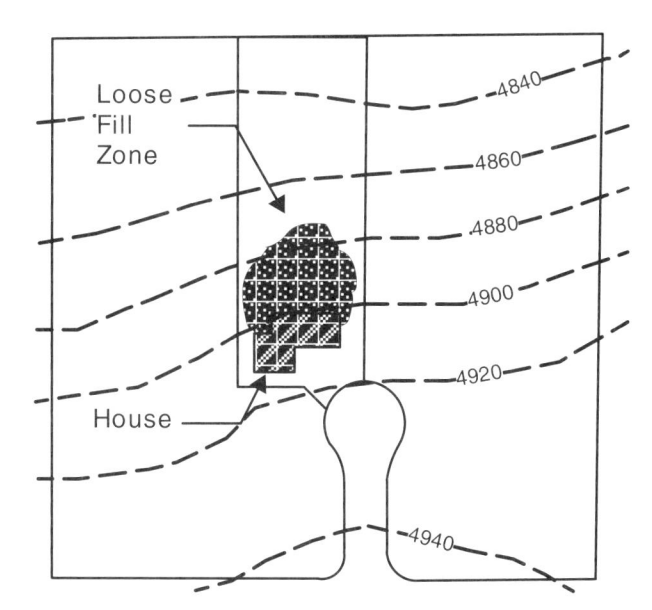

Figure 1-Original lot contours obtained from and courtesy of Yavapai Country GIS aerial photography database. The lot has approximately 100 feet of elevation differential from the front to the back of the lot. The front is at the cul-de-sac.

Figure 2 – View of rear of house showing the location and magnitude of the fill zone.

Photo 3 – View of eastern most column. This column exhibited the most settlement, rotation and lateral movement.

FIGURE 8.45

The cost of remediation included the following items:

- Engineering consultation and special inspection $3000.00
- Remediation construction and drainage control $100,000.00
- Landscaping $12,500.00
- Total cost of repairs to get the project done right $115,500.00
- Total estimated cost to have done it right the first time* $10,000.00

  *Includes the cost of the geotechnical report and grading and drainage plan

Risk to cost benefit ratio... 23 to 1

This is not a good ratio!!!

8.22 SELECT THE RIGHT CONTRACTOR OR YOU MAY PAY THE PRICE!

How often can one repeat "do it right or pay the price" before it sinks in? The following case is a sure reminder that just because contractors passed an exam and received a license from the state authority that licenses contractors doesn't mean that they are competent, and just because they can draw a floor plan doesn't mean they're an architect or engineer. All too often permits for the construction of residential homes are issued on minimal plans, or in some cases no plans at all, just a lot description. In some cases, the contractor or a friend prepares just enough plans to satisfy the permitting authority. The owners are elated that they didn't have to pay for the services of an architect or engineer, and the contractor was more than willing to make whatever changes the owners wanted with little concern that the local building inspector would even question any of them.

Couple the above scenario with an owner who has little, if any, knowledge of architecture, engineering, or construction, then throw in the complications associated with liability insurance, bonds, codes, standards, state statutes, OSHA, legal terminology, and so on, and you have a recipe for pending disaster.

Figures 8.46 to 8.50 reflect just the type of scenario outlined above. The construction agreement was two pages long and briefly laid out what was to be built. The plans were prepared by a "home designer" who had no formal training as an architect or engineer and knew nothing as far as what constituted minimal code standards. The owner was a retired medical professional who just wanted a "custom home" hidden away in the backcountry, away from the maddening crowd. Numerous changes were made during the course of the initial construction to take in the breathtaking views of the surrounding ambience. As the home began to take shape, the owner began to question some of the workmanship but continued to allow the work to proceed.

After a heavy rain proved that there were many problems with the roof, and the contractor was demanding more and more money for work that was obviously not completed, and certain items just didn't look like the picture provided to the house designer, it suddenly became obvious to the owner that there may be a problem with the construction. This concern prompted a call to the Registrar of Contractors, who came up with a list of defects and made a demand on the contractor to correct the problems. The contractor became irritated that the Registrar was brought in to check his work and became more demanding of the owner, delaying the completion of the home.

a   b   c

FIGURE 8.46 None of the girders were anchored to the pilasters with USP KST213 bent plates each side, as noted in the design documents, photo a. Long span girders were fit tight against the unreinforced stem wall, photo b. Code requires minimum $\frac{1}{2}$-inch clearance. Pilaster layout did not conform to city-approved plans. Girders were not attached to pilaster, photo c, and spanned 16 feet rater then the 8-foot designed span.

Care, Custody, & Control 127

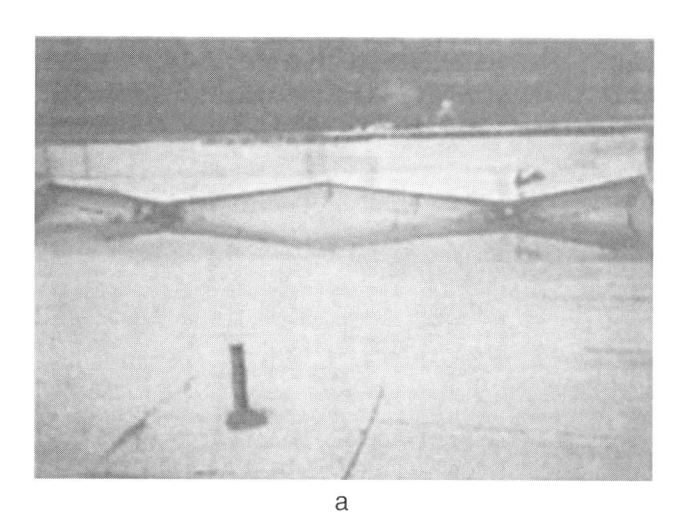

FIGURE 8.47 Photo a and b illustrate the scupper placement. Virtually every scupper was set 2 $\frac{1}{2}$ inches higher than the roof allowing ponding, which directed storm water runoff through various voids in the flashing and into the exterior wall framing.

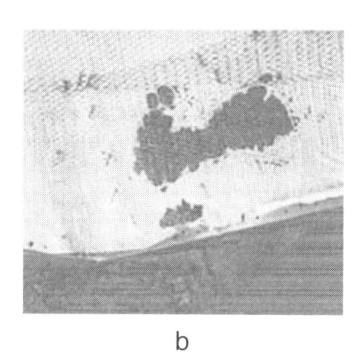

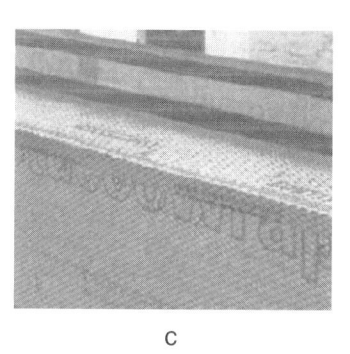

FIGURE 8.48 The exterior building wrap varied from Tyvek Home Wrap, to Tyvek Stucco Wrap, to unknown plastic sheeting, to no wrap or window surround flashing at all, as reflected in photos a, b and c.

FIGURE 8.49 Incorrect and modified joist hangers were found throughout the home, as shown in photo a, b and c. Many of these defects required shoring and disassembly of existing framing and mechanical systems to correct. The plans called for a 1$\frac{1}{8}$-inch first-floor plywood deck, glued and screwed. Contractor installed 5/8–inch OSB nailed.

In desperation, the owner retained another contractor to check out the project, who just so happened to be acquainted with qualified experts in the field of construction defects.

The photos cover only a partial listing of the construction defects discovered in this home.

Based on the above findings, the owner discharged the original contractor who promptly filed litigation to receive funds reportedly owed for completed work. This action prompted the owner into filing a counter claim for contractor negligence, at the recommendation of a reportedly

FIGURE 8.50 At the request of the owner, windows were added; however, the contractor did not address how such changes would affect the shear. In fact, he failed to follow any of the plans on addressing shear transfer from the roof to the foundations, requiring extensive modification and additional steel framework to correct the framing.

qualified construction attorney. At first look, this would seem like a great move, as the contractor did have general liability insurance. What happens next is a scenario all too often found in construction defect litigation:

- The contractor's insurance carrier retained counsel to defend the insured, the contractor, which helped the small contractor, who had limited finances, drag the case out for two plus years.
- The owner had to fund his defense and action against the contractor, starting with a $20,000 retainer for the attorney with the clock running at $250 per hour plus expenses. Total cost through interrogatories, disclosure statements, depositions, pleadings, briefs, meetings, and initial mediation was $65,000.
- The investigation of the construction defects required a construction expert and structural engineer at an hourly rate of $200 and $175 dollars per hour respectfully. Total cost through reports, calculations, remediation plans, building department approvals, and initial mediation was $28,000.
- Cost to remediate the construction defects caused by the initial contractor was $146,000.

Although there was more than substantial evidence to win a negligence case against the contractor, the home owner after receiving an award from the courts would have received nothing. To understand why this would have been the outcome, refer to Section 2 under Commercial General Liability Insurance. Unfortunately, the contractor's insurance carrier would only cover resultant damage. As virtually all of the remediation involved correcting construction defects, it became a contractor responsibility not an insurance company responsibility. The small contractor had no tangible assets, was incorporated, and would have quickly filed for bankruptcy, leaving nothing that would satisfy $350,000 judgment, leaving the home owner with nothing but bills and three years of frustration. Unfair, no question about it, totally unfair, which is the primary reason every client needs to know more about the contractor before entering into any kind of agreement or be faced with paying the price. One can find truckloads of judgments worth millions of dollars that were handed down but aren't worth the paper they are written on. Sadly enough, the home owner soon discovers that the contractor who created the problems is still in business, under another name, looking for the next fish to reel in.

SECTION NINE

SPECIALIZED & INVESTIGATIVE SERVICES

or

Finding the Needle in the Haystack

9.0 JUMP-STARTING SECTION 9

As the number of allegations of construction defects continues to grow, the role of specialized and investigative services takes on increased importance. In reading this section, one will want to note the following important points:

1. Like an architect or an engineer, the forensic investigator should not take on an investigative project that is not within his or her experience and competence, nor one lacking the time or staff to accomplish.

2. Not only does the investigation and report have to be scientific, demonstrable, and well organized, but also the claim of custody on tangible evidence has to be carefully established.

3. The forensic investigator has responsibilities to the justice system and the profession, not just to the party paying for the work.

4. The forensic investigator has an obligation to assess the situation so as to determine the appropriate testing, if needed.

5. When performing specialized testing, especially field testing, the forensic investigator needs to be conscious of who the client is; the forensic investigator is usually working for the owner and should not take direction from the contractor as to timing, placement, frequency, and so on of the test being performed.

In response to the overwhelming number of complaints that have surfaced over the past several years concerning design issues and/or construction defects, many new construction investigative providers have been created. There are two basic areas in which investigative services have developed, each in a different way. The residential area has seen the rise of home inspection firms. These are quite different from the forensic examiners one finds, ordinarily, in the commercial area. Some forensic examiners do some higher-end residential investigations (large expensive residences or class action situations), but the work in the residential field has been left mainly to home inspection firms.

This home inspection field has developed to service the needs of the home owner in dealing with (1) real estate industry issues prior to sale or (2) insurance industry problems related to loss and/or remediation costs due to property damage. While many of these services had been performed by licensed architects, engineers, or contractors in years past, the trend in minor disputes has shifted to using these new construction investigative providers who may or may not be licensed—or qualified.

As the involvement of home inspectors grew, it was discovered that many of these new investigative providers had limited knowledge of construction practices and very little knowledge of the architectural and engineering factors that go into developing a project. Although there are many high-quality inspection services in existence, it is becoming evident that some form of governmental regulation is required if there is to be consistency and a guaranteed minimum level of competency for home inspectors.

Some states have undertaken measures for the testing and licensing of home inspectors. However, as with testing and licensing procedures for contractors, the home inspector licensing schools are training applicants with a curriculum that may give little real knowledge and/or experience to the applicants, other than preparing them to pass a test and obtain a license as a home inspector. Under such conditions, the fact that someone is a licensed inspector does not by itself qualify that person to truly understand many of the construction defect issues affecting the industry today. A major problem in the field of investigative services is that there is no training designed to produce a good expert witness or forensic investigator. For further discussion, see Section 10, The Expert Witness, regarding the expectations for expert witnesses.

Because of this failure of adequate training in the home inspection area, we have prepared this section to better familiarize those in need of specialized inspection services related to construction problems as to what services are available to them. Good inspectors and good forensic examiners in the construction area are available, and this section will help you find them. This section will also introduce you to the different skills demanded of competent inspectors and building construction experts. Ordinarily, it is up to the consumer to find a home inspector, but it is up to the lawyer or a knowledgeable client to locate a competent forensic investigator for a construction problem.

9.1 FORENSIC INVESTIGATION IN THE CONSTRUCTION INDUSTRY

Forensic investigation services are considerably different from the investigation services performed by a home inspector. A forensic investigator examines physical evidence such as materials, products, structures, or components that fail or do not operate/function as intended and in so doing determines the cause or causes of such failure with an eye at correcting the failure. By correcting the deficiency, the item being investigated can be improved upon or, possibly, returned to service. A criminal aspect is possible in any investigation performed but is not always the case. Cases dealing with architectural, engineering, and construction most often are handled as a civil matter unless criminal intent is discovered during the investigation.

Forensic investigations include reverse engineering, which covers an examination of the failed component, an in-depth review of the component design, calculations, and witness statements, coupled with a working knowledge of current industry standards. The fractured surface of a failed component can reveal much information on how the item failed and the loading pattern prior to failure.

At one time, forensic investigations were almost exclusively associated with litigation, in which an investigator would be retained to present findings to a trier of fact. Engineering disasters (building collapses, out-of-design anomalies) are subject to forensic investigation by engineers and construction experts experienced in forensic methods of investigation. While those construction experts involved in forensic investigations still play a major role in construction litigation cases, they are becoming increasingly involved in a variety of activities that are improving design and construction techniques. The forensic investigator's primary responsibility in a construction-related issue that is being subjected to litigation is to determine the cause and effect of construction problems and to be able to testify to the findings. These problems can be as small as determining the risk of a minor load-bearing truss failure or more challenging investigations such as the following:

- Establishing the cause of a major mudslide that devastated hundreds of homes.
- Determining the safety of structures affected by catastrophic disasters such as the buildings left standing in proximity to the September 11, 2001, World Trade Center collapse.
- Determining the cause of major structural failures resulting in injury and death such as the July 18, 1981, Kansas City Hyatt walkway collapse.
- Determining the ability of existing structures to sustain further seismic activity, similar to the January 17, 1994, Northridge, California, earthquake.
- Determining the cause of the ceiling collapse in 2006 at the Ted Williams Tunnel at the "Big Dig" in Boston, Massachusetts.
- Determining the effects that design and/or construction defects had on failed or deficient construction projects throughout the world.

To accomplish this task, the forensic investigator must be sufficiently experienced in the applicable field and have the following attributes:

- Maintains an open and objective mind in the collection of evidence needed to properly develop a sound hypothesis necessary to arrive at an accurate conclusion related to the cause and/or causes of the defect or failure.
- Has a good working knowledge of local building codes, ordinances, specifications, and industry standards. This includes the understanding of the basis and intent of these documents, which is often included in their commentaries and historical records.
- Has an understanding of the scientific method and how to apply it to the issues being investigated.
- Has a firm understanding of accepted three-dimensional and other modeling, and the interaction of nature and construction to solve, as well as understand, the problem.
- Keeps informed on related science and technology.

These attributes, plus years of practical experience, are important in gathering evidence and deciphering it through sound scientific and engineering analysis in order to provide proof to the parties concerned that the findings are accurate and well supported by the evidence.

9.2 THE PRELIMINARY REPORT

The forensic investigator's findings are compiled in a Preliminary Report that will generally be presented in a format that is similar to the following:

Preliminary Report Outline:

1. **Letter of Transmittal.**
2. **Abstract or Executive Summary,** which describes in a brief and concise manner what the assignment was, what the report covers, and a summary of conclusions.
3. **Table of Contents**
4. **Introduction:**
   a. Objective of the investigation, covering why the investigation was conducted.
   b. Scope of services performed, including how and why they were undertaken to arrive at the objective.
   c. Background of the assignment, covering any history of the investigation, persons, places, data provided by others, and so on.

d. Team members and their assigned responsibilities in assisting in the investigation.
   e. Construction documents and/or other documents considered in the construction of the facility and/or item being investigated. See Section 10 regarding federal rules of civil procedure requirements as to what must be included in this portion of the report.
5. **Description,** of the accident, property, project, building, machine, and so on being investigated.
6. **Field Investigation,** including those persons involved in the investigation, written description of the process, location of the evidence, photographs of evidentiary findings, and a factual written description of the evidence.
7. **Photo Exhibit,** must be consecutively numbered and indexed to a plan and/or diagram that shows where the photos were taken and when. Provide complete descriptions of where the photo was taken using arrows or text to direct the reader to the key issue being shown. Reference and/or include important photos in the text of the report as needed to connect the field evidence to the findings.
8. **Laboratory Testing,** including those persons involved in the investigation, types of tests conducted, including photographic and/or graphically illustrated evidence of the results, and a factual written description of those results.
9. **Engineering Analysis,** including calculations and other supporting data.
10. **Discussion of the Findings,** covering both field and laboratory evidence and the interpretation of such.
11. **Conclusions,** should be concise and must be supported by the evidence and scientific facts as discussed in the findings.
12. **Proposed Remediation,** if applicable, should only be preliminary concepts, including a general cost of such remediation.

The Preliminary Report must present a convincing and logical argument that derives from the facts developed during the investigation. Facts are developed through collection of evidence, the careful study of that evidence, conduction of appropriate tests, study of applicable literature, and always passing all information through the filter of personal experience. As with any investigation, there is always the chance that new evidence may be found after the report is published. It is therefore recommended that the forensic investigator include a caveat similar to the following:

> The conclusions reached by Forensic Investigator, Inc., in the preparation of this report have been based on the known evidence. Should new evidence be discovered and/or provided, we reserve the right to modify our conclusions, accordingly.

The investigators must have impeccable credentials in their chosen field of investigation and will normally be affiliated with an organization that scrutinizes their practice through a peer review process and requires continuing educational credits. Some examples of such organizations include the following:

- The National Society of Professional Engineers (NSPE)
- The National Academy of Forensic Engineers (NAFE)
- American College of Forensic Examiners International (ACFEI)
- Association of Soils and Foundation Engineers (ASFE)

All of these organizations require continual education credits to remain in good standing and will normally conduct seminars to keep their constituents abreast of the latest scientific and technical findings, as well as the latest legal requirements for testifying experts.

9.3 PRE-INVESTIGATION CONSIDERATIONS

You have learned that a project has not performed well, and you are being asked to be an expert witness for one side (plaintiff or defendant). A number of issues need to be covered before any real investigative work, engineering, and/or document review can be accomplished. Sound business management practices, as discussed earlier in this book, still remain appropriate and the same procedures must be followed for defining what services you are being asked to provide. You must be prepared to ask a number of questions before you say yes to an assignment.

Client & Team Player Information

1. Who are all of the other players in the assignment? Attorneys, architect, contractors, engineers, and so on?
   a. Do you work well with other members of the team or have you had conflicts with one or more members of the proposed team?
   b. Do you have a conflict with any of the other players or attorneys involved with this project?
   c. Have you or are you currently working on any other project(s) for this client?
2. In litigation, the following questions should be asked:
   a. Who are the parties in the case and who wishes to retain you?
   b. Advise the attorney of your relationship with all the involved parties to determine if there may be a conflict of interest. Should a perceived conflict of interest arise after you have taken the assignment, you must immediately notify the attorney.

What is the Investigation About

1. What are the allegations or key issues in the investigation? You really need to know the specifics about the parts of the investigation that you are being asked to review and study.

2. How does your experience fit this investigation? Once you know the investigation specifics, you will then be better prepared to make an assessment of your experience and your comfort level for the assignment.

3. Based on your initial assessment, do you have the capability of performing an objective analysis free of any biases? Inasmuch as biases develop naturally in people in the course of living, it is important that you understand what biases you may have and then decide if you can still keep an even hand while doing the work. Your responsibility is similar to that of the jury members, who are not asked to have an empty head, but rather, an open mind. Recognizing that bias is the first step to freeing yourself from its effect, which will allow you to formulate sound opinions and defend those opinions from attacks from all sides, including your own client, as at times your opinions may be contrary to those of your client.

4. Will you be able to render a timely opinion? Many times, it is essential as well as beneficial that your client have a timely response in order to resolve an issue. Exposure to loss as well as possible failure could be at stake; therefore, the timeliness of providing the results of your investigation could be critical.

5. Do you have sufficient knowledge of the legal process to properly prepare your findings for the trier of fact, be it arbitration, mediation, mini trial, or court trial? Knowing how to do the investigation and arrive at an objective conclusion is one thing, presenting it to the trier of fact is another.

Project Fees

Forensic investigative services by their nature involve working with the unknowns. Because of this, it is virtually impossible to determine fees accurately in advance. Thus, most assignments are charged by the hour, while informing the client periodically of the progress and current costs. If a client demands a fee, this should be provided only in the form of a budget that will be continually revised as the investigation progresses. It is customary to receive a retainer that will be applied against the overall fee.

In the forensic profession, it is important that you understand how the billing and collections services work or you will become a very overworked and underpaid expert. Be specific; it is all right to be upfront about what your expectations are.

There are a number of questions that you need to be aware of and willing to ask and to receive answers to, including the following:

1. Who will be paying for your services and how will they be paid? It's always best to have a contract that sets forth the initial scope of services as you understand them, and most important how you will be paid. You can modify the scope as the investigation proceeds, but payment terms must be clearly understood and agreed to up front. The contract should cover specifics regarding payment, including the following:

   a. Your firm's fee schedule and those of any sub-consultants that may work for your firm.

   b. When you will invoice your services and what your expectations are for payment, that is, net 30 days after issuance of the invoice.

   c. Any interest assessment on overdue or unpaid fee balances in excess of 30 or 60 days.

   d. What is your client's turnaround time frame for receipt, approval, and payment of invoices?

   e. Will the invoices be shared by different clients (e.g., insurance companies)? This is a very difficult subject. Attorneys will often want to spread the cost of investigation over several different parties, which is not at all unreasonable. However, it is common that at least one of the other parties will balk at paying and will argue that it (1) did not agree to the work and/or (2) it had the right to participate and does not have to pay to secure its rights. The best way to proceed is to make the lead attorney responsible. Let the attorney collect the fees from the other participants.

2. What is the client's payment history?

   a. When does the client require monthly invoices to be received in-house so that you are assured that the current invoice will be paid in the next pay cycle?

   b. Will you be paid in a timely manner? You should know this information going into the investigation. Keep in mind some companies pay once a quarter or even once every six months. Can you live, or survive financially, with a long drawn out pay cycle? This is another "minefield" when working with attorneys, especially new attorneys without the financial resources for extensive investigation. The attempt is sometimes made, without your knowing it, to tie your payment to the time your attorney's client recovers money at the end of the case. THIS IS UNETHICAL! You cannot tie payment for your work and testimony to the outcome of a trial or hearing! Neither can you work on "contingency" the way many attorneys do. You work for a fee and the fee does not and cannot change with the

outcome of a trial. Remember the attorney is an advocate; you are not.

c. Who will be responsible for review and approval of your invoices? Many times, you will be instructed to send all correspondence, invoices, and so on to the client through an attorney's office. Be aware that this can significantly slow down the approval and payment process and result in your invoices going into the 90 to 120 day status.

3. How will you respond to urgent demands for additional work when you know that you have not been paid for past work in a timely manner? Are you prepared to say no to a request for additional work without benefit of being paid for past services?

Better to clear all of these questions in advance of starting the project rather than becoming angry or bitter about apparently working under rapid and sometimes inappropriate time demands and then apparently working for free since you don't get paid in a timely manner.

9.4 PROJECT & WORK TASK SCHEDULING

Once you have cleared up all of the administrative financial chores for initiating a new project, the real, and most challenging, work starts. You will need to know what the investigation schedule is at the time of project initiation. Again, a number of questions need to be answered so that you will have a thorough understanding of what will be expected of you and your firm.

1. If you cannot meet the time constraints imposed by the client, then it may be better to decline the project now rather than to be constantly rushed and not have enough time to do a thorough and professional job.
2. Do you have the staff to assist you in the endeavor to meet the schedules?
    a. What is your staff experience? Is it right for the investigation?
    b. Do you have enough staff to allocate the person-hours necessary to meet the schedule(s) without interfering with other work in progress?
3. Do you have the equipment to accomplish the work?
4. Will you have to engage the services of sub-consultants?
    a. Will their schedules match with the investigation schedules?
    b. Do they have the staff to meet the imposed schedules?
    c. Have you included their fees as part of your fees or will they bill separately?

d. Does the client understand that there will be fees from sub-consultants that you use?

5. In legal cases, what are the current drop-dead dates as mandated by the court, your client, other experts on your team, and/or the attorneys that you will be working for or with?
    a. How will these schedules fit with other work that you have in house?

Can you meet these schedules or will it push you beyond your comfort zone?

9.5 PROJECT-RELATED DOCUMENTS

Original Project Documents

In conducting any type of specialized investigation, it is essential that the forensic investigator is provided with all the necessary project documents that are in any way related to the issue being investigated. At a minimum, the following questions need to be asked:

1. Are the original project plans, specifications, reports, contracts, proposals, field memos, testing reports, communication documents, photographs, and any other documents that you deem necessary available for your review?
2. Where are these documents located?
3. Will you have free access to them?
4. Are they in digital or hard-copy form?
5. Will you be required to pay for reproduction and shipping costs?
6. If digital, do they require a special program for access?

Legal Case Documents

If the investigation has moved into litigation or some form of alternate dispute resolution, there may be other documents available from other parties who may have been involved in the issues surrounding the investigation. Should this be the case, the following questions need to be addressed:

1. What documents have already been produced by the other side? Are they appropriate to your area of expertise and work scope in this investigation?
2. Are they available for review, and can you get copies of them?
3. Where are these documents located?
4. Will you have free access to them?
5. Are they in digital or hard-copy form?
6. Will you be required to pay for reproduction and shipping costs?
7. If digital, do they require a special program for access?

9.6 WORK FLOW & DOCUMENT PROCESSING

Any forensic investigation consists of a multitude of tasks from client discussions, interviews and/or depositions of knowledgeable parties, reviewing and analyzing pertinent documents to conducting tests and developing models to accurately formulate final conclusions. As such, the forensic investigator can become inundated with volumes of documents, materials, and other items that become a part of the assignment. In view of this, it is good practice to develop a program on how best to segregate the information, as it is gathered, to ensure that it is cataloged into specific task categories that will fit into the overall investigative work plan.

Incoming Information

You need to have a thorough concept and understanding of what the scope of your services covers. From this scope of services, a listing of tasks to be performed that are needed to properly arrive at a conclusion can be developed. From this, one can develop a fairly comprehensive list of the information needed. In general, the initial information may consist of at least the following items:

- A review of initial project design documents. These documents may consist of preliminary and final design documents, engineering and/or architectural proposals, contracts, daily memos, and so on.
- A review of post-design and construction documents. These documents may consist of bid documents, bid supplements or addenda, pre-construction field meeting(s), contracts, daily field memos, weekly field meeting notes, requests for information (RFIs), change orders, photographs, and so on.
- A telephonic conference call with the client to discuss what your initial findings are, what you now understand your involvement to be, what the core issues of the investigation are, and what additional documents you are requesting for further review. Do not hesitate to ask for any additional reports, plans, specifications, contracts, field construction documents, memos, photographs, test results, RFIs, ASIs, change orders, responses, and any other information that you think will be needed to aid the process of discovery and evaluation of the investigation issues.
- Based on document discovery to date, it may be necessary to develop a scope of additional services that you think are warranted. This should be accomplished as early into the investigation as possible, so once approval is granted for the additional work, there will be adequate time to get it accomplished.
- Participation in field and/or office meetings with other experts.
- Participation in field observation of work being accomplished by either your own staff or consultants or by the opposing experts in legal cases.
- All of the above are discoverable activities and you may be requested to provide notes and/or recollections about any or all of what you have heard, found, recorded, or come to understand by any process. Further, an attorney, for strategic purposes, may decide to withhold some of this information from you. This is a dangerous thing to do and can result in the expert being in a very embarrassing situation. An attorney may reasonably limit the scope of your work, but such limitations need to be clearly stated and understood.

Incoming Materials

The day-to-day handling, processing, and filing of investigation documents are very important. The forensic investigator needs to identify and catalog the receipt of all documents used in the process of evaluating the issues being investigated. Set up an in-house system for cataloging all materials received, if one does not currently exist. See Section 10 for the negative consequences that occur when the materials you may have considered in your findings are lost or misplaced.

All incoming materials should be logged in and reflect the date received. Other important information includes the following:

1. Document author, date issued, project or job number
2. Document subject
3. Document addressee
4. Bates numbers (tracking numbers used to control documents), if they have not already been assigned in a legal case.

9.7 FORENSIC ANALYSIS, INTRUSIVE TESTING, & FAILURE ANALYSIS

It is not uncommon when conducting forensic investigations involving construction issues, particularly those related to structural and geotechnical failure, to begin with a preliminary analysis of the failure. The investigator often must review a large volume of evidence to find key issues to lead to the correct conclusions. A good investigator always practices the classic KISS (Keep It Simple Stupid) approach. Simplifying, instead of "complexifying," the issues is the hallmark of a good forensic investigator.

Engineering analysis is often necessary to determine the probable cause of a failure, be it a major structural member or distortion of other components of the structure, such as floor, wall, and roof assemblies. To assist in determining probable cause of a structural failure, calculations are available on failure analysis through computer programs such as NASTRAN, STRUDL, ANSYS, ADINA, and BOSOR.

When using such programs, it is essential that one does not become caught up in overreliance on such finite element analysis. These programs are oversensitive to distortion of the input for the model. The large number of input factors means that if one factor is erroneously stated, it may wildly unbalance the result. It is essential that simple hand calculations be performed to check the reasonableness of any result.

Other issues that have influenced failure in structures, and are often overlooked in the design, include temperature, humidity, creep, shrinkage, foundation settlement, and various stresses experienced during construction. Once the various forces affecting the structure have been calculated and the problem has been reduced to determining how and which structural or geotechnical element failed, the next step is to provide supporting proof of the finding. This effort is normally accomplished through previously performed research, finite element or finite difference analysis, or various types of testing programs. This testing may include the following:

- Intrusive testing
- Destructive testing (when and if required)
- Conformance and performance laboratory testing
- Constructing of models and conducting performance testing under laboratory conditions

It is essential that the forensic investigator be aware of potential legal consequences in conducting any type of testing. As explained in Section 10, the forensic engineer has an obligation to maintain and archive all parts of his or her file. Performing a test of one kind or another may place in the file, for all time, results related to aspects of the structure that are not in question. Most important, the forensic engineer must understand the intent and purpose of each test, including how it will properly relate to the investigation. The forensic engineer must thoroughly understand the relevant parameters of the test and any variables that may either negate or otherwise have an adverse effect on the results. The engineer must always ensure that the test is conducted by an experienced professional who can testify to the accuracy of the procedures and equipment used, the soundness of the chain of evidence, and the validity of the results.

Chain of Evidence & Protection Thereof

It is essential when conducting any forensic investigation that the handling and preservation of the evidence be both appropriate and accurately documented. The following discussion addresses the techniques and precautions that must be observed when conducting any type of testing program in support of the forensic investigation.

It is a legal requirement in any forensic investigation to identify, preserve, track, and properly protect any evidence discovered that might in any way be used in establishing the facts surrounding the investigation. This responsibility begins with the forensic investigator who first discovered the evidence. The first step after finding the evidence is to identify and photograph it by capturing the environment in which it was discovered. Each piece of evidence should have its own unique identification code that can be identified by an alpha/numeric symbol. That symbol can be, and often is, incorporated within the photograph taken of a specific piece of evidence.

Once the evidence has been located and identified with its own unique code, one must secure and protect it for a period of seven years unless released by the client. It is customary to charge the client for this storage. Obviously, in many construction investigations, this can range from placing the evidence in a small envelope to requiring special equipment and skills to remove, transport, and store it for further examination.

Some investigations may involve destructive testing or reconstruction of destroyed evidence. This happens because investigations by their nature occur after events, and much evidence may have been discarded, destroyed, or in some way altered significantly. In such investigations, it is essential that a detailed record of observed circumstances be maintained, including photographic, chronologic, and climatologic records; testing methodology; recordings of the investigation process; and results of the findings. If legal proceedings are expected to occur, all parties who have an interest in the investigation must be invited to observe the examination.

It is often essential to have another testifying witness who can verify the manner in which tests were conducted. In any investigation, it is the responsibility of the forensic investigator to properly identify, preserve, track, and protect any and all evidence discovered. Failure to do so can lead to unfortunate consequences in legal proceedings.

9.8 SPECIAL CONSIDERATIONS IN FORENSIC TESTING

When conducting examinations using standard test procedures, forensic investigators should keep in mind the following points to support findings and arrive at their conclusions:

- Although there are many instances when the results of a standard test will provide a clear answer to questions, there are other questions for which the answer is not so simple.

- It is often true that a failure, from something as simple as a loose floor tile to a complex system such as the engine mount on an aircraft, is not the result of a single item failure but is, rather, the result of a series of connected failures. With that in mind, always look for the "string" of failures and their sources. A single standard test will not find the string.

- A standard test, especially when developed over time by a reliable agency, is virtually always a good test. Almost by definition, however, it is not the "best" test. Standard tests are selected with many criteria in mind, including cost and availability of the test equipment, which will leave out some very good, but either expensive or difficult, testing procedures that are not generally available. Therefore, don't hesitate to use nonstandard testing *when helpful,* but you must justify the test methods. Oftentimes, a lawyer will consider a standard test as the best and only way to do the work. That is not correct.

- If you have a good understanding of the basics of how and why the test was constructed, you can modify it to meet your specific requirements. But, as always, you must be able to justify what you have done.

- A standard test is very much like a recipe for a chocolate cake; if you don't already know how to make one, the recipe is the way to go. You will get an acceptable result every time . . . so long as you already know how to cook and perform all of the little procedures that are implied by the recipe text before you can actually follow a recipe successfully. It's the same with standard tests; you must have the basic knowledge of how to conduct such work before your results are reliable.

- Keep in mind that standard tests are reviewed on a regular basis, usually several years apart. Be sure that you have used the most recent test, or possibly the test for the period of time in which the item being tested was manufactured or utilized.

- There is a connection between standard tests and building codes. The codes often reference specific tests. However, codes are often not *adopted* by various governmental agencies until well after they have been written. Cities often lag several years behind the code issue date. Be mindful of this difference when selecting which version to use.

9.9 EVALUATION OF FIELD-TESTING PROCEDURES

As field testing is a part of the quality assurance on a construction project, it is essential that the forensic investigator evaluate these testing procedures to determine if they were valid at the time the field testing occurred. Forensic testing may be desired or necessary for various reasons. It may simply be a question of whether the product has actually met project specifications. Such testing should conform as closely as possible to that defined by the project specifications. In some instances, forensic testing may be performed to determine the ability of a product to perform its function or meet the project specifications or industry standards.

In some cases, it may be necessary to go beyond the standard test methods to better define the product's properties or its acceptability for a particular application. In these situations, it is necessary that such modifications must have a scientific basis and should be well documented as a part of the test report. A client who requests forensic testing should be advised in advance that the information developed will become part of the project records, whether or not it shows expected (desired) results. See Section 10.

Forensic investigations often discover that field conditions were not in compliance with the requirements of applicable standards, faulty testing procedures were conducted, or there were erroneous interpretations of test results.

In the construction industry, ASTM specifications are the most commonly referenced and accepted. These specifications are detailed and need to be followed precisely. Some examples of field-testing programs used during construction or forensic investigations are presented.

Soils

Many construction-related failures are the direct result of unsuitable or improperly prepared soils. Since it is impossible to move the entire site into a laboratory for examination and testing, soils evidence must be properly identified, collected, and tested in the field. Field testing, like all testing, must be conducted in strict accordance with applicable industry standards.

Prior to the start of any field exploration and sampling services, it is imperative that the forensic soil investigator has a general understanding of the site conditions and probable soil composition that will be encountered during the course of the fieldwork. The investigator must then develop a plan for the field investigations and sampling techniques most appropriate for the anticipated field and soil conditions.

The suggested approach should include the following:

1. Review of available project plans relative to foundation and other structural element locations that have been or could be affected by the soil conditions.
2. Visual site reconnaissance to evaluate probable site and soil conditions.
3. Initial and subsequent photographic documentation to aid in the future development of a drilling/pit exploration and sampling plan.

And the following suggested questions should be answered:

4. Are there limitations as to when loud, noisy, or dusty exploration services can be accomplished?
5. Are any governmental permits required?
6. What are the requirements for site access?
7. Will some of the hardscape features require removal and replacement?
8. What types of soils are anticipated at the site?
    a. Granular soils will require sampling techniques and equipment that are totally different from what cohesive soils would require.
    b. Can the soils be evaluated using test borings and sampling or will pits and sand cone or nuclear density testing be required?
    c. Can test pits (either hand dug or backhoed) be accomplished around the facility?
9. What are the utility line services and off-site utility lines that exist at the project?
    a. Can the "as-built" plans, if available, be trusted?
    b. Can the services be located using a private or municipal utility locating service?
    c. Who will coordinate and verify the horizontal and vertical locations of the utilities?

10. What nationally and locally recognized and accepted exploration and sampling procedures will be used?
    a. Site characterization for engineering design and construction purposes can be found under ASTM D420.
    b. This standard lists a number of "Referenced Documents" that may or may not be appropriate to the tasks at hand. However, a thorough knowledge about these standards is necessary for the appropriate selection of the right standards.
    c. There may also be locally accepted standards to use where appropriate.

11. What nationally and locally recognized and acceptable laboratory testing standards are available for use?

Soil Construction Once design and specifications for soils are completed, the next task is to correctly construct the project. This requires the best efforts of the representatives of both the owner and the contractor who are in charge of the project. Quality control by the contractor's forces and quality assurance by the owner's representatives are necessary. After the responsible parties are familiar with the specification and testing requirements, both continual observation and periodic testing of the work in progress are essential.

Earthwork specifications generally are based upon achieving some minimum density and proper moisture content. It is not uncommon that specifications will include some other procedural requirements such as lift thickness and the manner in which the soil is to be prepared and handled during construction. Uniformity of density over the area being constructed is also essential, although it is not always included as a specification requirement. For instance, if a portion of the earthwork area is being used as a haul road for some portion of the work, repeated equipment travel may compact that location to a density well above adjacent areas. This nonuniformity in density may have an adverse effect on the soil structure's performance if not corrected; for example, differential settlement may occur.

Laboratory procedures upon which density and moisture requirements are based are performed by use of a standard compactive effort and procedure while compacting specimens at varying moisture contents. This procedure will yield a maximum density value and the moisture content (optimum moisture) at which it occurs. Similarly, it follows that the earthwork construction should obtain a certain density if uniform moisture content, placement method and compactive effort are maintained. Once procedures that achieve required results have been established, the effort then should be to maintain these procedures. Observation of the work while in progress plus a sufficient number of in-place tests to verify and document specification compliance should ensure that the required quality of work is being achieved. Trench backfill is a prime example of the need for both observation and testing.

An irresponsible contractor can place a significant depth of loose backfill very quickly in the lower portion of a trench. The defective backfill can then be covered with an adequately compacted lift or two, thereby obscuring the loose fill. Without continual observation of the work, and reliance solely upon density testing of the upper layer, the erroneous conclusion would be reached that the work was satisfactory.

It also is poor practice to give the contractor's forces the task of notifying the owner's testing agency when they are "ready" for testing. This allows the contractor to only test the locations that the contractor knows will pass. Giving the fox the key to the chicken coop is generally a bad idea. Further, testing by itself does not provide information on the manner in which the earthwork placement was performed. Removal of completed work in order to allow testing and verification of the quality of underlying soil structure is undesirable and costly. ***Doing it right the first time*** will avoid having to resort to such procedures to verify the quality of work.

Concrete

As with soils, improperly batched or placed concrete can result in serious structural failure. It is essential that appropriate field testing be conducted to determine that the concrete meets the specifications of the contract documents. Tests may be performed during delivery of materials or years after the installation. The type of test depends on the test objective.

Astm C31 Under this testing procedure, the concrete that is delivered to the site is tested prior to placement in order to verify that it meets the compressive strength called for in the contract documents. This test requires that after a thorough mixing of the concrete materials, a minimum of two cylinders be properly molded and protected from drying at a temperature between 60 and 80 degrees Fahrenheit for a period of 48 hours and then transported to the laboratory for testing. Additional cylinders are often collected for purposes explained below.

Burying cylinders in sand or allowing them to sit in the sun is a violation of the standards. To comply with the standards, the cylinders should be immersed in a tub of water where the water temperature is properly monitored. The old practice of molding three cylinders for a 7- and 28-day break, with one left to hold for a later break if the first two failed to meet the strength requirements was developed to allow work to continue as soon as the concrete met the compressive strength set forth for a 7-day break. Failure to protect the cylinders as previously recommended may produce invalid results. In such cases, the period of time necessary to reach the 7-day break requirements could vary.

Consistent test results are essential to proper evaluation of concrete. Companion cylinder breaks should be within approximately 200 psi for good control and no greater than 400 to 500 psi. A larger spread indicates poor testing procedures and can result in uncalled for conclusions by unknowledgeable inspectors. For example, the breaks for companion cylinders on a bridge project in Arizona had a spread of 1000 psi. The state project engineer penalized the contractor $13,000 because "that's what the test results showed."

If the 28-day strength is below that specified, ACI 318 must be read carefully before rejecting the concrete. Evaluation of the test results at 28 days (an average of three consecutive tests) must be in compliance with the appropriate specifications. If the 28-day test fails, ACI 318 provides many options to evaluate the suitability of the concrete. Such options may include obtaining and testing cores from in-place concrete in conformance with ASTM C42.

Astm C42 This procedure covers the drilling and testing of core samples when the cylinders tested under ASTM C31 fail to meet the required compressive strength in the specified amount of time. When failure occurs, the Building Code for Structural Concrete, ACI 318, Section 5.6.5.4, requires three representative cores be removed from the area in which suspect concrete was placed. Merely removing three cores from different locations in the structure does not constitute a proper test and the results would not be valid. A test is the average strength of the three representative cores. Cores may be damaged in the process of drilling or in some investigations in handling during delivery to the laboratory for testing.

ASTM C42 specifies the ratio of height to diameter of the cores to be two to one; however, if it is not possible to meet the two-to-one ratio, the standard allows shorter cores down to a minimum ratio of one to one and specifies reduction factors to be applied to the results. Although the standard requires a minimum diameter core based on the maximum size course aggregate, it is generally a good practice to use cores at least 3 inches in diameter. The standard requires the core to be trimmed in the laboratory to provide smooth and parallel tops and bottoms. With a 3-inch core, it is practically impossible to obtain a representative sample from a 4-inch concrete slab. Unfortunately, some laboratories will test cores with very small diameters taken from various locations. Results from such tests are not valid and only increase the probability of a dispute.

ASTM C42 contains a requirement that the cores be maintained in nonabsorbent containers in as close as feasible to the condition from which they were taken.

Masonry

It is not uncommon to have the grouting techniques and testing procedures used in reinforced masonry construction lead to a dispute. Grout requires large amounts of water in order to make it flow into the cells of the masonry units. It is well known that the strength of Portland cement products is controlled by the water-cement ratio. The more water in the mix, the lower the strength; however, in concrete masonry construction, the concrete units absorb the excess water, thus lowering the water-cement ratio. It is critical, and a requirement of the building code, that the grout be reconsolidated after the water is absorbed in the masonry. The net effect is a dense grout with an appropriate water-cement ratio.

Proper specifications require compression tests in accordance with ASTM C1019 to ensure that the grout has attained its proper strength. Inspection of the construction of the wall, as discussed below, is often more important than just testing the grout.

Astm C1019 ASTM C1019 requires a mold formed with four 8 x 8 x 16 concrete masonry units placed in a "pinwheel" arrangement to form a 4-inch square, 8-inch high void. The theory is that the masonry units will absorb water from the grout sample as the units laid in the wall. The problem is in reconsolidating the test sample in such a way as to be consistent with grout in the wall. To obtain a representative sample, it will require an experienced technician. Cardboard boxes are allowed by the standard under the assumption that the cardboard will absorb the excess water, but experience has shown such results can be erratic and consistently lower than those using the form of masonry units. The use of the cardboard box is not recommended. Once formed, the standard requires the molded cube be protected in the field for 48 hours before being transported to the laboratory. Leaving the sample exposed to the sun or adverse weather conditions is unacceptable and will almost guarantee a dispute.

Construction of a quality wall starts with control of the mortar during the time the masonry units are being laid. Under excessive dry heat, as can be experienced in the desert Southwest, the concrete masonry units at the end of the day, if not wet down, will cause abnormal shrinkage in the mortar, resulting in voids and poor bonds between the masonry units. The field supervisor should see that mortar is cleaned from the cells as the units are laid up; see that the reinforcing steel is placed and positioned properly in the cells; ensure that the cells containing reinforcing steel are fully grouted; see that the grout is reconsolidated after water has been absorbed by the masonry units; observe the batching of the mortar and limit re-tempering of the mortar on the board; and, most important, check the delivery ticket to see the that the proper mix was delivered to the project.

Problems in testing mortar are similar to those involving grout. Like grout, the masonry units absorb water from the mortar, making it difficult to obtain a representative sample. The value of field testing mortar is questionable.

Stucco

Stucco presents a challenge even greater than grout and mortar. There are no recognized tests for evaluating the stucco during application or after completion of the stucco product. There have been attempts made to develop a controlled penetration test to evaluate hardened stucco; however, these tests failed to produce a consistent result. It is not uncommon for the general contractor or the owner to demand tests be taken of the stucco to verify that it is of good quality, particularly if cracking starts to occur. This will usually require that samples of the stucco be taken to a laboratory and placed under either a scanning electronic microscope (SEM) or a petrographic microscope to determine the contents of the materials within the stucco. Some investigators want to know the sand content or the sand/lime/cement ratio. Petrographic examination will provide information on the mineral properties of the stucco. These microscopic examinations will be accurate but will they reflect a representative cross section of the overall surface? Further slight variations in sand content are not critical. In fact, "over-sanded stucco" may have a lower coefficient of shrinkage.

Experience has shown that properly mixed stucco, securely bonded to the reinforcing mesh, cured (protected from drying) for 24 to 48 hours, and with properly constructed and located expansion joints will minimize cracking.

In order to reduce the cost of the traditional three-coat stucco system, there has been an explosion of exterior cementitious wall-coating systems throughout the construction industry; these systems have been developed to seek a middle ground between traditional stucco and exterior insulation finish system (EIFS). These systems generally are composed of a weatherproof barrier attached to the substrate and covered with tongue and grooved insulation board covered with a wire lath, all of which are secured to the substrate framing members. Once in place, a base coat consisting of Portland Type I or II cement and a proprietary polymer additive are mixed with sand and water then either troweled or sprayed over the surface bonded to the wire mesh. Once the base coat has been properly cured, a color finish coat is then applied to the entire surface. Unfortunately, these exterior cementitious one-coat stucco wall-coating systems, when improperly prepared or applied, have been the cause of major water intrusion into both residential and commercial structures. In many instances, field testing is conducted with the use of moisture meters or is done in conjunction with the window/wall water testing. When done in conjunction with window testing, the two systems should be isolated from each other in order to properly determine if the water intrusion is the result of a product deficiency in the window, improper flashing around the window, or the improper application of the stucco system. See Windows & Doors below.

Exterior Insulation Finish System (EIFS)

Commonly called "synthetic stucco," EIFS is a finish system used for exterior surfaces on both commercial and residential construction. The EIFS system typically consists of four components: adhesive, insulation board (attached to the substrate with adhesive), a base coat into which fiberglass is embedded, and a decorative finish coat. This system is also called a face-sealed barrier that resists water intrusion at its outer surface. It differs from other types of cladding systems such as other "hybrid" stucco systems commonly called "one-coat" systems, which use a weatherproof barrier behind the cladding or may have air spaces between the cladding and the substrate. Because of these differences, it is essential that interfaces between dissimilar materials and the EIFS be properly addressed to ensure that water does not get behind the insulation board, not the EIFS lamina (base coat and finish coat). Water intrusion will occur when there is a void or opening between the EIFS and the dissimilar material, therefore testing of the EIFS will normally consist of the following operations:

- Bonding test that confirms the insulation board has been properly bonded to the substrate, which can be accomplished with either of the following methods:
  1. Applying a vacuum to randomly selected areas of the EIFS assembly
  2. Using a calibrated pull gauge in randomly selected areas of the EIFS assembly
- Moisture testing on wood frame construction should be done with a combination of two moisture meters in accordance with the latest edition of *Moisture Testing Guide for Wood Frame Construction Clad with Exterior Insulation and Finish Systems*, published by the New Hanover Inspections Department in Wilmington, NC.
  1. A noninvasive meter scans through the wall without penetrating the EIFS lamina.
  2. A probe-type meter that penetrates the EIFS lamina or between the EIFS and dissimilar materials and gives moisture readings of those materials in contact with the probe.

Windows & Doors

Windows and doors continue to be the predominant means from which various elements enter or leave a structure. Improperly manufactured or installed windows and doors pose problems for maintaining a balanced environment within habitable space, be it home, office, or other types of facilities. These problems affect the habitable space in many ways, including the following:

- Excessive operation of heating, ventilating, and air conditioning systems, leading to costly utility bills and increased maintenance costs.
- Allowing the outside elements, such as wind, rain, and odors, to enter the structure's substrate as well as the interior living space, resulting in mold, dry rot, mildew, and so on affecting the health and well-being of the inhabitants.

The standard field-testing method for determination of water penetration of installed exterior doors, windows, curtain walls, and doors is covered under ASTM E-1105 Modified and AAMA 502.02 Test Specification for Field Testing Windows. The testing procedures set forth in these standards are also quite effective on exterior wall systems as well. When utilizing an interior mounted vacuum chamber, air is exhausted from the interior surface as water is sprayed from a calibrated rack of nozzles against the exterior surface. These tests are excellent in determining how water infiltrates the various building components. Obviously, it is much more practical to accomplish these tests prior to the completion of finished interior wall assemblies. Unfortunately, many of these tests are performed after the fact, when water intrusion, mold, mildew, and dry rot have already taken place and prompted the necessity of conducting these tests, after the interior surfaces have been finished. When such is the case, suspect areas can usually be visually identified or located with a moisture meter or bore-scope. Once identified, selected areas can then be opened up so a proper test chamber can be set up and the appropriate testing performed.

It is essential when conducting these tests that each building system is isolated from the other; that is, the windows within a wood framed exterior stucco wall system would be isolated from the exterior stucco finish by masking off and covering the complete window unit from the exterior stucco with a water-repellant barrier. In this manner, the stucco system can be tested and observed to determine if the water-repellant membrane is doing its job and has not been breached by tears, staples, nails, or other such objects. Once the stucco has been tested and observed for possible water penetration, the water-repellant barrier can be removed or folded back so the window can be tested independently. By properly conducting and documenting the testing procedures as they are being performed, one can readily determine which assembly component is the primary problem, thus allowing a remediation plan to be developed.

Figures 9.1 to 9.4 are representative photos of typical water-testing methods and findings on a commercial facility and a residence:

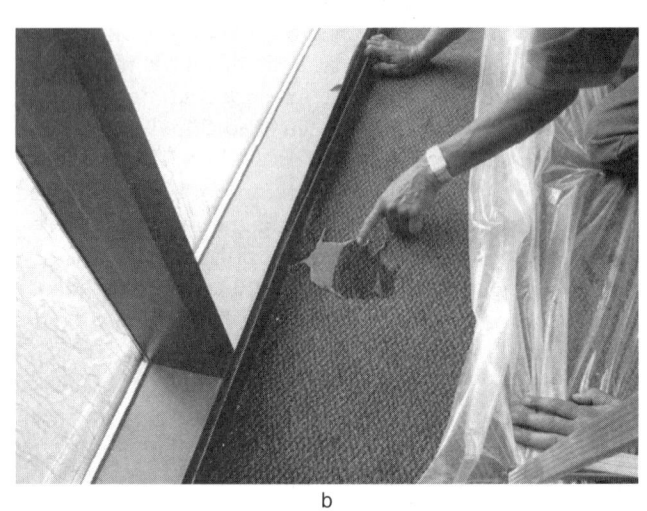

FIGURE 9.1 Step One: Identify and seal off test area and spray rack setup as reflected in photo a. Step Two: Start test; locate and mark the location of water intrusion, photo b intrusion points.

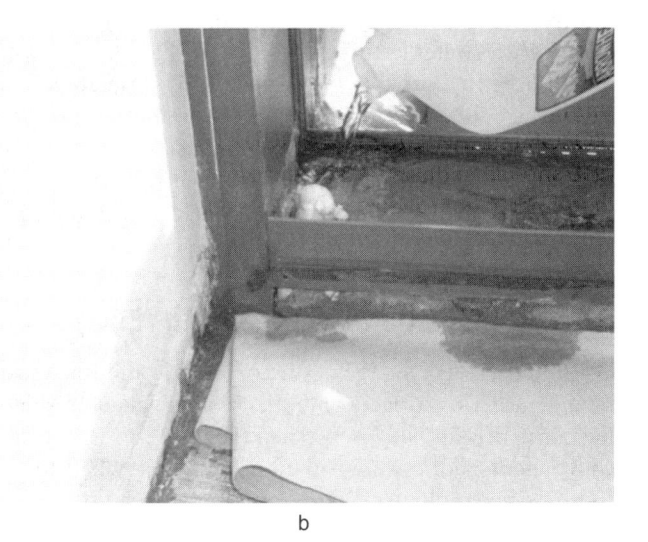

FIGURE 9.2 Step Three: Track and identify location of entry and cause as reflected in photo a and b.

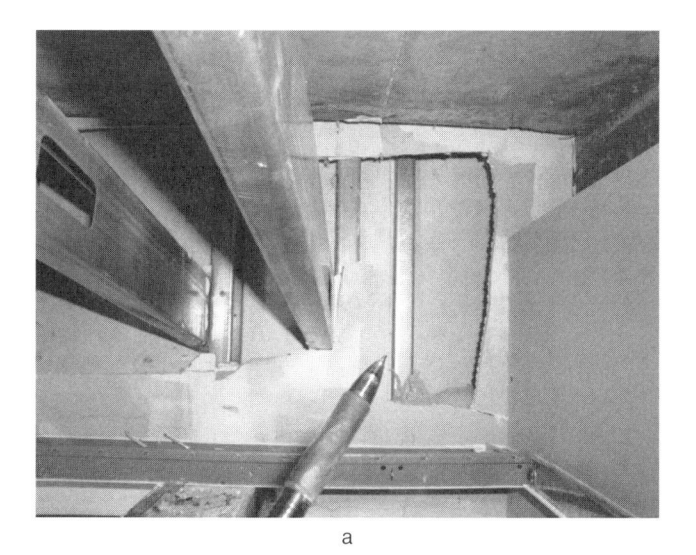

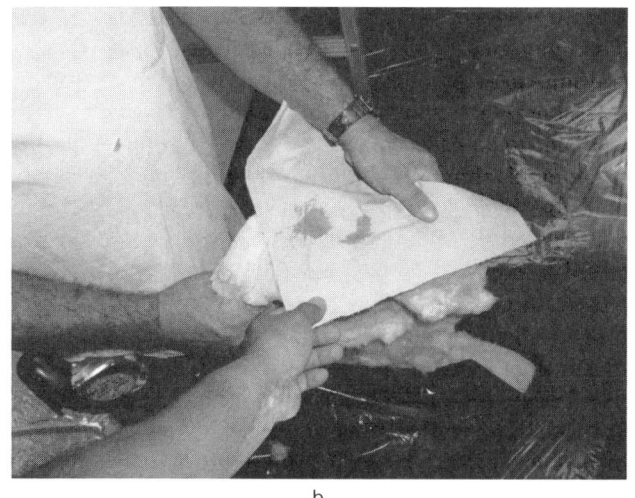

FIGURE 9.3 Following the water trail to an entry point can be complicated at times and require extensive destructive exploration, photo a, often leading to hidden microbial growth, photo b, and interesting points of entry. Photo c shows extensive mold forming on the outer wall. Photo d shows water entering the substrate via the screws used to anchor the window in place.

In addition to water penetration testing, the use of thermography to reveal variations of surface temperatures by electro-optical methods allows the trained professional to determine areas that are subject to thermal gain or loss.

Structural Investigations

Floors fit into two general categories: concrete slabs on-grade and suspended. Suspended floors include wood, steel, concrete framing systems, and combinations of these, for example, concrete floors supported by steel beams or steel joist. Load-carrying capacity, deflections, and vibrations are the design concerns of suspended floors. Level and flatness tolerances are construction considerations. Investigation of suspended floors may include visual observations to identify the obvious defects or deficiencies such as excessive deflections or structural distress. Structural analysis is used to establish load-carrying capacity and safety factors for the anticipated loads imposed in service.

Above Grade Floor Systems These systems structurally fall into three types based on the primary structural load-carrying members—wood, steel, or concrete. Regardless of the system, the first four steps are the same.

1. **Determine the total loading on the system.** To find the dead load, one must accurately field measure all the thicknesses of material and their weight. Sometimes this includes drilling holes to measure thickness and removing samples for weighing. Dead loads also include large permanently attached equipment, such as cooling towers, roof top A/C units, and elevator equipment rooms.
   - The prescribed live loads can be different.
2. **Determine the size and length of all members.**
   - If the original drawings are available they require checking (especially critical dimensions) in the field. Experienced Investigators do not trust "as-builts" as

142 SECTION NINE

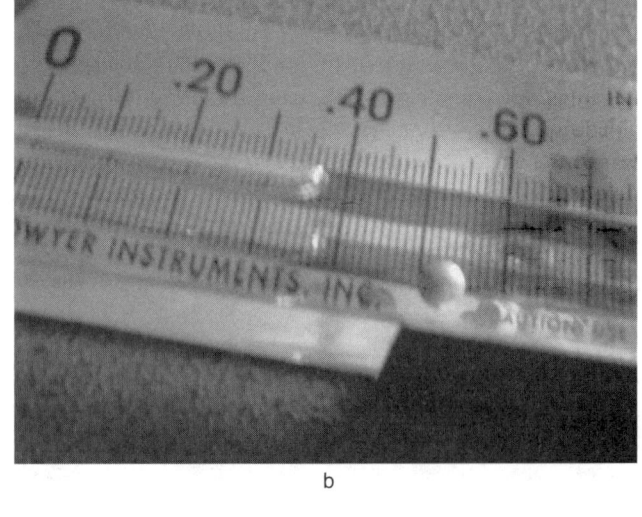

FIGURE 9.4 Vacuum tests are quite helpful in locating hard to detect leaks, particularly in custom-made storm windows as shown in photos a and b. Utilizing the same setup and tracking methodology, it doesn't take long to locate the leak as shown in photos c and d.

they often are wrong. Contractor substitution may not be documented.
- Without drawings, extensive field measuring, especially of critical dimensions, must be made. Scaled drawings should then be prepared to verify accuracy of the fieldwork in locating members and dimensions.

3. **Determine the strength of the materials of the members in place.** Again, if human lives are at risk, do not trust "as-built" drawings.
   - **Wood:**
     - In buildings built after 1960, field investigation may find grade stamps on the members. If a sufficient number of grade stamps are not visible, the investigator will need a qualified wood grader to determine the species and grade in place.
     - Buildings built before 1960 will not have grade stamps on the lumber and will require a wood grader depending upon the safety concerns.
     - The investigator must realize that wood is not a homogeneous material that it is graded by humans and machines with the expectation that some members will not qualify for the grade shown on the stamp.
     - The slope of grain and the size and location of knots are important to the grading and can adversely affect the wood strength.

- Grading rules allow up to 5 percent of the members to be one grade lower than marked.
- That the location of splits, checks, and twists that could have developed after the member was in place also must be evaluated.
• **Steel**
  - In buildings with small amounts of structural steel members, the fabricator would have purchased the steel from a warehouse that would have the most commonly available grade. The investigator then must determine what grade of steel was commonly used in the local area when the building was built.
  - In buildings with large amounts of structural steel members, the fabricator would have purchased as much steel as possible from a mill, thus all the same grade usually following the contract documents. This can easily be verified by taking a sample of steel (coupon) from a location of low stress and submitting the coupon to tensile testing to establish strength and ductility.
• **Concrete**
  - As builts are often reliable for strength of materials but never for placement of mild reinforcing (rebar) or prestress strand placement. Therefore, if the investigator has "as-builts," a few field verifications may be satisfactory.
  - For the investigator, the difference between prestressed or reinforced concrete is only in the analysis with one exception. The grade/strength of the prestressed strand can usually be determined by what was locally used when the member was manufactured.
  - Because the construction and manufacturing of concrete and its rebar have many variables, the in-place strengths can vary widely within a given structure. Thus, when investigating, you must check the material strength in several locations and use a standard strength for your analysis that would have been reasonable for the engineer who originally designed the structure. EXAMPLE: Concrete field test strengths are 3450, 3574, 3246, 2938 psi. One would use 3000 psi in the analysis. Rebar analysis would be the same.
  - Concrete must be cored vertically following the procedures outlined in the concrete testing section. Reinforcing steel must be located with a magnetic locator to make sure it is not cut during the coring.
  - Samples of rebar must be removed in low-strength areas and tested for tensile strength and ductility.
  - To locate rebar one must chip out concrete to determine size, location, and depth of cover. Magnetic locating devices can be used to determine where to chip but are not sufficiently reliable to establish size and depth.

Slabs On-Grade Floors Concrete slab on-grade floors (SOG) are widely used in both commercial and residential construction. Some of the more common issues related to SOG include cracking, moisture migration, flatness, level tolerances, and finish. Load-carrying capacity is rarely an issue except in very heavy-duty commercial and industrial facilities. When investigating cracking, the cause of the crack must be identified correctly for SOG. Too often, an inexperienced investigator will look at a crack and immediately conclude "the soil has moved up or down." In many cases, these novice investigators declare the crack is a "foundation failure." As noted elsewhere in this book, the SOG is not part of the foundation. (Interior and exterior foundations may be cast integrally with concrete slabs, but the slab portion is not part of the foundation system.) The SOG simply supports only nonstructural partitions and other elements.

It has been well documented that a very high percentage of all cracks in concrete (especially SOG) are caused by normal drying shrinkage and/or restrained thermal movement. Soil settlement or expansion in SOG makes up a small percentage of the cause of cracks observed. Furthermore, cracks caused by movement of the supporting soil will have specific characteristics that are readily identified by an *experienced* investigator.

Diagonal cracks across corners of SOG are usually caused by loads. Most of these cracks will occur at corners when shrinkage stresses lift the slab up off the subgrade at the corner. This is called curling (which is covered further in the discussion on flatness). Repeated loads on these curled corners will cause them to crack. In very few cases, these cracks may be caused by localized subgrade failure. Design of slabs for load capacity involves consideration of the slab and subgrade as a unit. Exposed corners in exterior SOG, such as driveways, often fail when the subgrade soil becomes saturated and is subjected to loading.

Water migration through SOG has become a major issue and the subject of numerous lawsuits. Many misconceptions exist concerning the properties of concrete and specifically in relation to normal moisture content in spite of the number of papers published on the subject in recent years. Further, there are many "magic" products on the market that the manufacturers claim will "waterproof" concrete. These are basically "snake oil." Some of the products will reduce the permeability of concrete, but there is no product on the market today that will make concrete waterproof.

All concrete, as all soil, contains water. Tests in arid Arizona show that samples of concrete brought to the laboratory contain 3 to 5 percent moisture. To put this in context, 4 percent moisture in a 4-inch-thick sample of SOG amounts to 2 pounds, or 2 pints, of water per square foot of surface area (or, 1/3 cubic foot) . This is the normal moisture in aged concrete in a desert environment.

To further complicate the issue, there are unrealistic specifications regarding moisture emissions and a lack of understanding of the results obtained from the moisture emissions test. It has been clearly shown that the moisture emissions test, ASTM F 1869, measures only the emissions at the time and under the conditions of the test. The investigator should apply these test results in a manner consistent with the structure's use. Retesting under different conditions will yield different results.

Floor-covering industry literature includes a disclaimer that relieves the manufacturer and installer of all responsibility for moisture issues if the emissions exceed 3 to 5 pounds per 1000 square feet per 24 hours. First, experience has shown that consistently meeting the 3 to 5 pound limit is practically impossible. Second, the installation of an impervious floor covering will alter the moisture content in the concrete. For example, retesting a SOG after an impervious floor covering has been removed will often yield moisture emissions on the order of 10 to 15 lbs/1000sq ft/24 hrs.

Moisture will flow into and out of concrete depending on the environmental exposure. This process is called "breathing" and is normal. Disrupting this normal process can, and most often does, create a moisture problem. Epoxy coatings, seamless vinyl flooring, rubber-backed carpet and porcelain tile are examples of impervious floor coverings. Generally carpet and ceramic tile will breathe. Placing plastic chair pads or flat-bottomed furniture on carpet may cause moisture accumulation in the carpet. Storing cardboard boxes on SOG will result in rotting of the cardboard. Inexperienced forensic investigators often claim the concrete is defective. This conclusion is wrong. The problem is that what was placed on the concrete disrupted the normal breathing of the concrete.

The floor covering industry also recommends moisture barriers, formally called vapor barriers, under SOG when impervious flooring will be installed. Unfortunately, the decision on the type of floor covering to be installed is not known before the slab is placed. Further, the floor covering may be changed during the life of the facility. This issue is further compounded by the fact that the moisture barrier may not guarantee that problems will not develop due to the normal moisture content of concrete. Further, current recommendations of the American Concrete Institute call for the SOG to be placed directly on the moisture barrier, which results in finishing problems and increases the risk of curling. This is another way the floor covering industry tries to avoid responsibility and liability for problems created by its products. Generally, moisture barriers are not recommended in arid desert areas due to the detrimental side effects.

In evaluating a finished SOG for flatness, an understanding of the governing rules and standards must be clear. For example, shining a light at a low angle over the surface will accentuate any imperfection and lead one to believe the slab does not meet tolerances. After evaluation, the slab in question may actually meet the required tolerance.

Specifications for Structural Concrete, ACI 301, and Guide for Concrete Slab and Floor Construction, ACI 302, published by the American Concrete Institute, define classes of floors and provide finishing tolerances for each. Standard Specifications for Tolerances for Concrete and Materials, ACI 117, provides additional information of tolerances and methods for evaluation. Project specifications should reference these standards and also include additional requirements. Unfortunately, some specifiers simply require the most stringent tolerance without regard to the requirements of the intended use of the facility. This practice will result in unnecessary cost or a dispute when the tolerance is not met. Finishing tolerances in all of the ACI standards require the measurement within 72 hours and recommends the measurements within 24 hours, after placement and prior to removing forms in suspended slabs. This requirement is intended to eliminate the effect of curling. Keep in mind that this is a finishing tolerance and should not be applied as a basis for determining the long-term performance of the concrete.

Curling must also be considered when evaluating concrete floors for flatness. Curling is caused by differential shrinkage between the top and bottom of a slab. The top is exposed to the atmosphere and shrinks and pulls together. The bottom is protected from drying by the subgrade, thus is wetter, and the concrete expands. The corners and edges of the slab lift off the subgrade due to this differential. Inexperienced investigators often equate the vertical displacement at a joint with soil movement when it is actually curling. It is very easy to identify curling using a 4-foot carpenter's level and a light hammer. Tapping (sounding) the surface with the hammer will detect the void where the edge of the slab is lifted off the subgrade.

In addition to the flatness tolerance, the standards also specify a levelness tolerance of $\frac{3}{4}$ of an inch. ACI requires that the surface of the slab should be level within $\frac{3}{4}$ of an inch throughout, regardless of the size of the floor. Local protocols and standards should be considered when evaluating SOG levelness.

For floor level surveys, a manometer will give more accurate results than survey instruments. It is essential to plot contours to properly evaluate floor levelness. Even then, careful attention to trends in the contours is required for proper evaluation. Unknowledgeable investigators may claim deferential movement when in fact the difference in elevation was the result of finishing tolerance. If a slab is out of level but essentially flat, it was probably built that way. Soil movement will not uniformly tilt a SOG.

Asphalt Pavement Mixtures

Production and placement of asphalt-aggregate pavement, often referred to as hot-mix asphalt (HMA), involves testing and analyses of both the component materials and the end product. The following is a brief summary of these processes, general testing that is utilized, and a few of the more commonly encountered problems. This discussion encompasses quality control and quality assurance processes during production and placement and forensic testing that becomes necessary when questions are raised regarding the acceptability of the end product.

HMA utilizes mineral aggregate and a bituminous binder in a heated condition that reduces the viscosity of the binder to allow mixing and placement to take place. Following placement, the pavement material cools and hardens to an extent that it can carry traffic and perform its intended function. Small quantities of admixtures such as Portland cement or hydrated lime are sometimes added during production to enhance properties of the completed pavement. Additives such as natural or ground crumb rubber and polymers are sometimes incorporated into the asphalt binder to modify mixture properties.

Quality control of aggregate production includes testing of the aggregate source for properties such as resistance to abrasion, soundness, and cleanliness. Once the quality of the aggregate has been established, further testing for control of gradation, fractured faces, and other properties is carried out during actual production. Proper handling of the aggregates is vitally important to prevent segregation, degradation, and contamination that might render the product defective for incorporation into the HMA. The total aggregate blend incorporated into the HMA is usually a combination of several aggregates, each with its unique gradation and properties.

Production of bituminous binder, usually referred to as asphalt cement, and mineral admixtures requires close control of raw materials and production to achieve a satisfactory end product. Manufacturer's quality control is often relied upon when accepting these materials for incorporation into the HMA; however, some additional testing of these components by both the producer and the buyer of the HMA is at times performed.

Prior to HMA production, a mixture design must be performed to establish the proportions of aggregates, asphalt cement, and admixture. This trial mix is performed in a laboratory by preparing and evaluating test specimens with various proportions of the component materials. This work must be performed under carefully controlled conditions using proper equipment and test procedures by qualified technicians. Verification of the acceptability of the final mixture design may be performed by further laboratory testing by the owner before actual production commences. All mixture design and verification testing and analyses must be under the supervision and direction of qualified professional personnel with technical knowledge of HMA and its application and properties.

Following are some test methods that have elements within their procedures that at times are not followed, resulting in erroneous or questionable test results. This list is by no means exhaustive and cites only a few examples of the problems that can occur during testing of HMA and its components.

ASTM D 6307 Asphalt Content of Hot-Mix Asphalt by Ignition Method This test method is used to measure asphalt content of a mixture by burning off the bituminous material and weighing back the remaining aggregate. The test procedure provides a means to test and calculate a calibration factor that is based upon the loss of mass of the aggregate during ignition, which is subtracted from the indicated asphalt content. The factor can be measured for a particular source and does not need to be determined each time a test is performed. However, it should be recalculated often enough to verify that it has not changed. All too often an erroneous calibration factor is used from another project or another aggregate source, not recognizing that the factor is source specific. Also, some aggregates have been found to break down under the high temperatures encountered in the ignition furnace. Consequently, gradation results using aggregate following ignition should not be solely relied upon without verification of gradation from samples prior to ignition testing.

ASTM D 2726 Bulk Specific Gravity and Density of Nonabsorptive Compacted Bituminous Mixtures This test method is used for testing both test specimens compacted and extruded from steel molds and for cores cut from compacted pavement. The test method has a provision that if a test specimen absorbs more than 2 percent of its weight in water while immersed in water during the volume determination portion of the test, it is not nonabsorptive and should be tested by **ASTM D 1188 Bulk Specific Gravity of Compacted Bituminous Mixtures Using Coated Samples**. This error is most often made when testing cores taken from compacted pavement for density determinations. When compaction is somewhat low and air voids high, absorption often exceeds 2 percent. Many times during testing, nonabsorption is assumed and the absorption, which is not necessary for final density computation, is not even calculated.

ASTM D 6926 Standard Practice for Preparation of Bituminous Specimens Using Marshall Apparatus This procedure is often used when performing laboratory mixture designs that incorporate asphalt cement modified by the addition of ground crumb rubber. Within this test method, specimens for determination of bulk specific gravity are compacted in steel molds. It is common laboratory practice to extrude these specimens from the molds shortly after compaction. However, when the binder includes ground crumb rubber, the rubber will rebound and increase the volume of the specimen if extruded while still in a heated condition. Consequently, specimens need to remain in the molds until cooled to ambient laboratory temperature prior to extrusion in order to obtain correct test results. Erroneous bulk specific gravity measurements will adversely affect final mixture design recommendations.

ASTM D 2041 Theoretical Maximum Specific Gravity and Density of Bituminous Paving Mixture When performing this test procedure, loose mixtures of aggregate and asphalt cement are immersed in water and subjected to a vacuum for a period of time to eliminate air from within the mixture and to subsequently measure the volume of the mixture. The test method includes a supplemental procedure for porous aggregate to allow for water that may be absorbed into the aggregate during immersion. All too often, it is assumed that the aggregate is nonporous, and this supplemental procedure, which involves weighing the sample in a surface dry condition, is never completed. No absorption limits are given to define nonporous aggregates at the completion of the test; however, the effect of absorption on the final test result should be examined by this procedure to see whether it is significant.

ASTM D 1856 Recovery of Asphalt from Solution by Abson Method and ASTM D 5404 Recovery of Asphalt from Solution Using the Rotavapor Apparatus These test methods are most often used for forensic testing to recover bituminous binder from samples of completed HMA pavement. The asphalt cement is recovered from the solvent at the completion of a bituminous extraction test performed under **ASTM D 2172 Quantitative Extraction of Bitumen from Bituminous Paving Mixtures**.

A few of the conditions of these test procedures that are sometimes neglected when asphalt recovery is being performed include the following:

- Extraction solvent must be from **ASTM D 2172, Method A**, the centrifuge method as the other methods within **ASTM D 2172** require heating of the sample during extraction. Hot extraction has been known to change the properties of the recovered asphalt. There also are precautions regarding the handling of samples during the extraction process that are not commonly adhered to if asphalt recovery is not intended.

- The extraction solvent used should be reagent grade trichloroethylene. Other common used solvents, such as methylene chloride, are not permissible under **ASTM D 1856** and **D 5404**.

- To prevent long periods of exposure of the asphalt to the solvent, the entire procedure from the start of the extraction to final recovery of the asphalt must be completed within 8 hours. There also are precautions for handling of the recovered asphalt until it is tested.

Examples of Other Recognized Construction Field Tests

The following are representative field-testing procedures used throughout the construction industry that will be noted in many of the specifications, project manuals, and other contract documents that become a part of the evidence trail used by the forensic investigator in arriving at a properly documented conclusion. When referring to any of these field-testing procedures in the preparation of the conclusions, the forensic investigator should not only understand the testing procedure but be prepared to respond to questions on why such a procedure was essential to arriving at the conclusion; that is, it was performed correctly but not properly documented or it was not properly performed and therefore the results of the test affected the issue being investigated.

Test Classification	Reference
Geotechnical Tests	
Hand, Power, and Hollow Stem Auger Drilling	ASTM D1452
Diamond Core Drilling	ASTM D2113
Vane Shear Test	ASTM D2573
Pile Load Test	ASTM D3966, D1143
Rock Bolt Pull Test	ASTM D4435
Drive Cylinder Test	ASTM D2937
Nuclear Method Test	ASTM D2922
Sand Cone Test	ASTM D1556
Sleeve Method Test	ASTM D4564
Seismic Test	ASTM D4428
Structural Component Load Tests	
Cladding Components	ASTM E997, E998
Beams and Girders	ASTM E529
Floors and Flat Roofs	ASTM E196, E695

Test Classification	Reference
Truss Assemblies	ASTM E73, E1080
General Practice	ASTM E575
Concrete Materials Tests	
Windsor Probe	ASTM C803
Swiss Hammer	ASTM C805
Metal Materials Tests	
Hardness Testing	ASTM A833, A370,
Weld Tests	
Radiographic Testing	ANSI/AWS B1.10
Ultrasonic Testing	ASNT/SNT-TC-1A
Dye Penetrate Test	ASTM E165
Magnetic Particle Test	A275/A275M
Masonry Tests	
Anchor Pull Out Tests	ASTM E754. E488
Water Permanence Test	ASTM E514
Water and Air Penetration Tests	
Window Flashing Test	ASTM D779
Window Wall Air Leakage	ASTM E283
Window Wall Water Leakage	ASTM E331, AAMA 501.3

These tests are fairly common throughout the industry, and they will often be involved in some type of destructive, contamination, load, strength, chemical analysis, or scanning electronic microscopic examination to assist in developing a failure hypothesis. That hypothesis, in conjunction with in-depth data analysis, can then be used to formulate a conclusion.

These tests can be done in the field, utilizing existing conditions, or in the laboratory, and they may entail the development of full-size mock-up models that will be subjected to various conditions to create a failure. The key to such testing is to determine the root cause of the defect or failure that prompted the need for a specialist in forensic examination. Without understanding what brought about the failure or defect, it becomes difficult to establish a foundation upon which the trier of fact can access liability and damages.

It is not, however, the purpose of this book to provide an exhaustive list of test methods. For one thing, there are simply too many of them. For another, they change over time. Most are reviewed on a regular basis, and then appropriate changes and/or additions are made. A few are discarded and new ones are being continuously proposed. The important point is to use the applicable testing procedure required by the project specifications. There are many useful standards available and one must know how to find them.

Perhaps the biggest single source for current testing information is the Internet. No matter how it may change, we can be certain that the Internet will be around in the future simply because it is so useful and people want it to exist.

If you don't know the specific test you need to have performed, or what group or organization wrote a test that you could use, then go to the Internet and type in key words and let the search engine do its work. "Key words" are words that directly relate to the subject of interest, including a few specifics. For example, during a recent search for test procedures useful for inspection of a failed fiberglass drainpipe, the key words were "Fiberglass Sewer Pipe." By not being any more specific than those three words, a large number of "hits" or possible matches were produced by the search engine. Hits included specifications, articles, manufacturers of such materials, and much more.

By following the logical thread to the makers of such materials, a series of specifications from four different test or specification-writing groups were quickly obtained. Those specifications were referenced by one manufacturer as being the relevant specs for its product. Checking other manufacturers produced similar results. From that short search, it was possible to obtain a list of specifications that related to the same type of pipe used for the same purpose as the case being investigated.

Further, information was obtained from the manufacturers regarding how the pipe should be backfilled, the nature of the ditch into which the pipe was placed, and other practical information, which reflected the "state of the art" as practiced by the people who make the product.

Also, by following the search engine "threads of information," discussions were found on the preferences of one pipe material over another. But, be careful— these discussions are NOT edited and reviewed the way technical articles are for publication. There is no "peer review" or other control. In other words, these are opinion pieces and may be very good or may be totally false. Read them with care.

After you collect a list of specifications or tests that you think may apply to the problem you are working on, you need to obtain copies to read. This can be a problem. You are also entering a very tricky legal area involving copyright laws.

Essentially, the group that wrote, tested, and published the specification owns it and desires to receive payment for its use. This is supported by the copyright laws.

Often, you can read a summary of the specification without charge but will have to pay a fee in order to read the full text.

There are two basic ways in which you can obtain the needed specification. One, you can contact the agency that owns it (e.g., ASTM) and purchase it. Two, you can contact a company that deals in specifications. There are several, and you can find them easily on the Internet. Again, try key words, for example, "purchase specifications."

With either source, you can download the full text using a credit card for payment. In minutes, you can have a first-class copy of virtually any specification written. Yes, they do cost money. They are a cost of doing business and can usually be passed on to the specific job you are working on.

There are certain limitations on what you can do with the specifications. Read the conditions that are supplied by the group that provides you with a copy. Just like software for your computer, these are often provided as a license to use rather than true ownership.

There is another way to approach the problem. Use the library! Large libraries, especially university libraries, will usually have large files of standard specifications and tests as provided by the major agencies. This source can be very useful, but keep in mind that it may not be a complete listing, it provides no data from the perspective of the manufactures, and copyright laws still apply. Just as you can't go to the library and copy a best seller to take home to read, you can't copy the specifications and use them.

Testing Caveats

When performing and interpreting the results of field or laboratory testing, the investigator needs to consider the reason for performing the test(s) and the application of the data obtained. If testing is performed to gain information regarding compliance with the contract documents, sampling and testing should be rigorously performed in accordance with methods specified within those documents. However, if the intent is to obtain information relative to the performance of a material or product, the tests stipulated by the contract documents may not be appropriate. If project conditions exist in which standard testing procedures are not applicable, specialized tests or variations from standard test procedures may be used. When used, these procedures must clearly define the scientific base used to obtain the test results.

Obtaining test data is only the beginning for the investigator. Without proper interpretation of these data, test results themselves are of little value. The manner in which samples were selected, obtained, and handled becomes important to the interpretation of the test results and must be thoroughly documented and reported. An investigator and the client must be aware that once testing has been performed and data obtained, these become part of the record and will be shared with all parties during litigation. It is inappropriate to perform tests and not reveal those results. The fact that they do not show what the investigator or their client had hoped for or expected does not justify concealing the results. These circumstances closely relate to discussion presented in Section 10, The Expert Witness.

9.10 LABORATORY TESTING

Laboratory testing is one of the primary means the forensic investigators use to assist in the search for truth while evaluating the evidence and separating fact from fiction in their quest to arrive at conclusions that can truthfully be defended. Laboratory testing, unlike field testing, is normally specialized and conducted in a controlled environment that is generally supported by various scientific equipment, which has been engineered and developed specifically to address defined testing procedures and/or assist in discovering unknowns through written protocols or other scientific methodology. In the construction industry, the forensic investigator may utilize one or more of the following laboratory procedures to assist in searching for the truth.

Microscopic Examination

The value of being able to examine materials microscopically cannot be overstated. A microscope, even of modest capability, is a direct and logical extension of our own eyes. In a very real sense, the purpose of a microscope is to allow one's eye to get very, very close to an object.

Most of us recall as children trying to get our eye right up against something to see just how big we could get it to look. A child's eye has a great range of focus, and we could get quite close indeed. The optical microscope, through its systems of light-bending lenses, allows us to extend that child's game on a scale almost beyond imagination. A good light microscope can easily magnify 1000x. A scanning electron microscope (SEM) can reach over a hundred thousand without difficulty. Other specialized imaging systems such as the scanning probe microscope and the electron tunneling microscope can approach atomic scale, but they are not usually of use to the forensic examination of construction problems.

To make a useful examination of most problems that come to the forensic examiner, the following types of optical microscopes are of high importance:

- **The Stereo Microscope** is actually two microscopes in one—one for the left eye and one for the right eye. Together, they provide an image of true 3D proportions. There is no training necessary in image interpretation, as every thing looks normal, just bigger. Magnification for such a microscope typically starts at about 10x and runs up to approximately 60x as a useful limit, often with the magnification being continuously adjustable over the entire range.

  The stereo microscope is an excellent choice for the examination of broken parts, bloom on the surface of a cementitious product, paint chips, and virtually anything that you need to "just get a good look at."

- **The Standard Binocular Microscope** with magnification that typically ranges from 50x up to 1000x. There is no stereoscopic capability on this type of microscope.

  If there is illumination from a light source in the base, passing light up through a transparent sample into the magnifying portion of the instrument, the microscope is often called a biological microscope because such lighting is necessary for the study of thin sections of tissue and so forth. They are also called transmitted-light microscopes, for obvious reasons.

  If the light source comes down onto the surface of the sample by passing through the lenses of the instrument and reflects back upward into the main portion of the microscope, it is then usually called a metallurgical microscope, or sometimes a reflected-light microscope.

  There are many variations on both basic designs depending upon the specific needs of the user, and, in fact, many such microscopes include at least some capabilities of both transmitted and reflected light designs.

  The reflected light microscope is useful for detailed examination of prepared samples such as polished cross sections of metal samples, rock samples, and other materials that are naturally very flat, such as microelectronic devices. This type of microscope has a very narrow depth of focus (i.e., the subject must be flat in order to have all of it in focus at any one time). The samples must usually be carefully prepared for observation, and a certain amount of training is necessary to fully understand the image produced by the instrument.

- **The Petrographic Microscope** is a very useful variation of the transmitted-light microscope. This microscope is used to examine very thin sections of solid materials that will transmit light. A very important sample for this type of inspection is cement and cementitious products. The petrographic microscope is fitted with a polarizing filter at the light source and a matched polarizing filter above the sample (the "analyzing filter"). Polarized light is affected to different degrees by different mineral composition and structures. A trained operator can recognize not only different mineral phases in the sectioned aggregate particles but also how the cement paste has developed, whether or not alkali-silica reactions have taken place, and a number of other very useful facts. Interpretation of the image to obtain these data requires extensive training. Anyone conducting examination of cementitious materials is well advised to find a good cement petrographer and keep his or her number at the ready.

- **The Scanning Electron Microscope (SEM)** is the third type of microscope used by the forensic examiner. This instrument can examine virtually any solid material that is stable in a vacuum chamber. There are limitations as to size, with quite a bit of variation among different models of SEMs produced.

  The SEM provides high resolution and tremendous depth of focus. Variations in surface height of many millimeters can be retained in functional focus at low magnification. These instruments are extremely useful for the examination of fracture faces to determine the mode of fracture, cut objects to determine what type of cutting device was used, fine sand particles to see what the surfaces look like, and virtually any solid object that requires detailed surface examination.

  Most SEMs are also equipped with an x-ray detector, usually of a type called an energy dispersive spectrometer, or EDS for short. The EDS allows determination of which elements, such as silicon, iron, or calcium, are present in the sample. X-rays are generated in the examined sample during routine SEM observation and the EDS takes advantage of that fact. The SEM and EDS together make up an extremely powerful analytical tool. With such an instrument, it is possible to examine the face of a fractured piece of concrete and look for mineral crystals that have grown in natural void spaces. When such crystals are found, they can be analyzed to determine if they are, for example, sulfates, which might be part of the evidence for sulfate attack.

  There are also specialized SEM instruments called environmental SEMs that can work at near-atmospheric pressure, making it possible to examine samples that are wet or otherwise unsuitable for the normal SEMs.

  When using a microscope of such high magnification capability, it is always important to keep in mind the scale at which you are working. The presence of a few sulfate crystals may look impressive at 5000x, but they do not by themselves constitute final evidence of sulfate attack.

  Although the images produced by SEMs look very much like optical images greatly expanded (but without color), the interpretation of the images does take some training.

These three microscope types, stereo optical, binocular optical in either transmitted or reflected light, and the scanning electron microscope, are basic instruments for the examination of material objects. It is difficult to imagine conducting any serious examination of failed materials and not making use of at least one of them.

With the quality of quick, secure delivery of materials available virtually anywhere in 24-hours or less, high-quality image transfer via the Internet and either text or voice messages by the Internet or telephone, the availability and utility of laboratories having all of the discussed microscopy are good. Even a small independent forensic investigator working in rural areas has ready access to the highest quality equipment.

Typical Examples of Microanalysis

The following four examples are indicative of utilizing a laboratory employing the techniques of microanalysis to assist interested parties involved in a construction dispute or litigation.

Example I—The Case of the House with Measles
The initial contact came from an attorney representing one of several home owners in an upscale home development, all of which had unsightly spots in the exterior finish. All of the homes in the development were of the same basic construction: concrete slab floor, single story, wood frame structure with painted two-coat stucco as the exterior finish. Information on the homes was provided by the attorney and arrangements made to visit the property, observe the problem in situ and take appropriate samples. The attorney also informed the laboratory that legal action was expected and that deposition and courtroom testimony should be anticipated.

The date of the initial contact was recorded, a work order started (an in-house document that traces all billable activity, location of evidence, client name, phone number, and other critical data), and a letter requested from the attorney that stated that the laboratory had been retained in the matter.

At the on-site meeting, a large number of people were present, including representatives, both legal and otherwise, of all interested parties.

An inspection of the residence showed numerous rust spots on all exterior exposed wall surfaces except for a wall that was shielded by a large patio cover. That wall was not only protected from the sun but was not exposed to the infrequent but often strong rainstorms that occur in the desert environment around Phoenix, Arizona. The builder stated that virtually all of the exterior stucco had been completed in a single day and would have been from a single batch of material, leaving no reason other than environmental exposure to account for the lack of spots in the protected area. Such information is extremely important to the work and is often lost if a site visit is not made by one or more of the people who will be conducting the tests.

The rust spots, or "measles," also showed vertical drip marks extending downward from the spot that displayed both rust and white colorations.

Figure 9.5 illustrates an area of one wall with several spots and associated drip trails marked.

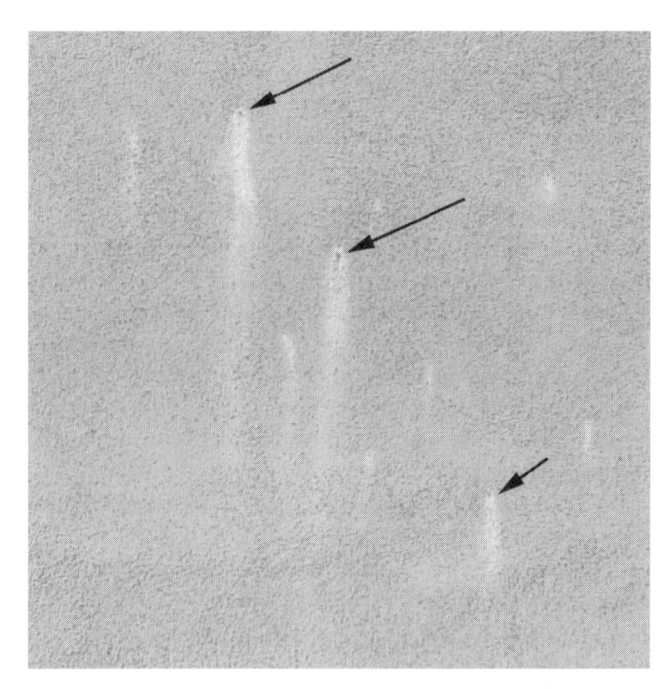

FIGURE 9.5 Spots in exterior stucco wall.

At least one area of all exterior walls was photographed for the record.

It was suggested that a sample of at least one spot be collected from each of the main compass-direction walls (i.e., north, south, east, and west), including the wall that did not display spots.

The parties were instructed that sample collection would be destructive and repairs would be necessary regardless of the eventual outcome of the dispute. Agreements as to the locations and manner for sampling were reached and samples were collected.

At each sample site, a spot was marked, identified, and photographed prior to sampling to properly document the location. Each sample was placed into an individually marked plastic bag for return to the laboratory.

Microscopic examinations at the laboratory using a stereo microscope having variable magnification of 10x to 60x were made of each sample. Gentle prodding of the full-thickness stucco sample (i.e., base coat and finish coat) quickly revealed a track of rusty discoloration extending from the surface spot toward the interior and directly to a relatively large area of soft, reddish-brown material in the finish coat. A portion of that material was recovered for analysis.

A sample was also taken of the surface paint below the spot that contained both unstained paint and paint that exhibited a portion of the vertical trail extending down from the spot.

Microanalysis of the reddish-brown source material using energy dispersive x-ray spectroscopy associated with a scanning electron microscope easily revealed a composition rich in iron and sulfur.

Analysis of the train on the exterior paint below the stop also confirmed sulfur and iron as well as calcium-rich deposits.

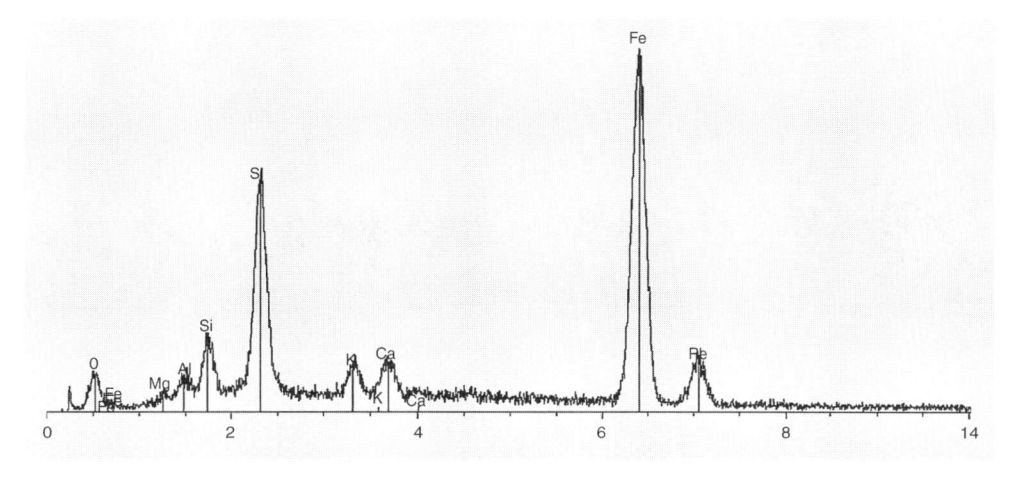

FIGURE 9.6 Typical spectrum obtained from the source material within the stucco finish coat layer, major iron (Fe) and sulfur (S).

These analyses, therefore, showed that the spots were the direct result of reaction of aggregate grains of marcasite (iron-sulfide) and moisture. The final product of the reaction is classic rust (iron oxide with water of hydration). Likewise, the drip mark was easily tied to the rust spot itself, along with a small amount of calcium carbonate from the cement.

Additional examinations of collected samples were also valuable in that they showed the marcasite grains to be in the finish coat rather than in the base coat. This simplified to some extent the possible mitigation strategies and also allowed the stucco company to review its sources of aggregate used to make the finish coat.

Figure 9.6 shows a typical spectrum obtained during the analysis of the marcasite grain. Such data are easily substantiated, if necessary, and can be reviewed by other experts without difficulty.

Except for the small amount of material lost during collection of the marcasite grains, the samples were essentially undamaged. The analytical technique itself, as separate from sample preparation, is nondestructive and the material remains available for others to analyze if needed.

The ability to nondestructively analyze evidence can be of considerable value, especially in legal procedures.

Although the Case of the House with Measles still had a number of interesting legal questions to settle, the ability of the laboratory to firmly identify the cause of the trouble and to present the findings in a clear and understandable way was critical to resolution of the problem.

This case also demonstrates the importance of keeping good records, including the preparation of a complete report, as the final resolution took in excess of two years. At any time during that period, the laboratory might have been required to provide competent testimony regarding the case, including the methods used for collection, analysis, and interpretation of the analysis results.

Example II—The Case of the Innocent Painter The laboratory was contacted to obtain assistance in making a determination of why both interior and exterior painted walls of newly constructed homes did not appear to be of high quality. The question was brought by the home builder, who had reservations regarding either the paint quality or the application of the paint. The short question was simply: Did he need to confront the paint supplier or the paint applicator?

The laboratory obtained the usual background information and letter of intent from the client and submitted a protocol specifying what would be done and how much it would cost. With the protocol accepted and a short contract in hand, the laboratory personnel obtained several samples of interior wallboard/paint and exterior stucco/paint.

The samples were mounted in epoxy, ground, and polished to allow a cross-sectional view of the support material (i.e., wallboard or stucco) and the applied paint. Figure 9.7 provides a microscopic view at low magnification of a typical exterior section.

Paint layers were measured and found to be well within specification; they displayed good coverage, appeared to have high solids content, showed good adherence, and were pliable as expected for latex paints.

Although specific tests were not conducted on the paint, it appeared as though the quality of the paint was very good and the application was also good.

Subsequent examination suggested that poor interior wall construction had produced an irregular surface to the wallboard that caused the paint to appear to have variable reflectivity when viewed at a low angle. The painter was not the problem. Neither was the paint quality.

In this case, a few relatively simple tests helped avoid a confrontation between the home builder and both a subcontractor and a supplier over an issue that, for them, did not exist. The home builder was then able to direct his attention to the more probable cause and determine how best to deal with it.

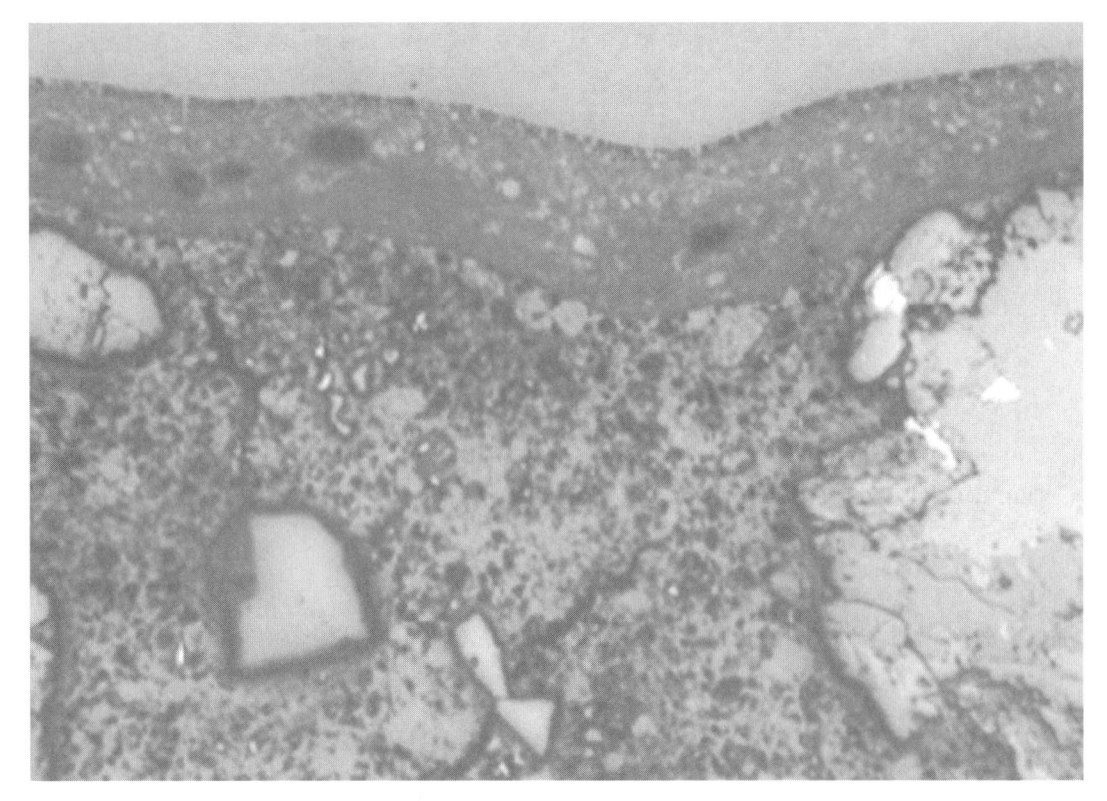

FIGURE 9.7 Approximately 100x. Layer of paint over stucco.

Example III—The Case with No Physical Evidence This case involves a young man who pushed his hand through a glass door at a hotel. The resulting injury might have killed him except for luck and quick action by those around him. Even with good care, however, he lost some motion of the hand and a blossoming basketball career came to an early halt.

Representatives of the hotel stated that all glass in interior doors was safety glass but did not deny that the young man had suffered the claimed injuries.

The emergency room doctor who performed the initial treatment had remarked in writing just how sharp the cuts to the man's arm were, even saying that they were as though "made by a sushi knife."

Unfortunately, all of the fractured glass had been thrown out.

When the hotel had been constructed, standard glass was "code" for interior doors. The building code at the time of the accident required safety glass, but the law does not require instant changes to updated codes. Just how much time you might have to make the changes is one of those interesting legal questions but was not in the area of interest of the laboratory.

In this case, the testimony of the expert from the laboratory was based upon the physical nature of safety and standard glass. Safety glass (other than having wire embedded within it) is prepared by carefully heat treating the glass such that the interior portion is in tension while the exterior surfaces are in compression. Glass in compression (the surface layers of the subject glass) is good at resisting breaking, but glass in tension (the core) has stored energy within it that will take it to failure instantly once a small fracture forms in the surface, compression, layer. The construction is such that when safety glass fails, it breaks into small blocky fragments that can cause minor scratches but is not able to produce serious cuts.

In other words, if the door had been safety glass, it could not have caused the observed injuries. In short, the injured man had carried out his own rather painful test of the glass in the door and had the evidence preserved in the scars in his arm.

The jury fully understood the "physical evidence" and awarded the client just what his attorney had requested.

This case not only shows that you can succeed without much evidence so long as you can describe what happened in terms of physics and common sense, but it also shows how personalities come into play.

The defense attorney was highly condescending to the plaintiff's expert as well as his attorney. There was no civility demonstrated. That attitude of essentially "how dare you sue my client" continued into the courtroom. Now, juries are collectively much smarter than they are sometimes credited and they pick up on attitudes as well as on evidence. They actually asked if they were limited by the amount requested!

Further, on appeal, the appellate judge criticized the defense counsel for his courtroom demeanor, including the way he treated witnesses.

The trial and appeal results were the main reward for the expert, who enjoyed them exceedingly and used the warm memory as encouragement to slog on when other cases were not going as well.

Example IV—The Case of the Fading Rock Garden In desert areas where plants require lots of water, many people use painted rocks for lawns. You may argue the aesthetics of painted rock lawns, but they do save water and work.

The client had purchased a business from an individual, which supplied the green rocks necessary for lawns. The product had a good reputation, and the new owner looked forward to a steady business.

However, not long after he had acquired the business and all the assets, including the green paint source, he started to get complaints that the green "lawns" were turning a sickly shade of yellow-green.

There was no question; the green paint was turning yellow. Actually, that's much like a tree leaf in the fall, but no one was amused by the comparison, although the knowledge of why leaves turn color turned out to be important to understanding the problem.

The laboratory was contacted by the attorney for the client with the fading green rocks business, and arrangements were made to acquire samples and background information.

Microscopic examination of the stones was useful in that it quickly showed that the undersides of the stones (i.e., the part shaded from the sun) retained their dark green color while the upper portions were turning yellow. That was an important observation that showed that the paint pigment was probably not stable in UV (sunlight) exposure.

Further background information developed that the actual formulation of the paint had probably been altered by the large, national paint company. Representatives of the paint company were not at all helpful.

A sample of the current paint was obtained with interesting results: After sitting for a time, the opened can clearly showed both blue and yellow paint not yet totally mixed to form the expected green! Samples of yellow and of blue were obtained and analyzed by a variety of methods.

The problem was quickly solved when the blue color was found to be a dye rather than a pigment, while the yellow color was a UV-stable pigment. In general, dyes are not stable in UV exposure, which is why exterior paints use different coloring agents than interior paints.

It turned out that the fading paints were very much like the tree leaves in the fall. The leaves don't actually turn yellow, they simply loose the overpowering green color, allowing the yellow that was there all the time to become dominant. In the case of the painted rocks, the UV-sensitive blue dye was fading out to leave the yellow portion to show up.

As soon as the true facts were known, the case was clear—at least from the point of view of the expert.

However, the case continued for another two to three years before settling in the plaintiff's favor. Why did it take so long? The lawyer for the defendant continued to believe his client's statements, which were incorrect. It was not until a much delayed deposition that the plaintiff's expert was able to directly demonstrate to the defendant's attorney just what had happened. When the attorney comprehended the evidence, things changed rapidly. Not to be swept away by evidence, however, the defendant's attorney did manage one last shot at the testifying expert: "This deposition will be continued. I'm not through with you yet." But, he was.

Computer-Generated Analysis

Computer-generated analysis has come of age in determining the effect various conditions have on one another. Obviously, in relying on computer-generated analysis, the forensic investigator must fully understand the software program that will be conducting the analysis, be assured that the information fed into the computer program being used is accurate and meets the criteria needs of the program, and be able to read and interpret the results.

Other Laboratory Tests

Other laboratory tests, for example, chemical tests, cover a multitude of issues in which the forensic investigator requires proof that the material or materials that are a part of the evidence being evaluated truly meet established design parameters, industry standards, and performance requirements for the purpose in which they were intended. Examples of some of the laboratory test standards used in construction follow:

Test Classification	Reference
Fire Tests of Building Construction and Materials	ASTM E119
Test Method for Surface Burning Characteristics of Building Materials	ASTM E84

Fire Resistance Testing Injuries and damage from fire claims have resulted in a continued rise in civil actions against those allegedly responsible for the loss. In view of the fact that many fires destroy crucial evidence, it is not uncommon to recreate the circumstances surrounding the issue that forensic investigators believed caused the fire. In so doing, it is essential that the model used to recreate the circumstances that led to the fire be as

154 SECTION NINE

accurate as possible to limit challenges from other experts on the accuracy of the test.[1] Figures 9.8 to 9.11, taken at University of California fire test laboratory in Richmond, California, demonstrate the fire intensity and spread difference between the foam insulation products without the thermal barrier applied versus the same product with the thermal barrier applied.

The next test was to determine the volatility of the product after the thermal barrier had been applied.

The following tests were conducted to demonstrate how the fire could have been contained, to some degree, had the contractor properly installed the fire stops as called for on the plans, in which case part, or all, of the structure could have been saved. This proved that had the thermal barrier been applied as required in the manufacturer's recommendations and constructed to meet NFPA Life Safety Code requirements, the fire would not have spread rapidly. By contrast, the project had been given a waiver by the local fire marshal on the pretence it would be encapsulated within the wall cavity. This waiver allowed the construction of the system without fire resistant coatings over the urethane insulation foam, resulting in the complete loss of the structure.

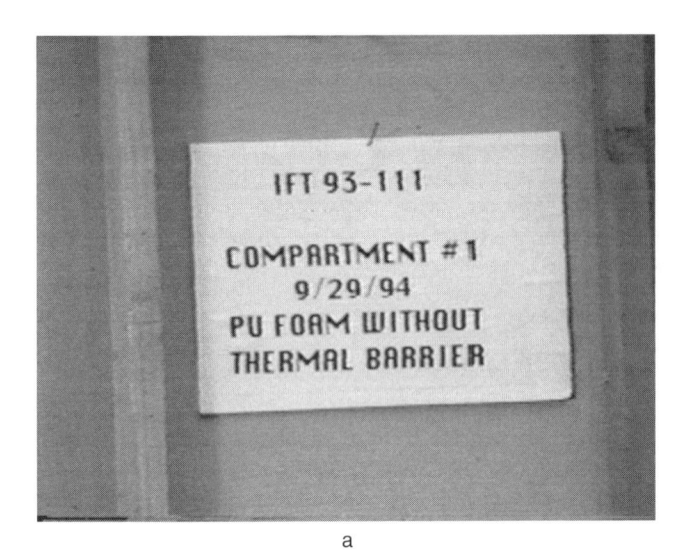

FIGURE 9.8 Photo a names the type of test being conducted. From the time of ignition to total involvement was less than 40 seconds, as noted in photo b. A clear indication that without a thermal barrier, this product was volatile and highly combustible.

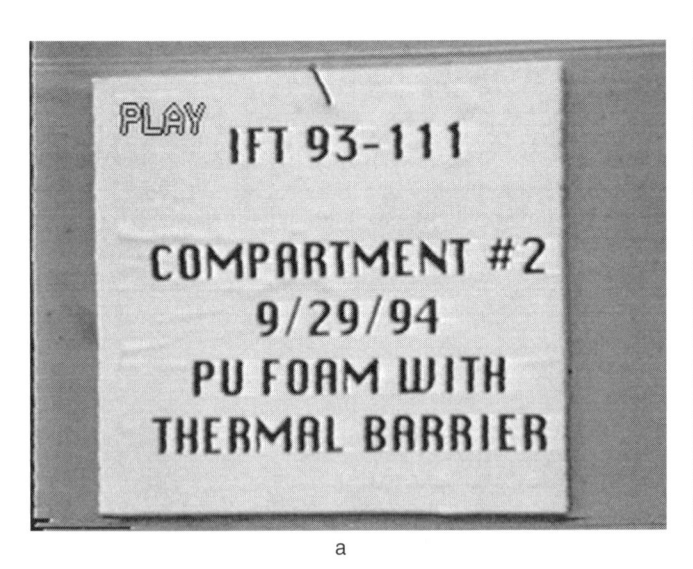

FIGURE 9.9 Photo a shows the type of test being taken. Seven minutes after ignition, the product had still not ignited, as reflected in photo b, even with the intense heat from the gasoline fire-box.

---

[1]The use of models in forensic work is helpful, but extremely risky if they are to be introduced in the courtroom. A lawyer will attack the use of the model, claiming that in some important respect, the model fails to comply with the real "on the ground" situation. If such is proved, the judge may refuse to allow any consideration of the model and may exclude from evidence a very expensive mock-up or demonstration model.

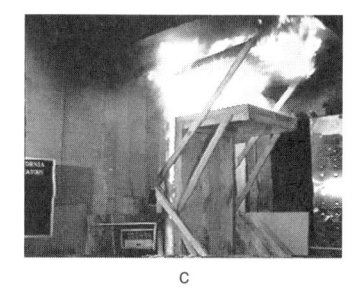

FIGURE 9.10 Full-sized test panel with no fire stops shown in photo a, goes from ignition in 24 seconds, photo b, to total involvement in 3.4 minutes as shown in photo c.

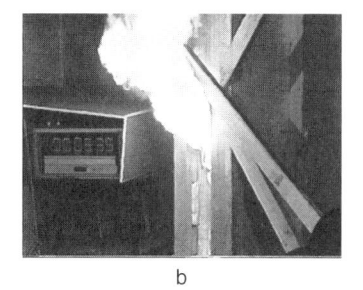

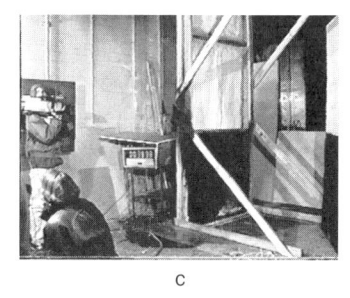

FIGURE 9.11 Full-sized test panel with fire stop is shown in photo a. Within 5.35 minutes after ignition as noted in photo b the fire has not bridged the fire stop. After 8 minutes, the fire was contained below fire stop, as shown in photo c.

9.11 FORENSIC PHOTOGRAPHY

For a forensic investigator, one of the most useful tools in his or her kit and one that will be needed during any investigation is the camera. It is often said that a picture is worth a thousand words. Truer words were never spoken, and the importance of high-quality photos cannot be stressed too strongly.

It is, therefore, essential that the forensic investigator, working in the field or in the laboratory, have a well-rounded knowledge of photography, particularly in meeting the challenges of opposing counsel when using photographs to assist in presenting one's case before the trier of facts. It is through photography that the forensic investigator brings the evidentiary findings for all to see. It is, therefore, imperative that each photo taken be properly logged and documented, which would include but not necessarily be limited to the following:

1. Description of the object, scene, or image captured in the photo.
2. When, where, and by whom the photo was taken. It is helpful to draw a sketch of the area or use a floor plan or elevation drawing to define the location where the photographs were taken. Oftentimes, it is quite helpful to take as many different photos of the area surrounding the scene as possible.
3. Care should be taken to define the scale of the evidence through the use of a ruler or a common object such as a pen, coin, or other definable object.
4. The direction in which the camera lens was pointed when the photo was taken.
5. The time of day and conditions under which the photo was taken, that is, bright sun, snow, rain, or smoke.
6. Type of camera and lens used to capture the picture.

Such documentation is needed to verify that the photos taken accurately represent the object, scene, or specific target depicted in the photograph. Remember that all of the surrounding conditions that could in any way have had a contributory effect on the photograph should be noted and recorded. In addition to defining an object, one must also give it dimension. This is accomplished in a number of ways, some of which include the following:

- Low-light-level photography used to permit photography under nighttime conditions without the use of supplementary light.
- Optical photomicrography to provide enlarged views of extremely small subject mater.
- Photoelastic recording in polarized light to reveal areas of stress in such things as gears, shafts, screws, and so on.
- Photogrammetry, which utilizes photographic methods to record distance or size of objects.
- Thermography is used to reveal variations of surface temperatures by the use of electro-optical methods.

The taking of photographs in today's world has drastically changed from even a few years ago. Only recently, it was a 35-mm SLR film camera body with a 50-mm f/1.2 or 1.4 lens that satisfied most fast lens requirements and was a standard tool in forensic photography. Today, digital photography as opposed to film photography has taken forensic photography to a new level through the use of electronic devices to record the image as binary data. These electronic devices facilitate storage and editing of the images on personal computers and have the ability to show and delete any unsuccessful images immediately on the camera itself. Other digital on-camera features not found in film cameras include the ability to record both video clips and audio data in addition to still photos.

The question arises: Do these digital cameras capture the same image quality as the film camera? Unlike the film camera, the quality of a digital image is the sum of various factors, many of which are similar to those of film cameras. Pixel count is one of the factors to be considered when selecting a digital camera; however, in forensic photography, the processing system inside the camera that turns the raw data into a color-balanced and well-detailed photograph is also very important. As with the film camera, the digital camera requires a good lens, which affects resolution, distortion, and dispersion.

Although not completely accurate, the film in a "normal" camera can be compared to the light-sensing device in the digital camera. Charge coupled device (CCD) arrays are currently the most common, but changes are always just ahead. Regardless of what the sensor unit might be, the basic concept is that of a large number of very small individual light sensitive units, referred to as Pixels (Pixel is an acronym for "picture elements"). These units, or pixels, are similar in concept to the rods and cones in the retina of our eyes. The more sensors available, the more detailed the image that can be recorded.

Several treatment processes can enhance the basic image. Some of these processes use multiple sequential images to make the apparent pixel count higher than the true number of units in the sensor device. Terms such as "apparent pixel count" or "functional pixel count" can be found in the advertising brochures and lead to confusion. For most applications, it is useful to simply relate the number of pixels to the sharpness of the image directly; that is, more pixels equal a sharper image.

That being said, how many pixels are enough? The functional answer is "enough so that you can't see the pixels in the final print." Roughly, that comes out to 300 pixels per inch of whatever you are looking at. A 4" x 5" print, for example, using the 300 ppi (pixels per inch) rule, would require 1200 by 1500 or over 1,800,000 pixels. That will produce a photo quality print, in color, when printed by a good printer using glossy "photo paper."

In the early development of digital recording photo equipment, 1,800,000, or nearly 2MP (mega-pixels), was an outrageous number of pixels and only cameras costing several thousand dollars could provide those types of images. With advances in production techniques, the pixel count of even inexpensive "point and shoot" cameras is well into the mega-pixel range. For forensic work, the pixel count should be at least 5 MP.

The size of the final print is not the whole story, however. The viewing distance must also be considered. Obviously, if the image is farther away, the eye will not distinguish loss of resolution resulting from enlargement. If, for example, you made a print that was 4 feet by 5 feet to use in a courtroom, you would need a huge number of pixels, probably well into the GP (giga-pixel) range, right? Not true, unless you wanted to look at the image close up. However, big pictures are usually displayed at a distance. If that 4-foot by 5-foot picture were at a distance that made it look like a 4-inch by 5-inch photo held comfortably at arm's length, then it would look fine even with a low pixel count. In fact, you could probably use about 2 MP and have it look just fine because it is "functionally" only a 4-inch by 5-inch image due to the distance it is placed from the viewer.

Because the image is stored as a series of electronic numbers, it can be processed by mathematical functions. The gain (contrast), brightness (zero level), color balance, and a long list of more subtle alterations are easily made and then reversed if the effect is not what was desired.

Once you take your important photos, don't forget how to save them. If you use a lot of pixel power to get high resolution, don't turn around and lose it by storing the image using a highly compressed storage format. In order to save memory space and to speed up the processing time, various techniques are employed to remove "unneeded" pixels. Although each format has its uses, the best advice is to use uncompressed image storage. TIFF format is very common, easily used by virtually all image processing and storage devices, and when employed as "uncompressed" mode, it will properly save all the data. Methods such as JPEG, possibly the most common image compression format, drop a lot of good data and you will not be able to get it back once the image has been compressed with this process.

A complete discussion of image storage formats is beyond the scope of this book, but the interested reader is encouraged to find information about the available formats for the camera he or she is thinking of purchasing. Fortunately for users, the availability of cheap memory devices makes compression of the images easily avoided, at least for the original, permanent "negatives."

Digital cameras also allow more complex metering because they have the whole CCD to meter with. Color balance of the image can be adjusted later on a computer. Digital cameras will soon be able to take multiple exposures of the same object at different focus ranges and different exposures to provide artificial depth of field and wider range than even the human eye. Added features available on digital cameras are voice recording, conversion to video, time dating, and other information directly recorded onto the image file.

Digital camera lenses, while small and fast, have a very deep depth of field, even when fast; however, future digitals may be able to simulate short depth of field, which would increase their adaptation to portrait photography. With new enhancements being developed on a continual basis, the digital camera may be the best camera for the forensic investigator.

Next to the importance of having a good camera to memorialize the investigative findings is the need to ensure that the camera kit has sufficient accessories to assist in properly documenting the evidence. The kit must include linear scales and other such devices that can dimensionally place the photographic evidence in a proper perspective long after the forensic investigator has left the scene of the investigation. This can be accomplished through the use of linear scales, L-shaped scales, folding 4-foot dimensional scale, 1-foot diameter circular templates, and so on. Each of these tools, when used properly, allows the investigator to develop grid patterns that accurately show the physical size, location, and positioning of the evidence being photographed within a given area. When dealing with most construction issues, copious points of reference will normally be available, such as doors, windows, concrete masonry units, and beams, of known dimension that can provide natural grid patterns. These patterns not only are useful in providing two-dimensional distance scales, but can also be converted to three-dimensional scales at the corners, which can become quite helpful in properly positioning the evidence for those who have never visited or seen the location in question.

The forensic investigator should ensure that the photos are taken of not only the evidence item and the immediate area being investigated, but also the surrounding area as well. A good standard practice when photographing the evidence scene is to step back from the primary object being photographed and take a shot, then take two steps to the left and take a shot, then two steps to the left for another shot. This results in capturing the evidence being photographed from three vantage points that are useful in producing a three-dimensional reconstruction of the evidence scene and also provides different views that will allow the selection of the view most representative of the issue that needs to be pointed out when developing a report of the findings.

It must be remembered that any photograph used at trial is viewed by the trier of fact as a graphic portrayal of what would have been oral testimony, and the photograph is admissible only when a witness has testified that it is an accurate portrayal of a relevant fact personally observed by the witness. This is called "laying the foundation." While the use of forensic photography is key to bringing the reality of the issue into focus for those who have limited or no knowledge of the facts surrounding the case, it must also be remembered there are certain limitations and caveats that must be addressed when using photography during trial.

With the advent of the digital camera, the potential for manipulation of photographic imagery has seen a major revolution and has raised numerous objections from opposing attorneys as to the validity of digital photos being used as evidence. This can become even more cumbersome when measurements and reconstructions are within the photo to better define the evidence. It is, therefore, essential to preserve the photographic evidence in its original form by immediately downloading from the digital camera to a non-erasable disc. This master disc should be copied and then archived in safe storage, remaining available should the authenticity of trial copies be challenged. Additionally, any copy of the master should also be logged and tracked as any other piece of evidence. In spite of these concerns, an expert witness familiar with the facts represented by scenes captured by digital photos can testify to the photograph's correct and accurate representation of the facts. When using the photo as a tool in helping better define the facts through reconstruction and measurement techniques applied to the photographic image, the admissibility rests on the expert's ability to demonstrate that a reliable scientific methodology was correctly implemented prior to offering the photo into evidence.

Photographs have been used in litigation to cover many issues other than images of the evidence found during a forensic investigation. It is not uncommon to use photographs to capture the time of day or year in order to demonstrate the effect such timing may have had on the scene. In some cases, when a person's vision may have had an influencing effect on the findings, experts who are knowledgeable in both photography and human vision use photography to demonstrate how shadows or lighting intensities may have played a role in the issue being investigated. In any event, it must be remembered that when venturing into areas such as psychophysics and the physics of photography, the expert must be able to prove that he or she is qualified to assist the trier of fact in understanding the facts.

In forensic photography, it is essential that the offering expert produce a complete and accurate analysis of the photographs used in a report or for demonstrative purposes at a trial, hearing, or arbitration.

SECTION TEN

THE EXPERT WITNESS

or

What to Look For—What to Expect

JUMP-STARTING SECTION 10

What is the correct definition of an "expert witness"? The term conjures up many misconceptions. If you or your business is in anyway associated with the fields of architecture, engineering or construction, there will probably come a time when you need an expert witness. With that in mind, let's try to clear up a few things.

Upon beginning this section, the following points need to be kept in mind to help you fully understand the role and importance of the expert witness. You may have seen the duties of an expert witness addressed in many different areas of this book, but this is certainly one place where the following needs to be clearly stated: ***The duty of the expert witness is to the truth. Although paid by one or another of the litigating parties, the expert witness is, in effect, an "officer of the court" and his or her duties are largely to the court only***. As such, the expert witness is *not* an advocate. These three points are paramount and must be understood before this section will make sense.

1. An expert witness must disclose all the information developed or reviewed in the course of the investigation. This is an important point and it does not leave room for attempts to designate part of the file as "work product" or similar contrivance made to shield any documents from discovery. Essentially, anything that the expert witness sees at any place or time relating to the case is discoverable,[1] whether or not it was used to form any opinions.

2. A report that an expert witness has prepared to be filed, providing the results of an investigation, should contain these seven elements:

    - It must be in writing and signed. If the expert witness is a registered professional it must also be sealed, as required in most states.
    - It must contain a COMPLETE statement of *all* opinions to be expressed by the expert.
    - It must contain the entire basis and all reasons for the opinions.
    - It must contain the data and all of the information **considered** by the witness in forming the opinions.
    - It must contain any exhibits meant to serve as support for or a summary of the opinions expressed.
    - Under the federal court rules, it must contain all the qualifications of the expert witness, including a list of *all* of the publications authored by him or her in the past 10 years and the identification of any and all cases where the witness has testified during the past 4 years (in deposition or in trial).
    - The compensation paid or promised to be paid for the report and for the testimony.

3. There are two types of expert witnesses: "testifying" and "consulting." The testifying expert witness is the person who writes the report and testifies in deposition and in court. A consulting expert witness is one who only advises the attorney, is not going to be used as a testifier, need not be identified, and cannot be deposed or questioned, and is *essentially invisible*. The expert witness, as an officer of the court, is held to a high level of ethical conduct. The expert witness is not a "hired gun," but owes duties to the court system, not just to the person paying for the expertise of the expert witness.

4. A testifying expert witness may be challenged using either the *Frye* test or the *Daubert* test. The much older (1923) *Frye* test essentially asks the question of whether the scientific basis of the witness's testimony is generally accepted in the scientific community. The *Daubert* test of admissibility is more rigorous and essentially requires the witness to show that the results were obtained using scientific methods. The *Daubert* test, especially, leaves wide discretion to the trial judge as the "gatekeeper" controlling what opinion evidence is allowed in and what is not. *Daubert* actually allows a broader range of information into the courtroom, but with tighter controls on how it was obtained.

5. In Arizona State Courts, *Daubert* has not been applied,[2] yielding instead to the *Frye* test. As Court decisions can quickly change the rules of acceptability, the reader is

---

[1] The American Bar Association has a committee that has for the past several years worked to undo the effect of this rule. The committee has just succeeded in amending Rule 26(a)(2)(B) of the Federal Rules effective December 1, 2010, to extend "work product protection" to the discovery of an expert's draft reports and to certain counsel-witness communications.

[2] A May 2010 statute (A.R.S. §12-2203) has just become effective, purporting to *require* the Arizona courts to follow the *Daubert* rule. The statute has been challenged saying that the legislature cannot control the courts' determination of what is acceptable evidence in a courtroom.

encouraged to check with an attorney on current local practice and to read carefully the acceptance conditions that apply for the jurisdiction in which he is appearing as an expert witness.

6. Although it is rare that such occurs, an expert witness may be sued in some states for negligence. Functionally, the person who retained the expert is the one most likely to sue. It will likely be alleged that the expert did not perform as well as someone of similar education, background, and so on, and that such poor performance materially and negatively affected the outcome of the legal action.

The initial reaction of some party who feels unfairly attacked by some allegation in a lawsuit is to look for an expert who will supply support for his or her opinion; to get someone who "thinks as I do." This is a dangerous road. A party should beware of hiring what amounts to a "yes man." A "yes man" will blind one to the facts and leave a party open to devastating surprises at trial. Remember, when properly presented, the truth is very powerful. A party may even find that his or her opinion is not so "bullet proof" as initially thought. A fair understanding of the facts (both good and bad) as developed by a qualified expert, often leads to a settlement between the involved parties rather than to an unpredictable and very costly trial. A fair settlement is a win-win situation.

If, as it happens in most instances, a party hasn't a clue on what to do in hiring an expert, this section will explain the importance of the expert witness and why it is essential that a litigant makes a correct selection when it comes time to employ one.

10.1 INTRODUCTION TO THE EXPERT'S ROLE

The expert is an officer of the court, even though he or she is *paid* by only one party. Federal and Arizona State Evidence Rule 701 states that an expert is any witness who, using professional expertise, attempts to assist the "trier of fact" (arbitrator, judge, or jury) to understand and evaluate the facts in the dispute.

If the expert can be portrayed as being one party's *advocate*, then the expert witness's central role and testimony will likely be disregarded by the trier of fact.

Not being an advocate, the expert's conclusions must be within a field of expertise and must be fully based upon, and supported by the evidence that was reviewed. Proving what the expert has "considered", or has not considered, is so important that Federal Civil Procedure Rule 26 and Arizona State Civil Procedure Rule 26.1 (to a lesser extent) require the expert to report in detail, all items considered. That requirement in the Federal Rule means that, even if an item was rejected after it was reviewed, it still must be disclosed in the expert's report as an item that was considered, but was not chosen as the basis for the report.

The Federal Rule requirement of disclosure and listing of all items "considered" by the expert effectively "trumps" any lawyer's claim of "work product privilege" as to any items that the lawyer may have shown to, or discussed with the expert. Some lawyers are poorly informed and don't realize that their showing memoranda, and other documents to an expert whom they have hired has "stripped the privilege away" as to the contents of those documents. An expert is not a member of the "advocacy team," but is instead an "outside" officer of the court, which means that, when he sees a document makes that document admissible. An expert has to be on guard and *careful* lest the lawyer makes a mistake and displays privileged "advocacy" material to the expert— thereby making it appear the expert was *not* an *independent* professional. To the extent the expert can be co-opted and made to look like an advocate, the expert has erred and allowed a careless lawyer to defeat the purpose of having an expert involved.

The expert needs to review *all* available relevant information. Some lawyers are reluctant to give the expert "the bad stuff" to evaluate as well as "the good stuff." To the extent that the expert does not probe and press to see all the relevant evidence, or to do the testing that is necessary, the expert runs the risk of being portrayed as a partially informed advocate, rather than a knowledgeable independent specialist assisting the court.

Perjury is a seven-letter word and we do not often see it coupled with the name of an expert. However, it is often the mindset of the opposing lawyer that the adversary's expert witness is shading the truth, concealing part of the truth, or deliberately not looking at crucial facts in order to present a *biased* view as if it were the truth. To the extent the expert witness fails to fully research the facts, to fully account for alternative theories, and to fully disclose caveats and countervailing considerations, the expert may be portrayed as an advocate. That portrayal will defeat or undermine the expert's role in the case.

Until December 1, 2010, anything considered by the expert and every contact with the counsel who employed him or her was examinable by the opposing lawyer. If notes, photos, sketches, memos, or drafts (which were published outside the expert's office) were destroyed or lost, then opposing counsel could ask the court for an "adverse inference" instruction to the jury. The instruction would require the jury to assume that those destroyed or lost materials would have supported the opposing party's case. That instruction defeats or undermines the expert's role in the case, harming the party whom the expert was employed to help. Loss of file items is a recipe for expert malpractice.

See *W.R. Grace & Co. – Conn. V. Zotos International, Inc.*, 2000 W.L. 18096 (W.D.N.Y. 2000); and *Residential Funding Corp. vs. DeGeorge Financial Corp.*, 306 F.3d 99, 53 Fed.R.Serv.3d 1105 (2d Cir. 2002), excerpts are attached as **Exhibit A**.

> **EXHIBIT A**
>
> **Summary of *W.R. Grace & Co.— Conn. vs. Zotos International, Inc.***
>
> **2000 W.L. 18096 (W.D.N.Y. 2000)**
>
> In an environmental case brought under CERCLA, the defendant accused of having polluted a certain site, Defendant Zotos, hired an expert (Barber), but when they presented Barber for his deposition, it came out that:
>
> - Barber's day diary with notes of contacts with defendant's counsel,
> - Correspondence between Barber and defendant's counsel;
> - Drafts of Barber's report sent to defendant's counsel and returned with notes by counsel before the finalization of Barber's report.
>
> were all missing and, except for the diary, had been destroyed pursuant to instructions (three weeks before the deposition) from defendant's counsel in order that Barber would "not confuse . . . things." Barber refused to produce various portions of his diary, including parts about notes of a meeting with defendant's counsel regarding preparation of his report.
>
> The magistrate writing this decision found that the Federal Rule 26 (A)(2)(B) does not permit withholding from production the lawyer/expert notes and correspondence, although the defendant claimed that such was work product showing the lawyer's strategies or impressions. In short, if *anything* is communicated to the expert, then it *can be* seen by the other side.
>
> The court found that the destruction of the draft reports by Barber pursuant to the lawyer's instructions was "spoliation" of evidence and was ready to enter an "adverse inference instruction to the jury," stating that such destruction supports the jury's inferring that the destroyed drafts *would have* supported plaintiff's position in the lawsuit. However, it was discovered that electronic versions of the reports *could be located* so the prejudice to the plaintiff was thereby avoided. The court awarded all of the costs and legal fees for the motion against the defendant whose expert had destroyed the hard copy reports.
>
> This case involved a *willful*, intentional destruction of hard-copy evidence. The attorneys and the expert came close to paying a dear price for having withheld from Plaintiff Grace evidence of how the lawyer had dealt with the expert and what had affected the expert and his report.
>
> **Summary of *Residential Funding Corp. vs. DeGeorge Financial Corp.***
>
> **306 F.3d 99 (2nd Cir., 2002)**
>
> This case involved a *non-intentional* "destruction" of evidence and obstruction of discovery involving an expert. Residential Funding sued DeGeorge. DeGeorge had lost the case (a $94 million verdict had been entered against it), but appealed the trial judge's denial of a motion that DeGeorge had filed for sanctions against Plaintiff Residential Funding's counsel for "dragging his feet" in order to conceal thousands of relevant e-mails until the very eve of trial. In response to an April 12, 2001 request for production of documents, Residential's lawyer said that he was locating them via "back-up tapes." In July, when the lawyer said that Residential lacked the technical personnel to restore the back-up tapes, DeGeorge took the matter to the trial judge and Residential agreed to hire a vendor. In August, Residential blamed technical problems and gave up only approximately 60 relevant e-mails. With court help, DeGeorge got a vendor involved and very quickly, thousands of e-mails claimed to not have existed came to light—three days before trial began. When DeGeorge moved for sanctions during trial, the trial judge said that there was not enough evidence to show "bad faith" or "gross negligence" by Residential. The case went to the jury without DeGeorge having had access to the withheld e-mails. The Second Circuit found this was not "spoliation" (destruction) of evidence, but said that this was "misconduct in discovery." The court said that mere *negligent conduct* by Residential's lawyers could have warranted the trial judge's granting an adverse inference jury instruction. The court found that there had been "purposeful sluggishness" by Residential and that such merited a sanction such as the adverse instruction. The court sent the motion for sanctions back down to the trial judge and found that the trial court, when it reviewed the facts, might well decide to undo the $94.5 million verdict that Residential had won—and stated that the trial court should presume that the evidence kept from DeGeorge would have supported DeGeorge's case. Obviously, the misdeeds related to the negligent delaying of the production of evidence was going to cost someone (who thought that he had won the case) a very dear price.

As of December 10, 2010, a change in the Federal Rules of Civil Procedure has occurred, which seeks to undo some of the effects of the *Grace* case. The new rule will give a "safe harbor" for certain revisions made to expert reports and, with three exceptions, the correspondence of counsel with the expert concerning the revising of an expert's report.

In new Rule 26 (a)(4), (b), and (c), any draft reports and any correspondence about a report draft will be deemed an "attorney work product" that does not need to be produced (unless the correspondence provides to the expert facts or assumptions to include in his or her opinion—and that sort of fact / assumption correspondence will have to be produced to the opposition).

10.2 WHAT TO LOOK FOR IN SELECTING AN EXPERT

Curriculum Vitae

A curriculum vitae should provide a broad brush look at the expert's education and experience. Keep in mind that having a Ph.D. in chemistry and teaching at the local university does not mean much if the "expert" person holding these credentials has had no practical experience in the field. Likewise, a person with years of field experience by working in the trades may have little ability when it comes to convincingly explaining to a judge or jury the chemistry or engineering aspects of an issue.

Specialization

What is the expert *specialized* in, such as general construction methods, structural steel, roofing, soils, microbial growth, and so on? Is the expert: certified or licensed in that field of specialization? Published in the field of specialization? Has field experience or only academic degrees? Has qualified as an expert in a court of law in earlier cases? Has failed to be qualified after attempting to do so in some earlier cases?

Case History

What types and size of cases has the expert worked on? In what part of the country? How many times has the expert been deposed? How many times has the expert testified at trial? What type of mediation and arbitration appearances experience does the expert have? What types of cases is the expert most comfortable with? How many cases has the expert worked on that are similar in nature to the issues confronting you now—and, if available, what were the results?

References

Who are the expert's references and how are they connected to the expert? How is the expert perceived by his or her peers?

Demeanor

Can you check on the expert's demeanor and poise in depositions and in courtroom testimony with other lawyers?

Where to Look

As most lawyers know, the best place from which to glean much of this essential information is from a report filed by the expert in a federal district court case. Under Federal Rules of Civil Procedure (Rule 26(a)(2)(B)), every report prepared by an expert in a federal district court case must meet the following requirements:

- It must be in writing and be signed.
- It must contain a **complete** statement of *all* opinions to be expressed by the expert in the case.
- It must contain the entire basis and all reasons for the opinions.
- It must contain the data and all of the information CONSIDERED by the witness in forming the experts opinions.
- It must contain any exhibits meant to serve as support for or a summary of the opinions expressed.
- It must contain all the qualifications of the expert witness, including a list of *all* the publications authored by the expert in the past 10 years and the identification of any and all cases that the witness has testified in during the past 4 years (in deposition or in trial).
- The compensation to be paid for the report and for the testimony.

This set of requirements is more stringent and exacting than the procedural rules of Arizona, or likely any other jurisdiction, so, if you can get the expert to give you the last such report (and it is one in the field in which you seek an expert), then you will have a lot of information from which to evaluate the expert's background and abilities.

One caveat: Providing the fourth item in the list of federal requirements (items considered) is onerous, but instructive as to the expert's role. Documents considered by an expert are no longer privileged and are discoverable. Case law has made it clear that "documents and information disclosed to a testifying expert in connection with testimony are discoverable by the opposing party, whether or not the expert relies upon the documents and information in preparing a report." Disclosure by the expert of what the expert checked on the Internet or was given or told is absolutely required. See *Trigon Insurance Company vs. U.S.*, 51 Fed. R. Serv. 3rd (Callaghan) 378, 57 Fed. R. Serv. (Callaghan) 664 (USDED Va. 2001), and *Aniero Concrete Company, Inc vs. New York City School Construction Authority*, 52 Fed. R. Serv. 3rd (Callaghan) 730 (USDSD NY 2002).

One last point: If one is an expert in a federal court case, one had better not do an abbreviated report. If an expert in a federal case were to commence testifying at trial to some aspect of that expert's field, which was not addressed in the expert's report, the judge would "rein the expert in" and not permit the expert to testify on that matter. The same thing can possibly occur in a state court case if the expert starts testifying on points that have not been covered in the disclosure statement, or in the expert's report, or in depositions.

10.3 THE NON-TESTIFYING "CONSULTING" EXPERT

Under the Rules of Civil Procedure, it is possible for a party to retain a "consulting" expert who will never testify. In fact, in this world of thorough and early, and thus onerous, disclosure statements, the identity of the consulting witness need not be divulged and such witness cannot be deposed. Such a witness can be a member of the "advocacy team," will never have to disclose what is said or given to the witness, and can advise the lawyer on strategy and deposition outlines without fear of their being made public. See Rule 26(b)(4)(B) of the Arizona Rules of Civil Procedure. Only in very exceptional circumstances can the contacts with a nontestifying expert become discoverable.

However, when a consulting witness later is *converted* to a testifying witness, essentially *all* of what was seen, heard, and received becomes discoverable by the other party. See *Arizona Independent Redistricting Commission v. Kenneth W. Fields, Judge of the Superior Court*, 206 Ariz. 130, 75 P.3d 1088, 1102 (App. 2003).

10.4 THE "HURRIED" EXPERT

In Arizona State Court practice, A.R.S. § 12-2602 requires that, in any case in which any licensed professional is made a defendant, certain procedures are to be followed. A party has to file at the outset of the case a statement as to whether expert opinion testimony is necessary to prove the licensed professional violated a standard of care. Thereafter, by the initial disclosure date (40 days after the answer is filed), the plaintiff must provide a "preliminary expert opinion *affidavit*." The affidavit has to contain the following:

1. The expert's qualifications to give an opinion on the standard of care
2. The factual basis for each claim against the licensed professional
3. The exact errors or omissions that the expert believes the licensed professional has committed, which constitute violations of the standard of care
4. The manner in which those acts, errors, and omissions have caused or contributed to the damages sought

Compliance with this rule can be postponed for a time with small consequences, but eventually, a court will force the plaintiff to comply. What seems to occur regularly is that a plaintiff's lawyer delays (the cost of) complying with this rule until the last moment, and then approaches an expert on an "emergency basis" asking for a preliminary affidavit to file quickly in court.

Hurried preparation of such an affidavit with insufficient fact investigation can result in a poorly devised affidavit, which can be subject to a devastating cross-examination by the defendant professional's lawyer in deposition or trial. If approached to prepare such an affidavit, a potential (competent) expert will want to attempt to get some time to do a cursory investigation, will state explicitly in his or her affidavit that the affidavit has been based on a less than full preparation or review of the evidence, and will highlight the affidavit with caveats saying that it is only a *preliminary* effort.

The difficulty posed by the statute is that this mini-report comes in the form of a *sworn* affidavit, and it is often prepared in haste and at a time when examination of the other side's documents, or perhaps investigation into the site, is not possible. It is not only a preliminary affidavit—it is one that may be well off the mark as to what the witness will later want to testify. These affidavits are troublesome creatures, and if one is needed, a party would do well to anticipate it, give an expert a lot of time to do a good job, and then file the affidavit with caveats stating that it is an early and preliminary effort done before disclosure.

10.5 THE "CARELESS" EXPERT

The "careless" expert and spoliation of documents or evidence can result in a negative inference jury instruction or sanctions for a party or his or her counsel.

Attached as **Exhibit B** is a short article by Gregory Joseph taken from the *ABA Section of Litigation Journal* (Vol. II, No. 2, Spring 2003).

While this article is written in the context of the four federal cases cited above, *Aniero*, *Trigon*, *W.R. Grace*, and *Residential Funding*, the guts of the Federal Rules can easily be made applicable to state court cases. All a party has to do is to issue a subpoena concerning, or ask about, subject matters and items listed in the Federal Rules, when taking a state court deposition.

"Spoliation" means the destruction of evidence. No expert should manage a file or conduct an investigation in such a way as to make it possible for the adversary to claim that information or documents have been destroyed or concealed. As Joseph states in his article, "We live in an era of spoliation. Parties long not so much for documentary *evidence* as for *evidence that documents have been destroyed*." (Emphasis added)

Given the Federal Rule (that the lawyer's contacts with the expert and anything said/shown between them may be discoverable), the expert who destroys field notes or notes of a chat with the employing lawyer has spoliated usable evidence. The expert who gives the employing lawyer or client or any third person a draft of the expert's report, gets it back with comments, changes the report but then destroys the annotated draft report, has spoliated usable evidence. For a lawyer to order or permit an expert to destroy such notes or emails or drafts is sanctionable conduct. See the *W.R. Grace* and *Residential Funding* cases cited above.

When that sort of conduct occurs, the trial judge is likely to give the jury an "adverse-inference" jury instruction. In effect, the jury is told by the judge that they should *assume* that the destroyed materials would have been damaging to the expert's client, and that the expert no longer appears to be an *independent* expert. In effect, by virtue of the expert having been caught up in a spoliation episode, the expert may become valueless for the task for which he or she was hired.

EXHIBIT B

Expert Spoliation

Gregory P. Joseph*

Can you properly instruct your experts to destroy drafts of their reports as they are working toward the final? Does it matter whether those drafts bear or reflect the comments of others? What if the comments reflected on the drafts are yours? Must communications with experts — including your emails — be preserved? Are your notes of conversations with your own experts discoverable?

We live in an era of spoliation. Parties long not so much for documentary evidence as for evidence that documents have been destroyed. This article explores the application of spoliation principles to expert-related materials.

The threshold question is whether the materials are discoverable. If so, there is necessarily a duty to preserve them since by definition there is a pending or reasonably foreseeable lawsuit. *West v. Goodyear Tire & Rubber Co.*, 167 F.3d 776, 779 (2d Cir. 1999).

Impact of Report Requirement

The discoverability of expert-related materials turns largely on an analysis of Fed. R. Civ. P. 26(a)(2)(B), the expert report requirement added in 1993. This Rule mandates disclosure not only of "a complete statement of all opinions" but also of *"the data or other information considered by the witness in forming the opinions."* The critical word is "considered." The 1991 draft of this rule originally proposed "relied," but that was deleted as too restrictive.

"'Considered,' which simply means 'to take into account,' clearly invokes a broader spectrum of thought than the phrase 'relied upon,' which requires dependence on the information." *Karn v. Ingersoll Rand*, 168 F.R.D. 633, 639 (N.D. Ind. 1996) ("considered" is satisfied where experts have "reviewed" documents "related to the subject matter of the litigation ... in connection with forming their opinions"). The 1993 Advisory Committee Note to Rule 26(a)(2)(B) observes that: "Given the obligation of disclosure, litigants should no longer be able to argue the materials furnished to their experts to be used in forming their opinions are protected from disclosure when such persons are testifying or being deposed."

Therefore, matters considered by experts are generally disclosable in their reports and, therefore, discoverable. This includes documents provided by counsel to the expert and the expert's draft reports and notes. *Corrigan v. Methodist Hosp.*, 158 F.R.D. 54, 58 (E.D. Pa. 1994); *Ladd Furniture v. Ernst & Young*, 1998 U.S. Dist. LEXIS 17345 at *34 (M.D.N.C. Aug. 27, 1998); *Hewlett-Packard v. Bausch & Lomb*, 116 F.R.D. 533, 537 (N.D. Cal. 1997).

Consequently, ordering experts to destroy drafts and notes is generally sanctionable. *W.R. Grace & Co. v. Zotos Int'l, Inc.*, 2000 WL 1843258 at *10-*11 (W.D.N.Y. Nov. 2, 2000). There are, however, a series of open issues — and a fundamental question whether this result is always the right one.

Comments of Consulting Experts

What if the drafts bear the comments of non-testifying, consulting experts, whose work product is generally non-discoverable, subject to the "exceptional circumstances" test of Rule 26(b)(4)?

An important 2001 opinion, *Trigon Ins. Co. v. United States*, 204 F.R.D. 277 (E.D. Va. 2001), holds this material discoverable. The defendant in *Trigon* retained a respected litigation consulting firm to supply experts (third-party academics) and to assist those experts in preparing their reports. The consulting firm and its principals remained non-testifying experts. The plaintiff sought all drafts worked up between the testifying experts and the consulting firm—and all communications (including email traffic) between them—much of which had not been preserved.

The *Trigon* Court held that since the drafts and substantive emails had been "considered" by the testifying experts in forming their opinions, the materials were discoverable. *Trigon* further ruled that the destruction of these materials was sanctionable because it was intentional, and that spoliation remedies attached regardless of whether the defendant acted in bad faith. The Court did not preclude the experts' testimony because that would have interposed a delay prejudicial to the plaintiff (the court would have permitted the defendant to engage new experts). Instead, the *Trigon* Court ordered the defendant to engage an outside technology consultant to retrieve as much of this data as possible— with the plaintiff's full participation in the process—and held it "appropriate to draw adverse inferences respecting the substantive testimony and credibility of the experts." *Id.* at 291.

In a late 2002 opinion, the *Trigon* Court also awarded the plaintiff more then $179,000 in fees and costs attributable to the spoliation. *Trigon Ins. Co. v. United States*, 2002 U.S. Dist. LEXIS 24782 at *7 (Dec. 17, 2002).

Interestingly, at the same time that it found sanctionable the destruction of drafts bearing the comments of other experts, the *Trigon* opinion stressed that it was not deciding "whether a testifying expert is required to retain, and a party is required to disclose, the drafts prepared solely by [the

(continued)

(continued)
testifying] expert while formulating the proper language in which to articulate that experts' own, ultimate opinion arrived at by the expert's own work or those working at the expert's personal direction" and that "*[t]here are cogent reasons which militate against such a requirement . . . .*" 204 F.R.D. at 283 n.8.

These cogent reasons were not specified, and, as noted above, other cases expressly allow discovery of draft reports and notes. At least one federal judge has issued a Standing Order requiring their production. *See Supplemental Order to Order Setting Case Management Conference in Civil Cases Before Judge William Alsup* at ¶15 (N.D. Cal. November 25, 2002).

However, there are cogent reasons why the Advisory Committee should reconsider whether this is the optimal result. Every carefully-drafted document has false starts. The quality of the final is not judged by the quantity or quality of the drafts. That is true of judicial opinions and briefs as well as expert reports. For the expert to formulate a reasoned opinion, he or she should be afforded the latitude to filter the facts through the prism of his or her expertise—using whatever process seems most appropriate—without intrusion and without the necessity of attempting to avoid committing matters to writing. If the concern is ghost-writing or undue influence by others, a party should be required to make a prima facie showing that validates that concern before piercing the report and opening underlying matters to discovery.

Regrettably, the proposed distinction in *Trigon* between the work-product generated by "those working at the expert's personal direction" and that of the outside consulting litigation firm is also difficult to sustain under Rule 26(a)(2)(B). Moreover, if it were sustained, the expert industry would no doubt be restructured so that experts relied only on "employees." But if the relevant concern is ghost-writing, there is no obvious reason why the courts should treat ghost-writing by employees differently from that of third-parties. The element of personal direction is really the key, and the question is always the same — whether the expert is giving the direction or receiving it. Is there a genuine issue as to just whose opinion the expert is espousing?

Counsel's Comments/Communications with Expert

The discoverability of communications between counsel and experts has split the courts since 1993. *See generally* 6 MOORE'S FEDERAL PRACTICE § 26.80[1][a] (3d ed. 2002). The technical issue is whether the protection for opinion work product set forth in Rule 26(b)(3) is trumped by the disclosure requirement of Rule 26(a)(2)(B). Many courts, like *Karn,* hold that it is and that all communications between counsel and the expert are discoverable. Others, following *Haworth, Inc. v. Herman Miller, Inc.*, 162 F.R.D. 289 (W.D. Mich. 1995), come to the opposite conclusion. I have advocated the latter position (*Emerging Expert Issues Under the 1993 Disclosure Amendments to the Federal Rules of Civil Procedure*, 164 F.R.D. 97 (1996)), but the trend of decisions appears now to favor the *Karn* approach. While that approach fairly addresses the perceived need to explore the basis of the expert's opinion, it is overly broad—capturing every exchange between counsel and the expert, regardless of the substance and regardless of whether there is any doubt that the opinion is in all respects that of the witness. This result operates to favor those litigants who can afford separate consulting experts off whom, for example, counsel may bounce ideas as to cross of opposing experts and trial strategy.

In those jurisdictions following the *Karn* approach, drafts of expert reports bearing counsel's comments are discoverable. *Weil v. Long Island Savings Bank*, 206 F.R.D. 38 (E.D.N.Y. 2001). There is the further question of the discoverability of counsel's notes reflecting oral communications with the expert. This is one step removed from the actual communications—assuming that the expert has never seen the notes—and necessarily implicates serious opinion work product concerns. The notes should be deemed immune from discovery, absent a prima facie showing of that (1) they reflect either misconduct or ghost-writing by counsel or form an important basis of the expert's opinion, and (2) cannot be recreated in any other way (*e.g.*, from testimony from the expert). Some courts have properly shown some reticence in ordering production of such notes. *See, e.g., B.C.F. Oil Refining v. Consol. Edison Co. of N.Y.*, 171 F.R.D. 57, 66-67 (S.D.N.Y. 1997). *W.R. Grace*, 2000 WL 1843258 at *5; *Amster v. Tiver Capital Int'l Group*, 2002 U.S. Dist. LEXIS 13669 (S.D.N.Y. July 26, 2002).

Practice Pointers

This discussion suggests the following practice pointers:

1. Each expert should, on retention, be made aware that everything he or she writes or receives, including every email, is potentially discoverable. Nothing should be discarded or purged (better yet, nothing written). This should be added to the expert retention letter, to show counsel's diligence in this regard. Special efforts must be undertaken by those experts working for organizations whose electronic documents are regularly purged to insure that potentially discoverable material is not destroyed.
2. Lawyers should curtail their written communications with experts, and those of others, like consulting experts (whose engagement letter should similarly afford notice of the preservation obligation). There is no duty to create exhibits for your adversary.
3. Lawyers should be conscious of the risk that notes of conversations with experts may be discoverable. For years you've urged your clients not to take notes. Now, it's your turn.

> 4. Even if draft expert reports are discoverable, there is no obligation to create them. There is no prohibition against having an expert work on a single version of a single electronic document. This will not prevent the adversary from requesting the hard drive of the expert's computer to see what can be electronically discerned. That, however, is expensive and less likely than a routine request for hard copies.
> 5. Be slow to request any of this discovery from your adversary. You, too, have an expert. It is effectively impossible to insure that no potentially responsive documents are lost, however hard you try. Mutual assured destruction worked for decades. It still has legs.
>
> *Gregory P. Joseph Law Offices LLC, New York. Fellow, American College of Trial Lawyers. Chair, American Bar Association Section of Litigation (1997–98). Member, U.S. Judicial Conference Advisory Committee on the Federal Rules of Evidence (1993–99). Author, MODERN VISUAL EVIDENCE (Supp. 2007); SANCTIONS: THE FEDERAL LAW OF LITIGATION ABUSE (3d ed. 2000; Supp. 2007); CIVIL RICO: A DEFINITIVE GUIDE (2d ed. 2000). Editorial Board, MOORE'S FEDERAL PRACTICE (3d ed.). © 2002–07 Gregory P. Joseph.

The case can be converted into a negative for the client and the lawyer who hired the expert, perhaps causing the client to lose the case—and to sue the careless expert.

10.6 EXPANDING RISKS FOR THE TESTIFYING EXPERT

Even though experts are increasingly seen as hired guns rather than as officers of the court, the legal system has been somewhat reluctant to expose expert witnesses to the level of exposure to suits of similar professionals. In a little more than a decade, however, the relatively new tort of "expert witness malpractice" has developed.

Expert witnesses owe their clients certain duties by virtue of their specialized knowledge, skills or training which, when breached, can be grounds for a cause of action by a client who has been damaged by negligent conduct of the witness. Negligent conduct can be an engineer's incorrect reasoning or unacceptable file management which draws the dreaded adverse-inference jury instruction, as was explained.

Presently, only eight state courts appear to have addressed the issue of expert witness immunity. So far no court has allowed an expert witness to be sued by an adverse party over testimony, but in six of the eight states, it seems clear *that lawsuits against "experts" by the employing party, while still relatively rare, are increasing*. Some courts are deciding that expert witnesses should be treated like other professionals who have to contend with the possibility of getting sued if they do something negligent—the experts are to be held to the prevailing standard of care of their profession, including file management as well as engineering abilities.

New Jersey appears to be the only state in which a court has held that even a court-appointed expert is not immune from liability for deviating from the applicable accepted professional standards. In *Levine vs. Wiss & Co.*, 97 N.J. 242, 478 A.2d 397 (1984), the defendants claimed that because they were court-appointed and the litigants agreed that the expert's determinations would be binding, they were effectively arbitrators, not expert witnesses. Therefore, the quasi-judicial capacity of their role should provide a judicial immunity shield against actions alleging negligence. The court disagreed, stating that the standard of reasonable care used for most professionals was applicable to expert testimony.

At the other extreme is the state of Washington, which has what is probably the broadest declaration of expert witness immunity to date. In 1989, the State Supreme Court in Washington decided a case in which the plaintiffs claimed that an engineer (their expert) had underestimated the cost of stabilizing the soil on some land adjacent to an excavation site. The expert's underestimation of the value of the plaintiff's case led to a loss for the plaintiff. The court held that the fact that the engineer had been retained and compensated by the plaintiffs, rather than appointed by the court, did not deprive the expert of witness immunity. The court also held that witness immunity applied not only to the engineer's *testimony*, but to any action the engineer took *before* trial that helped form the basis for the testimony (*Bruce v. Byrne-Stevens & Associates Engineers, Inc.*, 113 Wash.2d 123, 776 P.2d 666 [1989]). This decision cast the expert as being in the same mold as a lay witness, who is entitled to immunity from suit by testifying to what the layperson believes to be the truth.

With the exception of Washington, other state courts that have addressed this issue have recognized an exception for experts under the witness immunity doctrine. The nonexpert witnesses are immune from suit, but the experts *are not immune from being sued*. In California, the appeals court drew a distinction between neutral third-party witnesses and experts hired expressly to assist in the preparation of a case for trial. *The court said that witness immunity doctrine was established not to protect a litigant's own expert, but to protect experts from harassment claims by adverse litigants* (*Mattco Forge Inc. vs. Arthur Young & Co.*, 5 Cal. App. 4th 392 [1992]). The court indicated that applying the immunity doctrine "does not encourage witnesses to testify truthfully; and indeed, by shielding a *negligent* expert witness from liability, it has the opposite effect."

In *Murphy vs. A.A. Mathews*, 841 S.W.2d 671 (1992), the Missouri Supreme Court addressed an action brought against an engineering firm based on its alleged negligence in performing professional services involving preparation and documentation of a subcontractor's claim for additional compensation from a contractor. The court *held that privately retained professionals who negligently provide litigation-related services should not be covered by witness immunity.* That protection, the court said, should only cover defamation suits and retaliatory actions against adverse witnesses by opposing parties.

A Pennsylvania court decided that an expert hired by a developer to calculate lost profits from a failed venture could be sued by the developer for negligence. Judicial decisions in Connecticut and Texas have followed a similar pattern. Firms that provide expert witness services should know that while the essential status of the expert witness remains the same, the risk of being sued for negligent work as an expert may be increasing.

In Arizona, in *Jenkins vs. Bell Helicopter*, Cochise County CV97-000397 (2000), after the case was over, the trial judge found that the defendant's expert had falsely stated his credentials. The trial judge declared a mistrial and ordered the case retried—and charged the defendants the entire cost of $583,000 for the retrial. A misstatement of credentials could be framed as an act of malpractice by an expert.

10.7 PREVENTING EXPERT OPINION & TESTIMONY

Under certain circumstances the expert witness can be prevented from giving his or her opinion at all. Therefore, it is essential to the litigation team to understand how this can occur.

The decisions and rule changes of the past 15 years have been remarkable. The practice of requiring the expert to speak only in the context of a cumbersome, lengthy, "hypothetical question" has virtually disappeared. Modern decisions have expanded the subjects upon which experts may be permitted to testify, expanded the kind of facts upon which their testimony and opinion can be grounded, and then relaxed restrictions on the form in which their opinions can be stated.

One major development has been the elimination of the hurdle that used to exist requiring that an expert be used only regarding matters "beyond the ken of laypeople." See*McCormick on Evidence*, § 13, page 13 (3rd edition, 1984).

Rule 702 of the Federal Rules of Evidence has erased that hurdle and substituted the question of whether the witness *will be of assistance* to a trier of fact in determining any fact at issue. For a startling view of how far this can reach, see *Holiday Inns, Inc. vs. Rice*, 576 So.2d 322 (App. Fla. 1990) allowing an expert on bereavement to explain how the parents grief affected them after the death of their son. The new science of hedonics, that is, valuing a person's pain, such as the loss of use of an arm or leg, is another example of the relaxed standards as to the subjects on which experts may testify.

The last bulwark against the fall of all the hurdles regarding what subjects experts can testify to is the *Frye General Acceptance Test*, or the *Daubert/Kumho Tire/General Electric* test, depending on whether you are in the state courts of Arizona or the Federal District Court for Arizona.

The *Frye* test is 80 years old and stems from *Frye v. United States*, 293 F.2d 1013, 1014 (D.C. Cir. 1923). Since *Frye* is three-fourths of a century "old," there have been instances where, in one decade, a certain scientific theory is considered not "sufficiently established to have gained general acceptance in the particular field in which it belongs," only to have that science be fully embraced in the succeeding decade. See *Gianelli Symposium on Science and the Rules of Evidence*, 99 F.R.D. 187, 189–191 (1983).

The *Frye* test has been attacked as being conservative, "backward-looking," and fuzzy as it regards the definition of "the expert's field," and vague to the point of being useless. By the early 1990's, the *Frye* test was coming in for criticism in several states and circuits as being simply too restrictive.

The result was the *Daubert* opinion in 1993. In a nutshell, in *Daubert*, U.S. Supreme Court Justice Blackmun wrote that *Frye* had made "general" acceptance of a scientific theory a sine qua non of its being brought into the courtroom. Justice Blackmun termed the *Frye* test an "austere" standard of admissibility, noted that it did not appear in the Federal Rules of Evidence (Rule 702 as then written), and said that the too restrictive *Frye* test was incompatible with the Federal Rules of Evidence and should not be applied in the federal courts (*Daubert v. Merrell Dow Pharmaceuticals, Inc.*, 509 U.S. 579, 113 S.C. 2786, 125 L.Ed.2d 469 [1993]).

Daubert heightened the role of the federal district court judge as a gatekeeper. The trial judge will make a "preliminary assessment of whether the reasoning or methodology underlying testimony is scientifically valid and whether that reasoning or methodology can be properly applied to the facts at issue in the case before the court. 113 S.Ct. at 2796.

Daubert was followed by *General Electric v. Joiner*, 522 U.S. 303, 118 S.Ct. 512, 139 L.Ed. 508 (1997) holding that the admissibility of scientific evidence is an issue subject to the trial judge's discretion. The trial judge's decision will be overturned on appeal only for an abuse of discretion. The *Kumho Tire* decision two years later extended *Daubert* even further. The duty of the District Court judge to serve as a "gatekeeper" extends not only to *scientific* evidence, but also to *all expert testimony*. The trial judge is to assess relevance and reliability in regard to all expert testimony, applying the 4 key factors (testing, peer-review, error rates, and "acceptability") mentioned in *Daubert* and other factors as well. For a recent Arizona District Court decision following *Daubert* and *Kumho Tires*, see *United States v. Hidalgo*, 229 F. Supp.2d 961, 965 (Nov. 2002) in which Judge Martone excluded certain aspects of handwriting evidence commenting that, as the gatekeeper of what was good science and what was bad science, he had wide latitude to do so under *Kumho Tire*.

Meanwhile in Arizona, our State Supreme Court has simply not embraced *Daubert* or its progeny. In *State v. Bible*, 175 Ariz. 549, 858 P.2d 1152 (1993), cert. den. 1145 Ct. 1578, Justice Feldman authored one of the longest opinions among the Supreme Court's decisions (70 pages). The 129 headnotes contain a dozen which, in essence, reject *Daubert* (as to DNA probability evidence) and insist that Arizona

Courts will continue to follow the *Frye General Acceptability* test for admission of expert testimony.

In *State v. Johnson*, 186 Ariz. 329, 922 P.2d 294 (1996), the Supreme Court (via Justice Feldman) repeated that for the reasons stated in *Bible*, the *Frye* rule, which has been followed without causing significant problems since it was first adopted in 1962, remains the rule in Arizona. A year later, with Justice Feldman again the author, the Supreme Court issued its decision in *State vs. Hummert*, 188 Ariz. 199, 933 P.2d 1187 (1997), again sticking with the *Frye* test of *General Acceptance* and calling it the law in Arizona.

The last important opinion that bears on *Frye* versus *Daubert* as the standard for admissibility in Arizona is *Logerquist vs. Danforth*, 196 Ariz. 470, 1 P.3d 113 (2000), another Justice Feldman opinion. Here Justice Feldman said that *Frye* was still the law, but it was not applicable to certain psychological evidence where "general acceptance" does not need to be applied in explaining behavior by a scientist using *his or her own* observation and experimentation.

Against this backdrop of the long struggle in Arizona to stick with the older, simpler *Frye* test and reject *Daubert* and its amplification of the trial judge's gatekeeper role over allowing experts to testify, there is a new statute in Arizona. In August 2010, the Arizona Legislature's new A.R.S. §12-2302 adopted the *Daubert* Rule and announced that such would thereafter control the admission of scientific and other expert testimony. The statute carefully adopted the *Daubert* Rule's five principles and four factors. However, on September 23, 2010, the statute was promptly declared unconstitutional by Judge Douglas Rayes in *State vs. Joel Randu Escalante-Orozco* (CR2007-008288-001 DT). That case never got appealed, but another case by another judge is up to the court of appeals. Trial lawyers expect that the statute will be again declared unconstitutional as a legislative invasion of the court's power under Rule 702 of the Rules of Evidence to determine the admissibility of courtroom evidence.

Accordingly, in state court (unless the new *Daubert* Rule statute (A.R.S. §12-2203) is allowed to go into effect), Arizona lawyers and their experts can expect to find greater resistance to introducing novel science than they might find in the federal courts. In either jurisdiction, practitioners will be well advised to establish the expert's qualifications under Rule 702 as a witness whose opinions and expertise will be of significant assistance to the trier of fact in weighing the evidence on a fact at issue.

10.8 SHOOTING DOWN THE EXPERT

There are four factors (testing, peer review, error rates, and acceptability) that can be used to hold a *Daubert* hearing to exclude the expert or some part of his or her testimony.

In truth, there are several different *Daubert* arguments that can be advanced to knock out an expert. First, one can argue that the proponent has failed to produce sufficient evidence to permit the judge to rationally find that its expert's hypothesis has been empirically validated.

Initially, although the expert states that there is an empirical study of the hypothesis, the expert's own testimony tells the judge next to nothing about the design of the study. One can attack the expert as having failed to demonstrate peer testing and acceptability.

Next, the expert may disclose that the study entailed only a small database. In a pre-*Daubert* decision, the North Dakota Supreme Court held that a trial judge may bar a scientific opinion resting on a very small database. In the words of the North Dakota court, quantitatively there had "been too little research" (*Nelson v. Trinity Medical Center*, 419 N.W.2d 886, 892 [N.D. 1988]).

Assume now that the expert testifies that the database included 1000 subjects, but that all the subjects were infant animals rather than adult human beings. Such testimony about 1000 subjects might allay the *quantitative* concern about the size of the study mentioned in the preceding paragraph, but now there are *qualitative* concerns. Is the database *representative of the subject about which the expert proposes to testify*? If the expert ultimately contemplates testifying to a hypothesis about medical causation in human beings, the issue is whether, standing alone, the animal study is sufficient to carry the proponent's burden.

In short, the conditions during the experiments may not approximate the conditions involved in the case. In part for that reason, the North Dakota court ruled the experimental verification inadequate. If the test conditions do not match the case before the court, then the requisite "fit" between the research and the facts of the instant case will be lacking. *Daubert vs. Merrell Dow Pharmaceuticals, Inc.*, 509 U.S. 579, 591 (1993). As a matter of logic, it may simply be too great an extrapolation from the research data presented to the ultimate inference that the expert contemplates drawing from the data.

Suppose that the proponent of expert opinion presents barely enough evidence to allow the judge to find that its expert's hypothesis has been empirically validated, but the opponent's rebuttal *convinces the trial judge by a preponderance of the foundational testimony that the proponent's hypothesis has not been validated by* sound *scientific methodology*.

In *Daubert*, Blackmun made it clear that Federal Rule 104(a) governs the judge's ruling on the foundational question of whether the proponent's theory rests on sound scientific reasoning. Under 104(a), the judge acts as a true finder of fact. The judge may be convinced that an expert is an expert on the point in question, but the judge's opinion may then be changed when he or she hears the opposing party's expert, and the first expert may be disqualified under *Daubert*.

On hearing the second expert, the judge would resolve the foundational fact against the proponent, sustain the objection to the first expert, and bar the proponent's expert testimony.

A *Daubert* hearing is triggered by a motion to strike, or a motion to exclude, if raised in advance of trial. If it is to be raised at the commencement of trial, then a motion in limine is filed. Motions must call into question the experts:

i. Factual basis
ii. Data

iii. Principles
iv. Methods
v. Application

Federal Courts have yet to address clearly the burden on the movant seeking to strike expert testimony. The Fifth Circuit recently concluded that the issue was sufficiently raised "by providing conflicting medical literature and expert testimony."

It is the judge's decision whether to hold a *Daubert* evidentiary hearing or resolve the matter through arguments and materials presented to the court.

Once an expert's testimony is sufficiently called into question, the party who was seeking to have the expert testimony admitted will bear the burden of showing the expert's methodology is sound. The evidential requirements of reliability are lower than the proof needed by the jury.

The rules of evidence are not applicable during a *Daubert* hearing and counsel may use *otherwise inadmissible evidence* to attack an expert.

A "voir dire" may be conducted of the expert witness. Pursuant to *Daubert*, the following is a list of subject areas that may be covered:

i. Whether the scientific theory or technique can be or has been tested.
ii. Whether the theory or technique has been subject to peer review.
iii. The known or potential error rate.
iv. Is the same intellectual rigor applied in the courtroom testimony as required in the laboratory or field? Are the methods utilized solely for litigation or also for nonjudicial settings?
v. Did the expert rule out other possible causes? Was differential diagnosis performed in accordance with appropriate methodological standards?
vi. Has the expert established the existence of an alternative feasible design for his or her study of the situation?
vii. Is there too great an analytical gap between the data and the opinion?
viii. Is the expert qualified?

To survive a *Daubert* challenge:

i. The initial evaluation of an expert must go beyond his/her opinions; one must evaluate and demonstrate, even more than in the report, how the expert arrived at those opinions.
ii. One can use discovery of the opposing expert to garner information needed to support one's expert's methodology including test data, field data, analytical methods and research.
iii. Experts should be prepared to support their opinions through comprehensive written reports, not just by explanations given after the fact in deposition.
iv. It may be prudent to test one's own expert's hypothesis, although this may be costly.
v. An expert must be prepared not only to support his or her *conclusions* in deposition, but also his or her *methodologies* used to reach those conclusions.
vi. One can attack the other side's expert's methodologies and conclusions to buttress one's own expert. One can use the other side's expert to support one's own expert's analytical approach. Have the opposing expert acknowledge the use of your expert's approach and techniques.

Attached as **Exhibit C** is an index of court opinions in which an expert in a particular field (civil engineer, mechanical engineer, construction claims expert, appraiser, etc.) has been rejected by the court as *not qualified* to testify as an expert. These cases demonstrate that the *Frye/Daubert* disqualification hearings have to be taken seriously. Further, it is a "black mark" on the record of a professional if a court designates the professional as "unqualified." Such a ruling would likely "dog the footsteps" of that expert thereafter, blighting his or her ability to serve as an expert. It is probably well known to most professional experts these days that Lexis-Nexis has a database called "*Daubert* Tracker." That database contains summaries of all state court and federal court instances in which *Daubert* or *Frye* challenges to experts are attempted, and the success or failure of such motions. Becoming an entry in that database is not a positive occurrence, so experts should "get it right" in their reports.

10.9 CONCLUSION

The expert is an officer of the court with a duty to "tell it straight" and to serve the truth. To the extent one deviates from that path and appears to be the "paid mouthpiece" for an advocate, the expert runs the risk of being discarded as an aid to the judge or jury reaching a decision on what the facts are in the circumstances.

Daubert and *Frye* are two very important tests to be applied to the testimony of an expert and the witness should be familiar with both and prepared to beat back an attack on his or her report, its conclusions, and its methodologies.

The expert who is portrayed as an advocate for a party has thereby done a disservice to the party paying for expert services.

10.10 EXHIBITS

Incorporated in this article are three exhibits. **Exhibit A** summarizes the *Grace* case in which a federal district court in New York penalized a party for the destruction of evidence, draft reports circulated outside the expert's office, and lawyer-expert communications that had been destroyed two weeks before the expert's deposition. The strict Federal Rule led to an award of substantial expenses against the parties.

Exhibit A also summarizes the *Residential Funding* case in which a federal court sanctioned the attorney who destroyed and concealed e-mail and responded to discovery with "a somewhat purposeful sluggish" attitude. State courts tend to give significant weight to these federal decisions even though the state court rules are somewhat less strict.

Expert	Rejected Testimony	Case Citation
Appraiser	Not permitted to testify about impact of prior engineering report.	(LA) *Nesbit v. Dunn*, 672 So.2d 226 (La.App.2 Cir.1996)
CPA	Not qualified to testify on projected lost income.	(IL) *Maher v. Continental Cas. Co.*, 76 F.3d 535 (4th Cir. 1996)
Chemist	Testimony on cause of defective tire did not meet *Daubert* standard.	(LA) *Mitchell v. Uniroyal Goodrich Tire Co.*, 666 So.2d 727 (La.App.4 Cir.1995)
Civil Engineer	Not permitted to offer opinion on cause of escalator accident.	(NJ) *Jimenez v. GNOC Corp.*, 670 A.2d 24 (NJ Super. A.D. 1996)
Civil Engineer	Not qualified to testify about the standard of escalator accident.	(MO) *IMR Corp. v. Hemphill*, 926 S. W. 2d 542 (MO ED 1996)
Construction Expert	Not permitted to testify as to the cause and effect of delays.	(OH) *Jurgen Real Estate v. R.E.D. Constr.*, 659 N.e.2d 353 (Ohio App.12 Dist. 1995)
Economics Professor	Not permitted to testify as to extent of investors' damages.	(NY) *Three Crown Ltd. Partnership v. Salomon Bros., Inc.*, 906 F.Supp. 876 (S.D.N.Y. 1995)
Economist	Not permitted to testify on enjoyment of life.	(NE) *Talle v. Nebraska Dept. of Social Serv.*, 541 N.W.2d 30 (Neb. 1995)
Engineer	Not permitted to testify on defective tire.	(AZ) *Diviero v. Uniroyal Goodrich Tire Co.*, 919 F.Supp. 1353 (D.Ariz. 1996)
Engineer	Not permitted under *Daubert* to testify on alternative design theory.	(IN) *Cummins v. Lyle Industries*, 93 F.3d 362 (7th Cir. 1996)
Engineer	Testimony of cause of spinout insufficient under *Daubert*.	(MI) *Pomella v. Regency Coach Lines, Ltd.*, 899 F. Supp. 335 (E.D. Mich. 1995)
Friction Expert	Not permitted to testify on bathtub fall.	(NY) *Fedorczyk v. Caribbean Cruise Lines, Ltd.*, 82 F.3d 69 (3rd Cir. 1996)
Forensic Consulting Engineer	Not permitted to testify about effect of degreaser on parking lot.	(MO) *Scheerer v. Hardee's Food Systems, Inc.*, 92 F.3d 702 (8th Cir. 1996)
Mechanical Engineer	Not permitted to testify on sudden acceleration of auto.	(LA) *Lawrence v. General Motors Corp.*, 73 F.3d 587 (5th Cir. 1996)
Metallurgy Engineer	Not permitted to testify about sewer grate.	(PA) *Colston v. Southeastern Pa. Transp.*, 679 A.2d 299 (Pa. Cmwlth. 1996)
Occupational Physician	Not permitted to testify on state-of-the-art of asbestos hazards.	(OH) *Owens-corning Fiberglas v. A.M. Centennial*, 660 N.E.2d 819 (Ohio Com Pl. 1995)
Planning Regulator	Not qualified to testify on value of landowner's development rights.	(NV) *Suitum v. Tahoe Regional Planning Agency*, 80 F.3d 359 (9th cir. 1996)
Radiation Expert	Not permitted to testify in Three Mile Island case.	(PA) *In Re TMI Litigation Cases Consolidated II*, 910 F.Supp. 200 (M.D. Pa. 1996)
Roofing Expert	Not permitted to testify about negligence.	(SD) *Zens v. Hon*, 538 N.W.2d 502 (Neb. 1996)
Traffic Control Devices	Expert not qualified to testify about portable speed bumps.	(FL) *Goodyear Tire & Rubber Co., Inc. v. Ross*, 660 So.2d 1109 (Fla. App.4 Dist. 1995)
Transportation Consulting Engineer	Not permitted to testify as to who was driving vehicle in question.	(FL) *State Farm Mut. Auto. Ins. Co. v. Penland*, 668 So.2d 200 (Fla. App. 4 Dist. 1995)
Tile Abstractor	Not qualified to testify on land ownership.	(UT) *Butler Crockett v. Pinecrest Pipeline*, 909 P.2d 225 9Utah 1995)
Tire Designer	Not permitted to testify on defective tire.	(AL) *Charmichael v. Samyang Tires, Inc.*, 923 F.Supp. 1514 (S.D. Ala. 1996)

EXHIBIT C Experts Rejected by Courts as not Being Qualified to Testify

Although there have been important changes to the Federal Rule, making some draft reports and some counsel-expert communications "disappear from view" those rule changes have not yet been made in Arizona State Court procedures and the risk of spoliation and a devastating adverse jury instruction exists in both court systems if the file is not handled correctly.

Exhibit B is a 2003 article by Gregory Josephs exploring experts' destruction of file materials.

Exhibit C is a list of court decisions declaring that certain sorts of witnesses did not qualify as experts after a *Daubert* hearing.

Dictionary of Key Construction Terms

A

ABC Aggregate base course, inert material consisting of fine sand and gravel materials that bind together and are easily compacted. Used to provide a stable and even surface under asphalt and concrete finishes.

Acidification The lowering of soil and water pH due to acid precipitation and deposition usually through precipitation.

ACT The Construction Health and Safety Act. Within the contents of this book, Act refers only to the Construction Health and Safety Act.

Actual Damages Costs that are definable and can be proven to have been sustained by the injured party and for which the injured party may reasonably expect to be compensated.

Activity A task or item of work contained in a critical path schedule that must be performed in order to complete a project.

Activity Duration The amount of time estimated as required to accomplish an activity.

Actual Cost of Work Performed Terminology used in critical path scheduling to indicate the expenditure to perform the work of an activity.

Addendum (Addenda) Written or graphic document released prior to the execution of a contract that modifies, clarifies, interprets, and/or corrects the bid documents and becomes a part of the construction contract when executed.

ADR (Alternate Dispute Resolution) Resolving disputes without resorting to courthouse litigation, by use of arbitration, mediation, partnering, and dispute resolution boards.

Affidavit A written statement made or taken under oath before a notary public, officer of the court, or other representative who by law is authorized to confirm that the party making the statement was, in fact, that party. This is different from a "declaration," which is a statement without a notary public but made "under pain of perjury."

Agreement An understanding reached between two or more parties with respect to the effect upon their relative rights and duties of certain past or future facts or performances.

Alligatoring Craze cracking of the surface layer of materials, such as concrete, stucco, plaster, paint, and so on normally caused by sudden change in temperature, lack of a binder, insufficient drying time between coats or finishing, or poor penetration or a hard film over a soft undercoat. Can also be rectangular patterns of char formed on burned wood.

Allowable Stress Maximum permissible force used in the engineering and design of members of a structure based on a factor of safety against rupture or yielding of any type.

Alluvial Soils Soil, sand, clay, or gravel deposited by flowing water, normally at times of heavy floods. Such materials have been found to be collapsible and otherwise unstable for supporting heavy structures. Alluvial soils may not always be found on the surface of the land. It is not uncommon to find these soils at various depths, particularly in hilly or mountainous terrain.

Alternate Bid An amount stated in the contractor's proposal to be added to or deducted from the base bid if the alternate bid item is accepted and made a part of the final contract agreement. (See *Base Bid*.)

Anti-Indemnification Statutes State laws that invalidate contract clauses relating to a party being held harmless for damages, or which limit the ways in which such contract clauses can be utilized.

Apparent Authority In the absence of actual authority, a party can by his or her conduct or words give another party reason to believe that authority exists. If reasonably relied upon, this appearance will have the same effect as the existence of actual authority.

Appellant The party who contests a decision handed down by the court or arbitrator by bringing the proceeding to a reviewing court or body for reconsideration.

Approval (Architect's or Engineer's) Architect's or engineer's written or imprinted acknowledgment that the materials, equipment, and methods of construction are consistent with the construction documents and are acceptable for use in the work, or that a contractor's claim is valid.

Approved Equal Material, equipment, or method permitted by the architect or engineer (usually before receipt of bids) for use in the work as being acceptable as an equivalent in essential attributes to the material, equipment or method specified in the contract documents.

Approved Plans & Specifications The final set of drawings and specifications that define how the project is to be constructed and that have been reviewed and accepted by the owner and any governmental authority that may have jurisdiction over the completed work.

Arbitration A method of settling claims or disputes between parties to a contract other than through the process of litigation, under which a referee or panel of referees, selected for their specialized knowledge in the field in question, hears the evidence and renders a decision with essentially no appeal.

Arbitrator An impartial person vested with the power to render a final decision concerning issues in dispute, that is, someone chosen by the parties to resolve their dispute.

Architect Designation reserved, usually by state law, for a person or organization qualified and duly licensed to perform design services, including analysis of project requirements; creation and development of the project design; preparation of drawings, specifications, and other such documents defining the scope of the project; materials; equipment and methodology of the work required to construct the project, including bidding requirements; and general administration of the construction contract.

ASI Architects Supplemental Instructions are additional information that may be added to drawings and or specifications that add to or change the original plan or specification.

Asphalt Cement A visco-elastic bitumen obtained as the residual product from the fractional distillation of crude oil. There also are some natural deposits of asphalt cement referred to as lake asphalt. Asphalt cement is used as a binder for hot mixed asphalt concrete and also as base material for cutback and emulsified asphalts.

Assembly Occupancy A place of assembly including but not limited to a building or portions of a building used for gathering together 50 or more persons for such purposes as deliberation, worship, entertainment, amusement, or awaiting transportation.

Assignment of Contract A transferring of rights and obligations to some party who was not an original executor of that agreement.

Assumption of Risk An affirmative defense used by defendants charged with negligence in which it is claimed that the plaintiffs had known of the hazards and dangers involved, yet voluntarily exposed themselves to the hazard that reportedly was created by the defendant, therefore relieving the defendant of any legal responsibility for the resultant injury.

Atterberg Limits Measured water contents (shrinkage limit, plastic limit, liquid limit) determined by established testing procedures to define the boundaries between the different states of consistency of plastic soils.

Attorney In Fact A person duly authorized to act for or in behalf of another person or organization to the extent described in a written document known as "Power of Attorney."

Auger Spectroscopy A means of identifying elements, by energy analysis of Auger electrons emitted by the sample, when suitably excited, often by a high-energy electron beam.

Authority Having Jurisdiction A term used in many standards and codes to refer to the office, organization, or individual responsible for "approving and/or interpreting" such standards and codes related to equipment, procedures, and construction in a town, county, city, or state.

Autoignition Temperature The lowest temperature at which a flammable gas or vapor-air mixture will ignite without a spark or flame.

Award An arbitration decision, as distinguished from a decision, order, or verdict resulting from a court action.

B

Balloon Construction Wood frame construction in which the studs run two or more stories from the foundation to the eve line. The floor joists rest on ribbon boards nailed to the studs, leaving an unprotected opening running from the foundation to the attic.

Bank Measure Volume of soil or rock in its native location before excavation.

Base A layer of aggregate, treated soil, or soil aggregate that rests upon a subbase or natural foundation. The materials of the base are normally held together with a binder; however, certain materials may sometimes combine without the addition of an outside binder. The purpose of the base is to provide a compact, stable, and uniform surface on which to place a wearing course.

Base Bid The amount of money stated in a contractor's proposal as the sum offered to perform the work, not including that work covered in alternate proposals.

Basic Services A term used to define the architect's or engineer's normal scope of services, usually consisting of the following phases: schematic, preliminary, design development, construction documents, bidding/negotiation, and contract administration.

Bates Numbers These are sequential numbers applied to documents by a sticker or a stamp (or an electronic addition to an image) when documents are being indexed for litigation purposes. Although commonly used in legal work, Bates Numbers are not unique to legal documents and can be usefully employed in any task involving relatively large numbers of documents. Usually these numbers will have a prefix, such as "FCL 0001695" so as to denote their source.

Bearing Capacity The maximum unit pressure that a soil or other material will withstand without failure or without settlement to an amount detrimental to the integrity or the function of the structure.

Bearing Wall A wall that supports floor or roof beams, girders, or other structural loads.

Bedding Ground or supports on which pipe is laid.

Benching A methodology in which a hillside is cut back in a series of horizontal ledges, usually to provide level surfaces upon which engineered fill may be placed.

Bench Mark A point of known or assumed elevation used as a reference in determining and recording other elevations in topographical surveys.

Bending Moment The buckling effect at any section of a beam; it is equal to the sum of all moments to the right or left of designated section and is represented by the symbol M.

Beneficial Occupancy Use of a project or portion thereof for the purpose for which it was intended.

Bent A group of two or more piles or posts that support a trestle, deck, or falsework; a transverse framework, to carry lateral as well as vertical loads.

Bentonite A clay composed principally of minerals of the montmorillonite group, characterized by high absorption and very large volume change with wetting or drying.

Bid The offer of the bidder submitted in the prescribed manner (written or verbal) to furnish labor, material, equipment, and services to perform the specified scope of work within the prescribed time for the consideration of payment at the prices stated in the bid schedule.

Bid Bond A form of security executed by the bidder as principal and with a surety that the bid will remain capable of acceptance for a certain period of time and may not be withdrawn by the bidder without penalty.

Bidding Documents Those documents that define the requirements associated with bidding a project, normally consisting of the following: advertisement or invitation to bid; instructions, conditions, and procedures for submitting the bid; and the bid form and proposed contract documents, including any addenda issued prior to the receipt of bids.

Binder The fine soil (clay) fraction inherent or added to sands and gravels for cohesion; soil fraction passing a No. 40 sieve; also admixture for stabilizing soils such as Portland cement, clay, ground limestone, ground shells, and other such materials that tend to cement together the various particles into a cohesive mass.

Binding Arbitration This term is a misnomer. There is no such thing as nonbinding arbitration. Arbitration is a process by which the parties agree to be bound to the decision of the arbitrator or the arbitration panel. Any arbitration award may be enforced by the courts. Even though all arbitration is binding, there are very limited circumstances in which one of the parties can appeal the award through the court system (e.g., bribery of the arbitrator or the refusal of the arbitrator to grant reasonable continuances and hear offered relevant evidence).

Bituminous Material A liquid or semisolid material obtained during the fractional distillation of crude oil. It may be asphalt cement, which is the residue from the distillation process or it may be a liquid material created by modifying asphalt cement with a more volatile hydrocarbon (cutback asphalt) or emulsifying the asphalt cement in water by use of an emulsifying agent (asphalt emulsion). Asphalt cement, in some instances, is found in natural deposits (lake asphalts) that do not require distillation.

Boundary Survey The locating of existing property corners in the field of a parcel of land and recording this information on a drawing. This drawing must include the recorded legal description, the relationship of the found to recorded property corners and the traverse of both. The traverse is a mathematically closed diagram of the complete peripheral boundary of the site, showing dimensions and bearings. It must bear the seal and signature of a state licensed surveyor.

Breach of Contract The failure of one party to another to fulfill an obligation created by a written or oral agreement. The breach of a contract can be intentional or inadvertent or caused by the negligence of the party breaching the contract.

Breach of Duty Failure to perform an obligation created by law or by contract.

Bridging Lateral bracing fixed between roof or floor structural members.

Brief A written argument, submitted in connection with an application, motion, arbitration, trial, or appeal, normally prepared by an attorney, concentrating on legal points of law and/or authority affecting the issues of the case for the court or arbitrator to consider.

Brown Coat A coat of plaster that is applied with a fairly rough finish to receive the finish coat; in two-coat work, the term refers to the base coat of plaster that is applied over lath.

Building A structure enclosed within a roof and exterior walls or firewalls designed for the housing, shelter, enclosure, and support of individuals, animals, or property of any kind.

Building Envelope The elements of a building system that protect and enclose conditioned spaces through which thermal energy may be transferred to or from the exterior.

Building Inspector A representative of a governmental authority employed to inspect construction for compliance with applicable codes, regulations, and ordinances.

Building Occupancy Occupancies used for the transaction of business other than that covered under mercantile occupancy; includes doctors' offices, dentists' offices, city halls, general offices, court houses, libraries, and outpatient clinics among others.

Building official The officer or other designated authority charged with the administration of applicable building codes, regulations, and ordinances, or a duly authorized representative.

Build Out An area within a structure that has been completed although the entire structure may still have additional work required to complete the structure.

Burden of Proof The duty of a party to substantiate an allegation or issue in order to convince the trier of facts that the merits of a claim are valid.

C

Caisson A foundation system in which holes are drilled in the earth to a bearing strata and filled with reinforced concrete; a wood, steel, concrete, or reinforced concrete air- and water-tight chamber in which it is possible to work under air pressure greater than atmospheric pressure to excavate material below water level.

Calcium Chloride An admixture often added to cement during the winter months to accelerate the set of the cement and early strength gain. It reacts chemically with the cement to accelerate set. Due to the accelerated rate of hydration, it may increase heat gain.

Calibrate To check the graduations of an instrument or machine, and to graduate it correctly as necessary.

Capillary Action The action in soil by which water rises in a channel in any direction above the horizontal plane of the supply of free water.

Carbon Cycle The term used to describe the exchange of carbon (in various forms, e.g., as carbon dioxide) between the atmosphere, ocean, terrestrial biosphere, and geological deposits.

Cast-In-Place A concrete system that is formed on site and in its final position.

Cement, Portland The product obtained by pulverizing clinker, consisting essentially of hydraulic calcium silicates; usually containing calcium sulfates as an interground addition. The federal government and the American Society of Testing Materials recognize five major types of Portland cement: Type I Standard Portland Cement used as a general purpose cement in construction; Type II a modification to Type I that is resistant to sulfate attack and decreases the rate of heat evolution; Type III a high early strength of quick hardening cement; Type IV designed for a low heat of hydration; and Type V designed for extreme sulfate resistance.

Certificate of Occupancy A document issued by governmental authority certifying that all or a designated portion of a building complies with applicable statutes, codes, regulations, and ordinances and permitting occupancy for its designated use.

Certificate of Payment A statement from the architect or engineer to the owner confirming the amount of money due the contractor for the work accomplished or materials and equipment suitably stored or both.

Change Order A written document issued and signed by the owner and architect or engineer or contractor issued after the execution of a contract authorizing a change in the scope of work, which may or may not adjust the contract sum or contract time.

Chromatography Chemical procedure that allows the separation of compounds based on differences in their chemical affinities for two materials in different physical states, that is, gas/liquid, liquid/solid.

Civil Action An action filed in court, seeking enforcement or protection of private rights, which may include an award of money damages.

Claim A demand for money, services, or property based upon rights created by contract or law.

Class Action A lawsuit brought about by a large group of people or a representative group of people on behalf of all members of the group. This is quite common with home owner associations in construction defect cases.

Clay The finest of the particles found in soil, usually of less than 0.002 mm in size and possessing significant cohesive properties and a risk of expansion when wetted.

Closed Specification Specification stipulating the use of a specific product or process without providing for substitution.

Collateral Estoppel A legal doctrine that bars further litigation of issues in a case previously decided by other courts.

Collateral Source Rule A rule that prevents a defendant from introducing into evidence the fact that a plaintiff has received benefits for injury or damages from other sources, such as disability insurance, worker's compensation, and so on.

Combustible Capable of reacting with oxygen and burning if ignited; a material that will burn in air; a material or structure that will burn.

Compaction The process of inducing a closer packing of solid particles such as soils, aggregates, concrete, or mortar during placement by the reduction of the volume of voids, usually by tamping, vibration, rolling, or some combination of these actions.

Comparative Negligence The proportional share of liability between a plaintiff and defendant for damages based upon the percentage of negligence of each. Not all states allow a sharing of liability based on comparative negligence.

Compartmentation A type of building design in which the building is divided into sections that can be closed off so that there is resistance to fire spread beyond the area of origin; found in most hospitals and high-rise structures.

Compensatory Damages Damages awarded to pay a plaintiff for his or her injuries; includes direct out-of-pocket losses as well as compensation for pain and suffering.

Completion Bond A form of security executed by the contractor as principal and a surety to the owner and lending institution to guarantee the work will be completed, and that funds will be provided for that purpose.

Compressive Strength The strength of a sample of building material at its point of failure under a compressive load.

Concrete (Asphalt) Asphalt concrete is a mixture of asphalt cement and mineral aggregate that is mixed, placed, and compacted at an elevated temperature that temporarily reduces the viscosity of the asphalt binder. Small quantities of admixtures, such as Portland cement or hydrated lime, are often added to modify the asphalt concrete properties. Asphalt concrete is sometimes referred to as hot mixed asphalt (HMA) or, in some instances, asphaltic concrete.

Concrete (Portland Cement) A mixture of Portland cement, coarse aggregate, fine aggregate, and water. Some mixtures contain additives, including fly ash and admixtures.

Condensation The change in state of matter from vapor to liquid that occurs with cooling. This process releases latent heat energy to the environment.

Consequential Loss An indirect loss occurring as a result of some other loss.

Construction Defect A component or components within the matrix of a building system not built in accordance with the contract documents and which negatively affected the integrity, value, or performance of the building system.

Construction Documents Generally include the contract agreement between the parties and the working drawings and specifications, including any approved revisions and change orders issued by the owner and/or a designated representative that define how the project is to be built.

Consulting Expert There are two separate definitions of this term depending on the context in which it is used. A consulting expert is a person who through education, training, and experience has superior knowledge of a subject. In the legal system, if an attorney hires a forensic consulting expert, that person will be retained to assist in the development of a case but will not testify or provide opinions at trial or arbitration. If an owner or other professional hires a consulting expert, that person will simply advise the owner or professional on a matter within the consultant's expertise.

Contract An agreement between two or more persons that creates an obligation to do or not to do a particular thing. Its essentials are competent parties, subject matter, a legal consideration, mutuality of agreement, and mutuality of obligation.

Contract Bond An approved form of security executed by a contractor or subcontractor and his or her surety, or sureties, guaranteeing complete execution of the work and all supplemental agreements pertaining thereto and for the payment of all legal debts pertaining to construction of the project.

Contributory Negligence The plaintiff or claimant, by not exercising ordinary care, contributed to the injury; in some states, a plaintiff's contributory negligence will bar the plaintiff from recovering damages.

Copyright Exclusive right to control the making of copies of a work or authorship, such as granted by federal statute to the author for a limited period of time.

Counterclaim An independent cause of action or demand made by a defendant against a plaintiff. This occurs during a case when a defendant files a claim against the plaintiff.

Covenant A written, signed agreement by two or more parties pledging that something is done, or shall be done as, for example, a covenant not to sue.

Critical Path Method (CPM) A graphic method showing the relationship of work tasks (activities) on a project, and how they interrelate. The critical path or paths of a project will be the combination of activities between the start and completion of a project in which the successor activity depends on the completion of the predecessor activity before it can start. When joined together, these dependent activities formulate the critical path and determine the time length of the project.

Cross Examination The questioning of a witness by the opposing party regarding matters in which the witness under direct examination testified to or gave opinions in an effort to clarify, weaken, or refute the direct testimony.

Curtain Wall An exterior non-load-bearing prefabricated wall, usually more than one-story high supported by the structural frame, which protects the building interior from weather, noise, or fire.

D

Damages The amount claimed or allowed as compensation for injuries sustained or property damaged through wrongful acts, negligence, breach of contract, or other cause.

Damproofing Treatment of concrete or mortar to retard the passage or absorption of water or water vapor, either by the application of a suitable membrane coating to the finished surface or through the use of various admixtures or treated cement.

Date of Commencement The date established in the Notice to Proceed or spelled out in the contractual agreement between parties setting forth the date that the work shall begin.

Date of Substantial Completion The date certified by the architect or engineer when the work or a designated portion thereof is sufficiently complete in accordance with the contract documents so that the owner may occupy the work area or portion thereof for the intended use.

Dead Load An inert, inactive load, as in structures composed of the weight of all of its members, permanent attachments, and building code required weight resulting in the design constant requirement.

Declaration A substitute for an affidavit. This is a written statement of facts, signed under penalty of perjury by a declarant.

Declaratory Judgment The order of a court that establishes the rights of the parties on a question of law or on a contract.

Default An omission on that which ought to be done. Specifically, the omission, or failure, to perform a legal or contractual duty.

Defect Any characteristic or condition that negatively affects the integrity, value, or performance of the building system.

Defendant The party against whom relief and/or recovery is sought by a plaintiff in litigation.

Defense A response to the plaintiff's allegations, which introduces factual findings and evidence that refute all or a portion of such allegations.

Defraud To make a misrepresentation of an existing material fact, knowing it to be false or making it recklessly without regard to whether it is true or false, intending one to rely on the false statement, and under such circumstances the person does rely to his or her damage.

Degree of Compaction The measure of compaction of a material such as soil embankment, base course, or asphalt concrete pavement. It is normally expressed as a percentage of a density determined by a laboratory test of the material. Depending upon the reference density used, the percentage of compaction will be greater than the actual percentage by weight of material within a given volume. In some instances, asphalt concrete pavement compaction is referenced to the theoretical maximum specific gravity of the material and the percentage of compaction will yield the actual percentage by weight of material.

Deposition The sworn testimony of a party, witness, and/or expert involved in the case, in which the attorneys for the parties involved ask questions regarding the case. The testimony is recorded by a court reporter and/or on videotape. Transcripts of the deposition are available for the deponent to review, edit, and attest to. The transcript can be used in the courtroom to impeach a deponent who deviates from his or her deposition testimony. As the transcript will become a part of the permanent record, it is essential that the deponents review their testimony.

Derivative Tort An action in tort based on criminal conduct of the defendant that resulted in injury to the plaintiff and for which the plaintiff seeks compensation.

Design To form plan, scheme, conceive, and arrange in mind, originate mentally, plan out, contrive, more often with the intent to execute the preliminary conception of an idea in accordance with a preconceived plan. As a term of art, the giving of a visible form to the conceptions of the mind.

Design Build A term defining the coming together of a team of architects, engineers, and contractors all working together to design and construct a project.

Design Defect A flaw within the matrix of a building system that negatively affects the integrity, value, or performance of a building system, which flaw results from the failure on the part of the design professionals to meet the applicable established standards of care.

Design Strength The load-bearing capacity of a member computed on the basis of the allowable stresses assumed in the design; the assumed values for the strength of the concrete and the yield stress of the steel on whose basis the theoretical ultimate strength of a section is computed.

Design Team Normally, conceived as architects and engineers working together to formulate a concept with proper direction for a contractor to build.

Dimensional Stability In describing building materials, material is dimensionally stable if it has no moisture movement and little to no temperature movement and does not shrink or expand.

Directed Verdict A situation in which a trial judge may order the entry of a verdict without allowing the jury to consider the evidence because the party with the burden of proof failed to provide prima facia evidence for the jury to consider; therefore, as a matter of law, there can be only one verdict.

Direct Examination The initial questioning of a witness by the party on whose behalf the witness is being called.

Discovery The process by which the opposing parties collect information and facts from each other and third parties in order to assist in preparing for litigation, be it arbitration or trial.

Dispute A conflict or controversy; a conflict of claims or rights. The subject of litigation, the matter for which a suit is brought and upon which issue is joined, and in relation to which jurors are called and witnesses examined.

Doctrine of Objective Impossibility If something is impossible to do, it need not be done, and the failure to do it cannot be a breach of contract.

Doctrine of Respondeat Superior Or Vicarious Liability Employers are responsible for the torts of their employees.

Due Care The degree of care that a person of ordinary prudence and diligence would exercise under the same or similar circumstances; negligence is the failure to use due care.

Duty An obligation imposed by law or contract.

Duty of Good Faith Exists in almost all contracts and requires the parties not to engage in acts that prevent the other party from receiving the benefit of the contract.

Dwelling A building occupied with no more than two living units, each of which is occupied by a family; no more than three outsiders may rent rooms.

Dynamic Compaction A process that densifies soil by applying a series of impact blows to it.

Dynamic Loading Loading from units such as machinery start up or vibration-imposed stresses in excess of those imposed by the dead load of the machinery itself.

E

Easement A legally created restriction on the unlimited use of all or part of one's land.

Ecosystems Ecological units composed of complex communities of organisms and their specific environments.

Educational Occupancy A building in which six or more persons gather to receive instruction, such as a school, college, university, or academy.

Efflorescence A deposit of salt, usually white, formed on a surface, the substance having emerged in solution from within either concrete or masonry and having subsequently been precipitated by evaporation. Refer to ACI 116R for a more detailed definition.

Effluents Waste materials, such as smoke, sewage, or industrial waste, that are released into the environment.

Eichleay Damages Recoverable damages that include the amount of overhead that must be transferred to or borne by other projects when a construction contract is wrongfully terminated or delayed.

Engineered Fill Soil embankment that has been processed and compacted to a specific standard during placement. This is in contrast to placement of soils with no control over the manner in which it is placed. The standard to which engineered fill has been placed must be considered when evaluating the properties of the end product.

Entrance An opening into a building.

Estoppel A legal bar preventing a person from asserting a legal position because of his or her own conduct, which has been relied upon by others, or because of some other reason created by operation of law.

Exculpatory Clause A clause in a contract that attempts to transfer liability from a party with power or control of a situation to one who has little or no power to control.

Exhibits Physical or tangible items offered as evidence in support of the case.

Exit The portion of the means of egress that leads from the interior of a building or structure to the outside ground level.

Exit Access Any portion of an evacuation path that leads to an exit.

Expert Witness A person with specialized knowledge in a certain field gained through education, personal experience, or both who is expected to testify regarding the facts. In order to be qualified as an expert, the witness must have specialized knowledge that will assist the trier of fact in understanding the evidence. (See also Consulting Expert and Testifying Expert.)

Express Warranty A representation, usually in writing, that a fact in relation to a transaction is as stated or promised; a guarantee.

Extra Hazard Occupancy Properties in which flash fires can be immediate and result in the opening up of all fire sprinklers within the area affected. In such occupancies, sprinkler spacing must be closer and pipe sizes larger than in ordinary hazard occupancies.

F

Fair Labor Standards Act of 1938 (FLSA) Requires employers to pay employees a minimum wage and to pay certain employees overtime pay for work in excess of 40 hours per week.

False Light Offensive publicity that attributes to the plaintiff characteristics, conduct, or beliefs that are false, such that the plaintiff is placed before the public in a false position.

Family Medical Leave Act (FMLA) Requires employers to give employees up to 12 weeks of unpaid leave per year for various family-related problems such as birth of a child or serious illness of a family member.

Federal Rules of Civil Procedure A series of procedural requirements that governs all civil actions in U.S. district courts and has been modified and/or adopted by most states for use in their respective jurisdictions.

Fiduciary A person having a duty, created by an undertaking, to act primarily for another's benefit in matters connected with such undertaking.

Fines General category of silts and clays, the smaller particles.

Fire Rapid self-sustaining oxidation accompanied by the production of varying intensities of heat and light. Class A Fire is one of ordinary combustibles; Class B Fire is one in flammable liquids, gases, or grease; Class C Fire is one that involves energized electrical equipment; Class D Fire is one that occurs in combustible metals.

Fire Code A set of rules and requirements whose purpose is to establish levels of fire protection considered adequate for procedures, practices, and equipment within a specified area.

Fire Resistive Construction Construction in which the structural members, including walls, columns, beams, floors, and roofs, are of noncombustible or limited combustible materials and have fire resistance ratings not less than those specified in NFPA 220.

Fire Stop An obstruction across an air passage, or concealed space, in a building to prevent fire from spreading.

Firestopping The blocking off of concealed spaces in structures to prevent fire spread through the ceilings and walls.

Flatness A finishing tolerance for concrete specifying the limits of the deviation from a plane (not necessarily level). (Two acceptable methods for measuring flatness specified in ACI 117 are measuring the deviation from a 10-foot straight edge or using an instrument that measures a profile and reports the results in terms of "F-Numbers.")

Forensic Expert A professional in a specific field, or discipline, who conducts an in-depth investigation to determine how and why a specific situation occurred and who appears in legal proceedings to give testimony.

Fraud An intentional misstatement or failure to disclose information that leads to damage.

Freshwater Water with very low soluble mineral content; sources include lakes, streams, rivers, glaciers, and underground aquifers.

Frivolous Suit A suit that is so totally without merit on its face, as to show bad faith or other improper motive on the part of the plaintiff.

G

Galvanic Corrosion Accelerated corrosive action brought about by two different metals touching each other while wet with a solution capable of conducting an electrical current that creates a galvanic voltage that tends to accelerate corrosion in one of the metals and protect the other.

General Conditions That part of the contract documents that sets forth many of the rights, responsibilities, and relationships of the parties involved or of the contract.

General Contractor One who is licensed by the governing authority and contracts for the construction of the entire project, rather than for a portion of the work. The general contractor is in privity with the owner/developer and will normally hire subcontractors for specialty work, such as electrical, mechanical, and plumbing but is responsible for coordination of the work as well as the care, custody, and control of the work site and payment to the various subcontractors.

Geotechnical Engineer A civil engineer who has studied extensively and is concerned with the engineering behavior of earth materials.

Girt A horizontal member between columns or bents, acting as a stiffener.

Goods All forms of tangible personal property including specifically manufactured goods, supplies, mobile homes, and materials.

Governmental Immunity Federal, state, and local governments are not amenable to actions in tort, except in cases in which they have consented to be sued. Most jurisdictions, however, have abandoned this doctrine in favor of permitting tort actions with certain limitations and restrictions.

Grade To establish a level by mathematical points and lines and to bring to the level by manipulating the elevation or depression of the natural surface to meet established level.

Grade Beam A horizontal load-bearing foundation member, supported like a standard beam by the use of caissons, piers, and so on, unlike a ground-supported foundation stem wall.

Groundwater Water sources found below the surface of the earth often in naturally occurring reservoirs in permeable rock strata; the source for wells and natural springs.

Guarantee A contract that something shall be done exactly as it is agreed to be done.

H

Habitability The condition of residential or other premises being reasonably fit for occupancy, and which does not impair the health, safety, and well-being of the occupants. If the conditions are not met, due to a failure to provide heat or potable water, for example, the occupant may be eligible for rent abatement or may in some circumstances be just cause for the occupant to vacate the premises.

Harmless Error Error that is trivial or formal or merely academic and not sufficiently prejudicial to the party that it affects the outcome of the case.

Hearing A proceeding wherein evidence is taken for the purpose of determining an issue of fact and reaching a decision on the basis of the evidence.

Hearsay Rule A rule that declares not admissible as evidence any statement other than that by a witness while testifying at a hearing and offered into evidence to prove the truth of the matter stated.

Heavy Timber Construction Construction in which nonbearing exterior walls, bearing walls, bearing portions of walls, columns, beams, and girders are heavy timber and noncombustible. Floors and roofs are of wood and there are no concealed spaces. Also called mill, plank-on-timber, and slow-burning construction.

Hidden Defect Defect not recognizable upon a reasonable inspection of a good product, or which is not readily apparent for which a seller is generally liable and which would give rise to a right to revoke a prior acceptance.

"Hired Gun" A common term used to describe an expert witness who will provide opinion in favor of the party who retained him or her without regard for the application of truth.

Horizontal Exit A protected way of travel from one area of a building to another area in the same building or in an adjoining building on approximately the same level.

Hypothetical Question A question that assumes facts that the evidence tends to show and calls for an opinion based on the hypothesis. In trials, hypothetical questions can only be posed to an expert witness who is qualified to give an opinion on the matter in issue.

I

IBC International Building Code is the new standard developed by the International Congress of Building Officials (ICBO), which has the responsibility of establishing life safety standards for the construction industry. The IBC has for the most part replaced the UBC (Uniform Building Code) in 2000 as well as many other building codes developed in various parts of the country. All such codes are adopted by states, counties, and cities and may have revisions to meet the specific needs of the location.

Ignorance Lack of knowledge or ignorance of the law does not justify an act, since every person is presumed to know the law; however, mistake of fact may provide a legal excuse.

Immunity (Governmental or Sovereign) A right of exemption in which the federal, and derivatively, the state and local governments are free from tort liability arising from activities that were governmental in nature; however, most jurisdictions have abandoned this doctrine in favor of permitting tort actions with certain limitations and restrictions.

Implied Authority That which is necessary, usual, and proper to accomplish or perform the main authority that has been expressly delegated to an agent.

Implied Warranty A representation, not expressly made in writing, that certain conditions exist or that a fact in relation to a transaction is as stated or promised.

Indemnification A contractual obligation by which one person or organization has agreed to secure another against loss or damage from specified liabilities.

Indemnify To protect against loss or damage, or to promise compensation for loss or damage. The duty to indemnify may be created by rule of common law, by statute, or by contract.

Indemnitee The party who is to be indemnified from loss or damage.

Indemnitor The party who is providing the indemnification.

Industrial Occupancy Includes factories of all kinds, including processing plants; assembly plants; and mixing, patching, finishing, and other related facilities.

Inexcusable Delay The contractor has no excuse for the delay in the completion of a defined segment of work and is responsible for paying the owner's damages. Normally occurs when the owner has committed to others to have a project up and running on a specific date or suffer financial loss.

Inflammable A material that will burn; easily set on fire.

Informed Consent A person's agreement to allow something to happen that is based on the full disclosure of the facts including all alternatives and risks involved. Constitutionally required in many areas where one may consent to what would otherwise be an unconstitutional violation of a right.

Ingress An entrance or the act of entering.

Injunction A court order ordering one party to do or not do something.

Injury Any wrong or damage done to another, either to person, rights, reputation, or property.

In Situ The Latin term for "in its original site or location." The term is generally used to describe a material such as a naturally deposited soil or mineral aggregate. However, the term is sometimes used to describe the properties of a material manufactured and placed by other than natural means to differentiate between in situ properties and properties measured from a sample of the material fabricated and tested by a standard laboratory method.

Inspection To analyze; examine; evaluate; scrutinize; investigate; look into; check over; view for the purpose of authenticity or condition of an item, product, document, residence, business, and so on.

Intent A state of mind in which the person knows and desires the consequences of his or her act, which, for the purpose of criminal liability, must exist at the time the offense is committed.

Interior Finish The surface material of walls, fixed or movable partitions, ceilings, and other exposed interior surfaces; includes plaster, paneling, wood, paint, wallpaper, carpeting, tile, and the like.

Internal Friction Resistance by a soil particle to sliding within the soil mass. For sand, the internal friction is dependant on the gradation, density, and shape of the grain and is relatively independent of the moisture content, while for clay, internal friction will vary with the moisture content.

Interrogatories A discovery process of written questions allowed under the Rules of Civil Procedure, which are put forward to a party having information of interest in the case.

Invert Elevation The lower inside of a pipe or sewer, at a given location in reference to an established benchmark.

Issue of Fact A question dealing with what happened at some point in the past; asks "who, what, when, where, why, or how?"; decided by a jury in a jury trial, by a trial court judge in a bench trial, or an arbitrator in arbitration.

Issue of Law A question dealing with the meaning of a law or which laws apply to a situation; always decided by a judge.

J

Joinder Uniting two or more elements into one, such as the joinder of parties as complaintiffs or codefendants in a lawsuit or as parties in arbitration.

Joint Liability Such shared liability as results in the right of any one party sued to insist that others be sued jointly with him or her.

Joint Venture A temporary business relationship between two entities and having the same legal effect as a partnership.

Judgment The determination of a court of competent jurisdiction upon matters submitted to it; a final determination of the rights of the parties to a lawsuit.

Jurgenson Formula Formula used to determine shear strength of most soils based on unit pressure and depth of stratum.

Jurisdiction A particular geographic area over which a particular government has power.

Jurisdictional Dispute Disagreement between two or more labor unions over which union shall do certain work.

Jurisprudence The philosophy of law or the science that treats of the principles of positive law and legal relations (i.e., of the form as distinguished from the content of the systems of law). Also, a collective term denoting the course of judicial decision (i.e., case law as opposed to legislation, sometimes used simply as a synonym for "law").

Jury Trial A trial in which a panel of citizens, called jurors, decides the factual issues of the parties in dispute; the legal issues are decided by the judge.

L

Latent Defect A hidden or concealed defect; one that could not be discovered by reasonable and customary inspection or would not be apparent to a buyer/owner on normal observation.

Lateral Soil Pressure The horizontal component of the force due to a wedge of earth and surcharge, if any, moving or tending to move downward along its natural cleavage plane.

Legal Liability A legal obligation that arises out of contract or by operation of law.

Levelness ACI requires that the surface of the slab should be level within $3/4$ of an inch throughout regardless of the size of the floor. Local protocols and standards may relax this requirement and should be considered when determining levelness. (As it pertains to concrete slabs, it is the deviation from a level plane.)

Liability An obligation one is bound in law or justice to perform, an obligation to do or refrain from doing something. A duty that, eventually, must be performed, that is, an obligation to pay money. Signifies money owed, as opposed to assets. Also used to refer to one's responsibility for conduct, such as contractual liability, tort liability, or criminal liability.

Lien Right of claim or charge made upon property for payment or satisfaction of a debt.

Lien Laws Allow architects, engineers, contractors, material suppliers, and in some states laborers to file a claim on the owner's property for amounts owed to them for work done, which has enhanced on the owner's property.

Life Safety Code NFPA 101, a standard containing provisions for building design whose purpose is to ensure that lives are protected should fire occur; usually this entails providing a safe and accessible means of egress, but in health care occupancies, it involves designing buildings that can be defended with occupants in place.

Limited Liability The limitation placed on the amount an investor of a corporation can lose resulting from a lawsuit against the corporation or other loss suffered by the corporation.

Liquidated Damages Amounts stipulated in advance for damages to be recovered for breach of contract.

Litigation Legal action/proceedings including all proceedings therein.

Live Load The nonpermanent load to which a structure is subjected, in addition to its own weight. It includes occupants of the building, operating machinery forces, free standing materials, and mobile equipment.

M

Malfeasance The doing of an act that is wrongful and unlawful. A wrongful act the actor ought not to do, or the unjust performance of some act that the party had no right to do, or that the party had contracted not to do.

Malpractice A breach of a professional duty by rendering professional services below the standard of care required, which then are the proximate cause of injury, loss, or damage to other person or persons.

Material Breach A serious major infraction of a law or contract.

Materialman A supplier of material to a project.

Mediation An effort by an independent and impartial party retained to assist parties in conflict in reaching a settlement of a controversy or claim voluntarily.

Meritless Claim A claim that is so obviously insufficient that it should be rejected on its face without argument or proof.

Mini-Trial An effort by an independent and impartial party retained to assist parties in conflict in reaching a settlement of a controversy or claim voluntarily. Usually this involves a brief recitation of evidence and the presentation of just one or two witnesses per side. The decision is often advisory only.

Misfeasance The doing of an act in a wrongful or injurious manner. The improper performance of an act that might have been lawfully done.

Misrepresentation Any manifestation by words or other conduct by one person to another, under the circumstances, amounts to an assertion not in accordance with the facts. An untrue statement of fact. An incorrect or false representation. That which, if accepted, leads the mind to an apprehension of a condition other and different from that which exists. Colloquially, it is understood to be a statement purposely made to deceive or mislead.

Mitigation of Damages The doctrine of "mitigation of damages" is sometimes referred to as the doctrine of avoidable consequences and imposes a duty that the one injured by reason of another's tort or breach of an agreement exercise reasonable diligence and ordinary care to avoid aggravating the injury or increasing the damages. The term also is an affirmative defense and applies when the plaintiff fails to take reasonable actions that would tend to mitigate the injuries.

Motion For Summary Judgment Federal Rule of Civil Procedure 56 permits any party to a civil action to move for a summary judgment on a claim, counterclaim, or cross claim when believing that there is no genuine issue of material fact and that one is entitled to prevail as a matter of law.

Motion In Limine A written motion that is usually made before or after the beginning of a jury trial for protective order against prejudicial questions and statements. The purpose of such a motion is to avoid the injection into the trial of matters that are irrelevant, inadmissible, and prejudicial.

Motion For Protection Order A motion to obtain an order or decree of a court (protective order), to protect the person from further harassment in discovery. Often used to avoid a deponent or witness from being asked about privileged or irrelevant matters

Movant The moving party; applicant for an order by way of motion before a court.

N

Negligence Failure to exercise such care as a reasonable, prudent, and careful person would use under similar circumstances. The term refers to conduct that falls below the standard established by law for the protection of others against unreasonable risk or harm.

Negligence Per Se An act or omission regarded as negligence without argument or proof because it violates a standard of care defined by statute.

Neutral A position of disengagement in which a party or parties do not take any position concerning the matter in question.

NFPA (National Fire Protection Association) An association that has established extensive reference material, used throughout the design and construction industry, covering virtually all phases of fire protection.

Nonconforming Use A particular landowner's use that continues after zoning regulations require a different use.

Nonfeasance The total omission or failure of an agent to enter upon the performance of some distinct duty or undertaking that was agreed to in principle.

Notice Information concerning a fact, actually communicated to a person, or actually derived from a proper source, such as constructive notice that is presumed by law to have been acquired; implied notice that may be inferred from facts that a person had means of knowing, and so on.

O

Obligee A person in favor of whom some obligation is contracted, whether such obligation be to pay money or to do or not to do something. The party to whom someone else is obligated under contract. The party to whom a bond is given.

Onerous Clause A clause in a contract that attempts to transfer liability from one party with power or control of a situation to one who has little power or control.

OSB Oriented Strand Board, also known as waterboard, sterling board or exterior board; an engineered wood product formed by compressing layers of wood strands (flakes) together with wax and resin, (95% wood 5% wax and resin). The resins may vary depending on whether the end product is for exterior of interior use.

P

Payment Bond A bond posted by a surety for a contractor guaranteeing payment of labor and materials, should the contractor be found in default, or unable to pay such costs due to insolvency. Many times the payment bond will be written as one bond referred to as a performance and payment bond.

Performance Bond A bond of the contractor in which a surety guarantees the owner that the work will be performed in accordance with the contract documents. Except in some states where such is prohibited by statute, the performance bond is frequently combined with the labor and material payment bond.

Performance Specification A more general specification detailing the desired result and leaving it to the contractor and/or specialty subcontractor to design and develop the system to meet the desired result.

Perjury The willful assertion as to a matter of fact, opinion, belief, or knowledge, made by a witness in a judicial proceeding, as part of the evidence, either by oath or in any form allowed by law, and known by the witness to be false.

Permeability The quality of a substance that permits a liquid or gaseous substance to travel through a material.

Personal Injury Physical or mental injury to a human being.

Pier A word with many meanings, including a solid support on which the arches of a bridge rest, a solid support for masonry, a solid support of a wall or beam, or a wharf projecting from the shore.

Plaintiff A person or entity that brings action, complains, or sues in a civil action.

Pleadings The formal allegations by the parties of their respective claims and defenses in litigation.

Pollution The contamination of a healthy environment by human-made waste.

Porosity A measure of the void volume as a percentage of the total material volume.

Post-Tensioned Normally refers to the method of prestressing reinforced concrete in which steel tendons are tightened after the concrete has attained its initial set, or hardening.

Potable Water Water that is drinkable, safe to be consumed.

Precedent An adjudicated case or decision of a court that is recognized as authority for the disposition of future cases.

Predecessor Activity That work activity within a critical path schedule that must be started and/or completed before the next work activity can commence.

Preemptory Challenge A challenge to a juror at the time the jury is being empanelled for which no reason need be given. In various jurisdictions, each party is entitled to a certain number of such challenges as well as to challenges for cause. The challenge removes the juror.

Premises Liability If the owner violates in his or her duty to maintain a safe structure and a user of the structure is injured, the owner is liable.

Prestressed Concrete Concrete in which internal stresses of such a magnitude and distribution are introduced that the tensile stresses resulting from the service loads are counteracted to a desired degree. In reinforced concrete, the prestress is commonly introduced by stressing the tendons.

Pretensioning A method of prestressing reinforced concrete in which the tendons are tightened before the concrete has hardened.

Pretrial Motions Legal documents made prior to the beginning of the trial that seek judicial remedies. Such documents primarily deal with discovery disputes and positioning one's case, such as arguing to withhold certain facts from the jury.

Pretrial Order An order embodying the terms and stipulations agreed upon at a pretrial conference or hearing that governs the conduct of the trial, including discovery deadlines and disclosure of witnesses and expert witnesses.

Privity The connection or relationship that exists between two or more contracting parties. It was traditionally essential to the maintenance of an action on any contract that there should subsist such privity between the plaintiff and defendant in respect to the matter sued on. However, the absence of privity as a defense in actions for damages in contract and tort actions is generally no longer viable with the enactment of warranty statutes, acceptance by states of the doctrine of strict liability, and court decisions.

Probability Likelihood; appearance of reality or truth; more likely than not likely to occur. A condition or state created where there is more evidence in favor of the existence of a given proposition than there is against it. A greater than 50 percent chance of it reasonably being likely to occur.

Pro Bono For the good; used to describe work or services done or performed free of charge.

Proctor A commonly used name for a standard laboratory test of maximum density and optimum moisture for a soil or base course material using a specified compactive effort. The name is derived from R. R. Proctor, who developed the test procedure in 1932. A similar laboratory test using a greater compactive effort is often referred to as a modified

Proctor. The term Proctor does not appear in these test procedures. Maximum density is a relative term to indicate the highest density occurring during the test. However, there will always be air voids within the laboratory compacted specimens such that a theoretical maximum density is never achieved by these tests.

Proctor Modified A moisture-density test of more rigid specifications than the standard Proctor. The primary difference between the modified Proctor and the standard Proctor is the use of heavier weight being dropped from a greater distance in laboratory determinations. It is used primarily by the Corp of Engineers and state highway departments.

Product Liability A tort that dictates that a manufacturer is strictly liable for placing a manufactured article in the market, knowing that it is to be used without inspection for defects, which article proves to have a defect that causes injury to a human being.

Professional A person who is deemed to have specialized knowledge and skills, acquired through education and experience, which are used in advising or providing services to others.

Promissory Fraud A promise to perform made at a time when the promisor has a present intention not to perform. A misrepresentation of the promisor's frame of mind is, for that reason, a fact that makes the basis for an action for deceit, also known as *common law fraud*.

Proximate Cause A cause that leads directly and naturally to a given result without any intervening causes.

Punching Shear Shear stress calculated by dividing the load on a column by the product of its perimeter and the thickness of the base or cap, or the product of the perimeter taken at one-half the slab thickness away from the column and the thickness of the base or cap.

Punitive Damages Normally awarded separately as punishment in addition to compensatory (actual) damages due to findings of malicious or wanton misconduct on the part of the defendant.

Q

Quality Assurance The act of attesting that the work performed was in fact done in accordance with established industry standards, contracts, specifications.

Quality Control The act of authority to properly manage, direct, superintend, restrict, regulate, govern, administer, or oversee that work will be done in accordance with established industry standards, contracts, specifications, or detailed account of the various elements, materials, dimensions, and the like involved.

Quantum Meruit An equitable doctrine, based on the concept that no one who benefits by the labor and materials of another should be justly enriched thereby; under those circumstances, the law implies a promise to pay a reasonable amount for the labor and materials, even absent a specific contract thereof.

R

Redirect Examination The examination of a witness by the direct examiner subsequent to the cross examination of the witness. Sometimes also referred to as damage control.

Reflective Cracking Cracking occurring within a layer of material that has been placed over another cracked layer of material. Cracks that occur at the same location as the underlying cracks are referred to as reflective cracks, being a repeat of the underlying crack pattern.

Release A document that memorializes the fact that claimant has abandoned his or her claim because it has been satisfied or deemed invalid.

Relief/Remedy Terms used in addressing a list of things, such as attorney fees, punitive damages, and injunctions that a court can award to the winning party.

RFI(Request for Information) A written document prepared by the contractor and addressed to the architect or engineer of record requesting clarification of the design intent or other related issue.

Retainage Amount due the contractor from a particular invoice but held by the owner and not paid until the end of the project after final acceptance.

S

Sacrificial Coating A coating applied over a finish coat that is designed specifically to take on damage from the elements, such as wind, rain, snow, and so on, in order to protect the finish coat from damage.

Sand A mixture of rock particles ranging from 0.06 mm to 2 mm in diameter.

Settlement A voluntary agreement in which the parties in dispute have agreed to resolve their differences.

Several Liability Each entity is individually liable for the entire debt should any of the other liable entities not have sufficient assets to cover their portion of the debt.

Shear The strain upon, or the failure of, a structural member at a point where the lines of force and resistance are perpendicular to the member.

Shear Modulus The modulus of rigidity, which is equal to the shear stress divided by the shear strain.

Shear Wall A wall that in its own plane carries and resists shear resulting from wind, blast, or earthquake.

Sheathing The first layer of wall or roof covering attached to the studs, joists, or trusses.

Shop Drawings Incidental drawings furnished by material suppliers, equipment suppliers, manufacturers, and fabricators defining specific portions of the work and how it will be installed and/or function to be in compliance with the contract documents.

Shrinkage Shrinking, *per se*, is simply a reduction in volume of a material. It is generally related to solid materials; however, liquids also shrink or lose volume with environmental changes such as temperature. Shrinkage may result from several phenomena. Drying shrinkage is shrinkage resulting from a loss of moisture. Thermal shrinkage is the result of a reduction in temperature. Other shrinkages, such as chemical or molecular shrinkage, are the result of changes in the structure of a solid resulting in a reduction in volume. If the shrinkage is restrained, shrinkage or tensile stress will be produced. When these stresses exceed the material's tensile strength, cracks will result.

Shrinkage Cracks   Cracking of a structure or member due to failure in tension caused by a reduction in volume.

Silt   Moderately fine particles of rock from 0.002 mm to 0.06 mm in size.

Slump   A measure of consistency of freshly mixed concrete, mortar, or stucco equal to the subsidence measured to the nearest ¼ inch (6 mm) of the molded truncated cone immediately after removal of the slump cone. The results of this test have little if any correlation to *strength* of the cementitious material being tested.

Soil   Material found on the surface of the earth not bigger than 20 mm in size, not including rocks and boulders and predominantly nonorganic. If soil is to be used for building material, it must not contain any organic material, and it can be a natural selection of particles or a mixture of different soils to attain a more suitable particle distribution.

Soil Erosion   The removal of soil by the action of water or wind.

Sovereign Immunity   A doctrine that precludes a litigant from asserting an otherwise meritorious cause of action against a sovereign or a party with sovereign attributes unless sovereign consents to suit. However, under certain circumstances, there are federal and state laws that will allow a suit to be filed. When dealing with any governmental body, U.S. or otherwise, it is a good practice to determine if the doctrine of sovereign immunity applies, or to what extent it has been waived.

Special Conditions   A section of the conditions of the contract, other than general conditions and supplemental conditions, that may be prepared to establish specific clauses setting forth conditions or requirements peculiar to the project to be constructed that are not normally considered.

Special Inspection   An inspection required by the governing authority that is not normally conducted by that authority but deemed necessary to protect the life and safety of the public.

Special Master   An impartial person chosen by the parties in conflict to perform services in their affairs, and who has the right to control the physical conduct of others in performance of services to resolve their dispute.

Specification   A description, for contract purposes, of the materials and workmanship required in a structure, as also shown in the working drawings. A written description containing the minimum quality standards required and necessary as they pertain to workmanship and materials that will be furnished under the contract.

Stabilized Soil   Soil that has been treated to improve its structural characteristics using one or more of the following stabilization techniques: mechanical, chemical, and/or physical.

Standard of Care   In the law of negligence, that degree of care that a reasonable prudent person should exercise under the same or similar circumstances. A person whose conduct falls below such standards may be liable for injuries or damages resulting from the conduct.

Statute   An act of legislature, adopted pursuant to its constitutional authority.

Statute of Limitations   A statute prescribing periods of time within which certain rights must be exercised.

Statute of Repose   The legal time limitation allowed by law for bringing a claim on a specific cause of action within a specified period. Such a period, which varies by state, starts to run from an event specified in the statute, such as the date of substantial completion of a project. A statute of repose differs from the statute of limitations, which runs from when the party bringing the claim knew or reasonably should have known of his or her injury or damages.

Stem Wall   That part of a wall that extends from the top of a foundation footing up to the floor, generally constructed of masonry or concrete.

Strain   The unit deformation caused by the application of a unit of stress. Axial strain is the total deformation divided by the length over which the deformation is distributed. Unit stress divided by unit strain is defined as the modulus of elasticity.

Stress   Intensity of internal force (i.e., force per unit area) exerted by either of two adjacent parts of a body on the other across an imagined plane of separation. When the forces are parallel to the plane, the stress is called **shear stress**; when the forces are normal to the plane, the stress is called **normal stress**; when the normal stress is directed toward the part on which it acts, it is called **compressive stress**; when the normal stress is directed away from the part on which it acts, it is called **tensile stress**.

Strict Liability   The concept of strict liability in tort is founded on the premise that one who engages in an activity that has an inherent risk of injury, such as those classified as ultra-hazardous activities, is liable for injuries proximately caused by such enterprise, even without a showing of negligence.

Subpoena   A command to appear at a certain time and place to give testimony upon a certain matter.

Subpoena *Duces Tecum*   A process by which the court, at the insistence of a party, commands a witness to produce all documents, books, papers, computer discs, and so on pertinent to a pending litigation that are in the witness's possession or control.

Subrogation   The lawful substitution of a third party in place of a party having a claim against another party. Insurance companies, guarantors, and bonding companies generally have the right to step into the shoes of the party whom they compensate and sue any party whom the compensated party could have sued.

Successor Activity That work activity within a critical path schedule that follows a predecessor activity and sustains the like part and character of the work, may or may not be critical, but is required to complete the work.

Suit A generic term, of comprehensive signification, refers to any proceeding by one person or persons against another or others in a court of justice by which an entity pursues such remedy that the law affords.

Summary Judgement A situation in which a judge hearing the case determines that there is no disagreement between the parties as to the applicable material facts. The judge then applies the law to the agreed facts and decides the case; thus, the evidence is never presented to a jury.

Summons A legal paper to be served on a person named as a defendant in a legal action notifying the party to answer the complaint or be found in default. Also used to require nonparty witnesses to appear for depositions or at trial or arbitration hearings.

Supplemental Instruction Instructions to the contractor from either the architect or engineer that clarify, modify, or change initial instructions or specifications.

Surcharge An excessive load or burden that has a resultant effect on the surface upon which it is placed. A methodology in which site soils are preloaded (surcharged) to preconsolidate soft compressible clays that would otherwise allow unacceptable settlement.

T

Testifying Expert Under the rules of civil procedure in any litigation, the lawyers may use testifying experts to explain aspects of the facts to the trier of fact. (See also Consulting Expert and Expert Witness.)

Testimony Evidence given by a witness under oath or affirmation.

Third Party Someone other than the original parties involved in a contract, claim or action, such as a subcontractor to a contractor who has been sued by an owner. If the contractor then sues the subcontractor then the subcontractor is a third party.

Third Party Benificiary Someone who is not a party to a contract but one who has a direct interest in some or all of the terms and conditions of the contract.

Third Party Complaint A complaint filed by a defendant against a third party that alleges that the third party is or may be liable for all or part of the damages that the plaintiff may win from the defendant.

Tort A private or civil wrong or injury, other than breach of contract, resulting from a breach of a legal duty, for which the court will provide a remedy in the form of an action for damages.

Tortfeasor One who commits or is guilty of a tort.

U

Ultimate Bearing Pressure The pressure at which a foundation sinks without increase in the load. In plate-bearing tests, the ultimate bearing pressure is the pressure over all of the plate at which the settlement amounts to one-fifth of the thickness of the plate.

Ultimate Compressive Strength The stress at which a material crushes, and the current way of defining the strength of stone, brick, and concrete.

Ultimate Load The maximum load that may be placed on a structure before it fails due to buckling of common members or failure of some component; the load at which a unit or structure fails.

Ultimate Strength The maximum resistance to load that a member or structure is capable of developing before failure occurs; or, with reference to cross sections of members, the largest moment, axial force, or shear a structural cross section will support.

Underpin To provide new permanent support beneath a foundation without removing the superstructure, in order to increase the load capacity of a preexisting structure, or providing temporary supports to contain an existing load while repairs are being made.

Under Protest The making of a payment or doing of an act under an obligation while reserving the right to object to the obligation at a later date.

Underwriters Laboratories Label A (UL) label affixed to material that shows that the material is regularly tested and complies with the minimum standards as set by an Underwriter's Laboratory specification for safety and quality.

Undisturbed Sample A soil sample of cohesive soil from a test hole that has been changed so little that it can safely be used for laboratory testing.

Unforeseen Site Conditions A contract clause that transfers the risk of an unforeseen site condition not predicted by the contract documents or reasonably expected by the parties onto the owner.

Uniform Commercial Code (UCC) A model statute, dealing with commercial transactions, that has been adopted by every state except Louisiana. UCC provisions do not normally apply to professional services but apply to the sale of goods.

Unilateral Change A change made by one of the parties to contract and under traditional contract law is not allowed. Unilateral changes are allowed in construction contracts as long as the owner pays the contractor a reasonable sum for the change.

Unjust Enrichment A legal equitable concept that prevents a party from a monetary benefit to which he or she is not entitled.

V

Value Engineering An organized creative approach by the architect, engineer, and contractor to apply the scientific method in identifying unnecessary initial and life-cycle items and costs, which provide neither quality, use, life, nor appearance to a project, thus producing an optimum building at a reasonable cost.

Verdict A formal decision or finding made by a jury, impaneled and sworn for the trial of a matter submitted to it, and reported to the court.

Vicarious Liability The imputation of liability upon one person, for the actions of another.

Voir Dire This French law term denotes the preliminary examination that the court may make of one presented as a witness or juror, where one's lack of competency, interest, and so on is objected to.

W

WAG (Slang) "Wild Ass Guess," a term often used when an estimator is attempting to put a figure together quickly without sufficient information.

Waiver The intentional and voluntary relinquishment or surrender of a known right.

Waiver of Subrogation The relinquishment by an insured of the right of its insurance carrier to collect damages paid on behalf of the policyholder.

Warranty A statement made that a fact is as stated or promised, which may be based on an explicit representation or inferred by law, depending on the circumstances.

Water Cement Ratio The ratio of the amount of water, exclusive only of that absorbed by the aggregates, to the amount of cement in a concrete or mortar mixture. The lower the water/cement ratio, the stronger the cement.

Waterproof Made secure against the flow or permeability of water by the use of materials or admixtures that repel and/or are impervious to water or dampness.

Weep Screed A leveling "L"-shaped strip of material, usually made of galvanized metal or vinyl, with a series of holes in the bottom leg. When properly installed at the bottom of a vertical wall surface that will be covered with an exterior finish system, it will provide a level surface guide for the finish material being applied, as well as a means to evacuate water that may enter the exterior finish.

Willful And Wanton Commonly referred to as an act knowingly and intentionally committed that is not only negligent but exhibits conscious disregard for the safety of others.

Wind Pressure The pressure, measured in pounds per square foot (psf), of wall or roof area, due to wind pressure that increases with wind velocity. On sloping roofs, low pressure suction occurs on the lee side, which may at times be nearly as much as the pressure on the windward side.

Witness One who gives evidence in a cause before a court or hearing, and who attests or swears to the facts or bears testimony under oath.

Index

A

ABC (aggregate base course), 112
above grade floor systems, 141–144
   slab on-grade floors, 143–144
adding water to concrete, 49
additional insured, 20–21
ADINA, 134
adjusting framing hangers, 50
affidavit, 162
AGC. *See* Associated General Contractors of America (AGC)
aggregate base (AB), 53
agreements. *See* contracts
AIA. *See* AIA Contract Documents; American Institute of Architects (AIA)
AIA Contract Documents, 138, 144. *See also* General Conditions of the Contract for Construction (AIA A201-2007/A202-1997)
   A-Series: Owner/Contractor Agreements
      A101-2007: Standard Form of Agreement Between Owner and Contractor where the Basis of Payment is a Stipulated Sum, 14–16
      A107-2007: Standard Form of Agreement Between Owner and Contractor for a Project of Limited Scope, 19
      A111-1997: Standard Form of Agreement Between Owner and Contractor where the basis of payment is the Cost of the Work Plus a Fee with a Guaranteed Maximum Price, 19
      A114-2001: Standard Form of Agreement Between Owner and Contractor where the basis of payment is the Cost of the Work Plus a Fee without a Guaranteed Maximum Price, 19
      A401-2007: Standard Form of Agreement Between Contractor and Subcontractor, 13, 19
   B-Series: Owner/Architect Agreements
      B101-2007: Standard Form of Agreement Between Owner and Architect without a Predefined Scope of Architects Services, 19
      B102-2007: Standard Form of Agreement Between Owner and Architect without a Predefined Scope of Architects Services, 19
      B141-1997: Standard Form of Agreement Between Owner and Architect without a Predefined Scope of Architects Services/Standard Form of Architects Services: Design and Construction Contract Administration, 14–17, 56, 58
      B151-1997: Standard Form of Agreement Between Owner and Architect, 19, 58
      B201-2007: Standard Form of Architects Services: Design and Construction Contract Administration, 19
   Comparatives and Comparisons chart for, 18–19
   C-Series: Architect-Consultant Documents
      C141-1997: Standard Form of Agreement Between Architect and Consultant, 58
   matrix of, understanding, 14–18
   numbering system for, 18–19
alkaline, level of, 52
allowable stress design (ASD), 78
all-weather access road, 103
Alpine Industries, Inc. vs. Gohl (1982), 86

alternate dispute resolution (ADR), 14, 26–27
American Arbitration Association (AAA), 27
American Architectural Manufacturers Association, 64
American Association of State Highway and Transportation Officials (AASHTO), 30, 94
American College of Forensic Examiners International (ACFEI), 3, 5, 131
American Concrete Institute (ACI), 48, 84, 144
American Institute of Architects (AIA), 3, 84. *See also* AIA Contract Documents
American Institute of Chemical Engineers (AIChE), 3
American Institute of Mining, Metallurgical and Petroleum Engineers (AIME), 3
American Institute of Steel Construction (AISC), 84
American Land Title Association (ALTA) survey, 10
American National Standards Institute (ANSI), 83
American Society for Testing and Materials (ASTM), 83
   ASTM-C Series
      C31: Standard Practice for Making and Curing Concrete Test Specimens in the Field, 137–138
      C42: Standard Test Method for Obtaining and Testing Drilled Cores and Sawed Beams of Concrete, 138
      C94: Standard Specifications for Ready-Mixed Concrete, 49
      C1019: Standard Test Method for Sampling and Testing Grout, 138
   ASTM-D Series
      D1188: Bulk Specific Gravity of Compacted Bituminous Mixtures Using Coated Samples, 145
      D1856: Recovery of Asphalt from Solution by Abson Method and ASTM D 5404 Recovery of Asphalt from Solution Using the Rotavapor Apparatus, 146
      D2041: Theoretical Maximum Specific Gravity and Density of Bituminous Paving Mixture, 145–146
      D2172: Quantitative Extraction of Bitumen from Bituminous Paving Mixtures, 146
      D2726: Bulk Specific Gravity and Density of Nonabsorptive Compacted Bituminous Mixtures, 145
      D6307: Asphalt Content of Hot-Mix Asphalt by Ignition Method, 145
      D6926: Standard Practice for Preparation of Bituminous Specimens Using Marshall Apparatus, 145
American Society of Home Inspectors (ASHI), 42
American Society of Mechanical Engineers (ASME), 3
American Society of Testing Materials (ASTM), 61, 93, 94
American Subcontractor Association (ASA), 6
Americans with Disabilities Act (ADA), 30, 60
Anderson, Arthur, 1
Aniero Concrete Company, Inc vs. New York City School Construction Authority (2002), 161
ANSYS, 134
approvals of work, allowing others to control, 87
arbitration, 27
   alternate dispute resolution, as type of, 27
   arbitrator, as dispute-resolver, 27
   consolidation of, 21–22
   expenses of, 27
   litigation *vs.*, 19, 21
   risks of, 27
arbitrator, as dispute-resolver, 27

architect. *See also* architectural issues & construction defects
  commissioned, 56
  contract, role in, 12
  contractual privity, 56
  "don't list" for, 67–68
  ethics of, 4
  General Conditions of the Contract for Construction (AIA A201-2007/A202-1997), 20
  lawsuits against, 56
  purpose of, 4
  role and responsibilities of, 55–58
  "to do list" for, 65–67
architectural drawings, knowledge of specified products on contract documents for, 60
architectural general notes of contract documents, 62
architectural issues & construction defects, 55–68. *See also* case examples of architect or engineers failure to maintain care, custody, and control of work
  code references, 63
  communication and coordination of contract documents, professional, 57–58
  contract drawings, clear and concise, 63–65
  cover sheets or title sheets of contract documents, 62
  early problems with of contract documents, 56–57
  environmental durability and maintenance of project site, 58–60
  general notes of contract documents, 62–63
  knowledge of specified products on of contract documents, lack of, 60–61
  specifications of contract documents, 61–62
architecture, 3–4. *See also* architect; architectural issues & construction defects
Arizona Independent Redistricting Commission v. Kenneth W. Fields, Judge of the Superior Court (2003), 162
Arizona State Civil Procedure Rules, 159
Arizona Revised Statutes (ARS), 23
as-built drawings, 91
asphalt concrete (AC), 53
asphalt pavement, 53
asphalt pavement mixtures, 144–146
ASPR (Armed Services Procurement Regulations), 95
Associated General Contractors of America (AGC), 3, 6, 11, 56
  AGC 200: Standard Form of Agreement and General Conditions Between Owner and Contractor, 11
  AGC 260: Performance Bond, 11
  AGC 655: Standard Form of Agreement Between Contractor and Subcontractor, 11
  industry advancement and opportunity, commitment to, 6
Association of Soils and Foundation Engineers (ASFE), 131
attorney's fees and interest for project, provisions to, 14
Automobile Liability insurance, 25

B
backfill behind retaining walls, 47
bench trial, 26
bent framing hangers, 50
biased view of expert witness, 159
bid bond, 22

Board of Ethical Conduct (NSPE), 4
Board of Supervisors, role of, 33
Board of Technical Registration, 31, 40
Boards of Licensure, 3
Boards of Registration, 3
"boilerplate" of contract documents, 61
borrowed specifications of contract documents, 62
BOSOR, 134
brittle floor coverings, cracks in, 49
broad form of indemnity, 23
Bruce v. Byrne-Stevens & Associates Engineers, Inc. (1989), 165
budget for project, 56
Builders Risk insurance, 25
Building Officials and Code Administrators International, Inc. (BOCA), 29, 31, 32, 33, 95
burden of proof, 7
Bureau of Yards and Docks, responsibility of construction work, 93
burial grounds, 20

C
CAD (computer aided design) drawings, 76
calcium hydroxide, 52
California, state construction codes of, 95
California Association of Window Manufacturers, 64
California Real Estate Inspection Association (CREIA), 42
camera kit, 156–157
canned specifications, 59
canned specifications of contract documents, 61
careless expert witness, 162–165
case examples of architect, engineers and contractors failure to maintain care, custody, and control of work, 98–128
  compaction of fill portion, 112–113
  concrete sports stadium, repairs to, 104–105
  condominium complex fire, 102–103
  contractors, 124–126, 126–128
  CPVC fire sprinkler piping, crack in, 118–119
  dormitory fire, 99–101
  expansive clays, 119–122
  landslides, 122–124
  local building officials, incompetence of, 109–112
  parking garage deferred maintenance, 103–104
  pipefitter welding fire, 98–99
  safety harness accident, 107–108
  sewerage treatment plants, construction of, 117–118
  sewer line trench, excavation of, 112
  shoring, use of engineer for, 105–106
  split responsibilities on construction project, consequences of, 113–114
  trenching, safety procedures for, 108–109
  underground storage tanks, compaction around, 114–117
  utility tunnel accident, 106–107
  wood bow string trusses in United States, use of, 105
  wood trusses, bracing of, 106
caulked joints, 59
causation, 26
caveats, testing, 148
certified home inspector, 42
Certified Public Accountant (CPA), ethical standards of, 1
chain of evidence, 135

Index

change order clause of contracts, 9
charge coupled device (CCD), 156
China Grove, Tennessee, 24
Chinle Clays, 119
claim restrictions in contract, provisions to, 14
claims made policies, 25
clay soils, 46, 47
Clean Air Act of 1970, 29
Clean Water Act of 1977, 30
client information, 131
clients, 57, 74
Coatesville Contractors & Engineers vs. Borough of Ridley Park (1986), 93
code noncompliance in plans, contractors obligation to catch, 20
code of conduct, for forensic investigator (ACFEI), 5
code of ethics, 4, 6
Code of Ethics and Professional Conduct (AIA), 3
code references, 63
Commercial General Liability Insurance (CGL), 24–26, 128
    causation, 26
    claims made policies, 25
    indemnity, 24
    occurrence policy, 25–26
    subrogation, 25
    types of, 25–26
commissioned architect, 56
common law indemnity, 24
communication and coordination
    architect responsibilities, 57–58
    client responsibilities, 57
    clients expectations, 74
    of contract documents, professional, 57–58
    design concepts, 73
    with others, lack of, 73–74
    sub-consultants responsibilities, 58
Communities Facilities Administration, 29
compaction, 46, 112–113
company philosophy (mission statement of company), 69–70
comparative form of indemnity, 23–24
Comparatives and Comparisons chart for AIA Contract Documents, 18–19
competence, 74
Competent Person Training, 109
complaint procedures, management of, 33, 40
compound cracking, 32
computer generated analysis, 153
computers, 76–77
concrete, 137–138
    adding water to, 49
    ASTM C31, 137–138
    ASTM C42, 138
    brittle floor coverings, cracks in, 49
    concrete slab on-grade floor, 47–48
    gypsum, 99
    myths & facts about, 47–49
    post-tensioned concrete slabs, 47–48
    spalling and deterioration of, 48–49
    in truck, leaving for extended periods of time, 49

concrete in truck, leaving for extended periods of time, 49
concrete slab on-grade, 47, 48
concrete sports stadium repairs, case example, 104–105
concurrent condition, 10
condition precedent, 10
condition subsequent, 10
condominium complex fire, case example, 102–103
conduct, AIA's principles of, 3
conformance and performance laboratory testing, 135
Consensus Documents of General Conditions of the Contract for Construction (AIA A201-2007/A202-1997), 22
consequential damages waiver, clarification of, 21
constructing of models and conducting performance testing, 135
construction claims, 92–93
construction contract, 11–13
    architect and engineer's role in, 12
    contractor's role in, 12–13
    cost plus a fixed fee with a guaranteed maximum, 11
    cost pus an incentive fee with a guaranteed maximum price, 11
    critical path scheduling, 11
    fixed price, 11
    general, 11
    owner or the owners representative, 12–14
    provisions to, 11–14
    stipulated sum, 11
construction defect litigation's, 42
construction defects, 45, 46. *See also* architectural issues & construction defects; engineering and construction defects
construction documents, 78
construction drawings, 77–78
Construction Engineering Research Laboratory, 30
construction field tests, 146–148
Construction Industry Standards, 78
construction issues, 81–97
    contractor, expectations of, 81
    documentation, lack of insufficient, 91–93
    field supervision, 81–86, 86–89
    job harmony, factors affecting, 89–91
    laws, federal, state and municipal, 95–96
    quality assurance/quality control, 93–95
    tradespeople, lack of skilled, 96
construction manager at risk (CMR), 56
construction phase of project, 79–80
construction services, establishment of fees for, 71–72
Construction Specifications Institute (CSI), 61
contract agreement forms. *See* AIA Contract Documents
Contract Change Directives (CCDs), 20
contract documents, 61, 63
contract drawings
    details of, clear and concise, 63–65
    general notes of, 62–63
    roofing and flashing details, 64
    waterproofing details, 64
    window and flashing details, 64–65
contract liability, 7
contractor, 6–7
    code noncompliance in plans, obligation of, 20
    code of ethics for, 6

contractor *(Continued)*
   contract, role in, 12
   equipment for, financing statement on, 22
   expectations of, 81
   general, timely paying to subs, 20
   licenses and requirements of, classifications of, 32
   mistakes of, covering up, 88–89
   performance standards of, minimum, 13
   qualifications of, case example, 124–126
   responsibilities of, 12–13
   selection of, case example, 126–128
   superintendent, proposed by, 20
Contractors State License Board, 32–33
contracts, 9–27
   of AIA, 14–22
   alternate dispute resolution, 26–27
   change order clause of, 9
   construction, 11–13
   contractor, role in, 12
   cost plus, 14
   cost plus a fixed fee with a guaranteed maximum, 11
   cost pus an incentive fee with a guaranteed maximum price, 11
   Covenant of Good Faith and Fair Dealing in, 93
   design, 10–11
   design-build, 56
   duties inferable from/consistent with, 19
   fixed price, 11
   general provisions to, 11, 12
   indemnity, 23–24
   insurance, 24–26
   No Damages for Delay Clause of, 92, 93
   subcontract agreements, 13
   surety bonds, 22–23
   terms and conditions of, 9–10
contract terms and conditions, 9–10
contractual privity, 56
control cost, 55
coordination of contract documents, 78. *See also* communication and coordination
coordination of work activities, inability for, 84–85
copper water piping, 102
corporate ethics, defined, 2
Corps of Engineers, 30, 93
correlation of contract, general provisions to, 12
correspondence files, 91
cost plus a fixed fee with a guaranteed maximum contract, 11
cost plus contracts, 14
cost pus an incentive fee with a guaranteed maximum price contract, 11
costs of project, provisions to, 14
county ordinances, 33–35
   Board of Supervisors, role of, 33
   on-site inspections, 34
Covenant of Good Faith and Fair Dealing in contracts, 93
covenants, conditions, and restrictions (CCRs), 42, 72
cover sheets of contract documents, 62
CPVC fire sprinkler piping crack, case example, 118–119
cracks/cracking
   in brittle floor coverings, 49
   of concrete slab on-grade, 48
   in masonry walls, 51
   of post-tensioned concrete slabs, 47–48
   in stucco, 51–52
criminal act and tort, distinguishing between, 7
critical path scheduling, 11
curriculum vitae for expert witness, 161

D

daily logs, 91, 92
daily progress reports, 91, 92
dampproofing materials, 54
Daubert Tracker, 168
Daubert vs. Merrell Dow Pharmaceuticals, Inc. (1993), 166, 167
Davis-Bacon Act of 1931, 29
dead load, 59
deductibles in insurance, 25
deficient thickness, 1 inch of, 53
delay damage in contract, provisions to, 14
delays in construction work, 92–93
Demand for Arbitration, 27
demeanor for expert witness, 161
dense soil, 47
density testing, 46–47
Department of Insurance, 43
design assignment, 74–75
design-build contracts, 56
design calculations for construction drawings, 77–78
design concepts, 73
design contracts, 10–11
design information about project, lack of adequate, 72–73
design professionals, in weekly progress meetings, 90
design services, establishment of fees for, 71–72
destructive testing, 135
digital camera, 156
digital recording photo equipment, 156
disasters, 44
discounts for project, provisions to, 14
dispute-resolver, arbitrator as, 27
documentation, lack of insufficient, 91–93
   daily progress reports, incomplete or lack of, 92
   documentation and construction claims, 92–93
   requests for information (RFIs), 92
   weekly progress meetings, no, 92
documentation claims, 92–93
document processing, work flow and, 134
documents, project-related, 133
Donnelly Construction vs. Oberg Hunt & Gilliland (1984), 56
"don't list"
   for architect, 67–68
   for engineer, 80
dormitory fire, case example, 99–101
drainage collection system, 46, 47
drawings, general provisions to, 12

E

elastomeric coating over one-coat stucco, use of, 52
electrical trench, 108
electronic data, instruments of service and, 19
electron tunneling microscope, 148

ell metal flashing, 53
energy dispersive spectrometer (EDS), 149
engineer. *See also* engineering and construction defects
   code of ethics for, by NSPE, 4
   construction phase, role of, 79
   contract, role of, 12
   "don't list" for, 80
   Fundamental Canons for, 4
   judgment of, lack of, 78
   post-construction and, 79–80
   role and responsibilities of, 5, 69–70
   State Board of Technical Registration, defined by, 31–32
   "to do list" for, 80
engineering, ethics in, 4–5. *See also* engineer; engineering and construction defects
engineering and construction defects, 69–80. *See also* case examples of architect or engineers failure to maintain care, custody, and control of work
   communication and coordination with others, lack of, 73–74
   company philosophy (mission statement of company), 69–70
   computers, 76–77
   construction documents, 78
   construction drawings, 77–78
   construction phase of project, 79–80
   design assignment, 74–75
   design information, lack of adequate, 72–73
   issues promoting, 75–78
   load and forces affecting structure, failure to properly determine, 75–76
   project information, 70–72
Engineering Founder Societies include the American Society of Civil Engineers (ASCE), 3
engineering notes, 63
Engineers Council for Professional Development, 2
Engineers Joint Contract Documents Committee (EJCDC), 11
English translation, 96
Enron, 1
environmental durability and maintenance at project site, effecting project performance, 58–60
   human conditions, 60
   maintenance manuals, 59
   natural conditions, 59
   performance expectations, proposed project, 59
   physical conditions, 59–60
Environmental Protection Agency (EPA), establishment of, 30
ethical behavior, defined, 1
ethics, corporate, 2. *See also* code of ethics; professional ethics
Ethics and the Practice of Architecture (publication), 4
execution of contract, general provisions to, 12
expansion joints and weep screeds, 51–52
expansive clays, case example, 119–122
experience in scheduling, lack of, 85
expert witness, 158–169. *See also* forensic investigator
   biased view of, 159
   careless, 162–165
   curriculum vitae for, 161
   defined, 158
   demeanor for, 161
   hurried, 162
   importance of, 158
   independent, 162
   negligent, 165
   non-testifying consulting, 161–162
   opinion of, preventing/exclusion of, 166–168
   references for, 161
   role of, 158, 159–160
   selection of, 161
   specialized, 161
   testifying by, expanding risks of, 165–168
exterior insulation finish system (EIFS), 54, 139

F

failure analysis, 134–135
failure to complete schedule, 85–86
FAR (Federal Accounting Regulations), 95
Federal-Aid Highway Act of 1952, 29, 30
Federal and State Evidence Rule 701, 5
Federal Arbitration Act (FAA), 21
Federal Civil Procedure Rule 26, 159, 160, 161
Federal Highway Administration (FHWA), 30, 93
Federal Housing Act of 1959, 29
Federal Housing Administration/Authority (FHA), 29, 33, 93
federal laws/legislations, 29–31
   construction issues and, 95–96
Federal National Mortgage Association, 29
federal or Native American projects, regulations for, 30–31
Federal Rules
   104(a), 167
   of Civil Procedure, 27, 160
   defined, 159
   of Evidence, 166
   requirements of, 159
fees
   for design and construction services, establishment of, 71–72
   for project, 132–133
field inspector, role of, 32–33
field supervision
   approvals of work, allowing others to control or, 87
   coordination of work activities, inability to, 84–85
   inspections and testing, 87–88, 90–91
   knowledge of work, insufficient, 82–83
   mistakes of contractor, covering up, 88–89
   poor, fallout from, 86–89
   professional assistance, failure to seek, 89
   quality standards, knowledge of, 83–84
   schedule/scheduling, 85–86
   specifications, inability to follow plans and, 86–87
   supervision, inexperienced, 83
   uneducated, unaware, or incapable, 81–86
   value engineering, lack of understanding, 84
field-testing procedures, evaluation of, 136–148
   above grade floor systems, 141–144
   asphalt pavement mixtures, 144–146
   caveats, testing, 148
   concrete, 137–138

field-testing procedures *(Continued)*
   construction field tests, 146–148
   exterior insulation finish system (EIFS), 139
   masonry, 138
   soils, 136–137
   structural investigations, 141
   stucco, 138–139
   windows & doors, 139–141
fill zones, dumped or spread, 46
Fireman's Fund, 22
fire resistance testing, 153–155
fire-stop, 99
fire-watch policy, 102
5% of construction documents, failure to complete last, 78
fixed price contract, 11
flashing, 64–65
floor coverings, 45
Florida, state construction codes of, 95
flow down clause of General Conditions of the Contract for Construction (AIA A201-2007/A202-1997), 13
footing depth, 47
forensic analysis, 134–135
forensic investigations. *See also* forensic investigator
   attributes of, 130
   challenges of, 130
   computers, limitation sof, 76–77
   in construction industry, 130
   ethics in, 5
   types of, 130
forensic investigator, 5, 130–131
forensic photography, 155–157
   camera kit, 156–157
   charge coupled device (CCD), 156
   digital camera, 156
   digital recording photo equipment, 156
   dimensions of, 155
   documentation and logging of, 155
   JPEG format, 156
   pixels, 156
   TIFF format, 156
   treatment processes for, 156
   in trial, use of, 157
forensic testing, special considerations in, 135–136
foundation, 47
Foundation of Real Estate Appraisers (FREA), 42
four-ply built-up roofing system, 54
framing, myths & facts about, 49–50
framing hangers, 49–50
fraud, 93
Frye General Acceptance Test, 166, 167
Frye v. United States (1923), 166
full-length standing seam metal, 53
Fundamental Canons for engineer, 4

G

galvanized steel connection plates, 49
general building contractors, role of, 32
General Conditions of the Contract for Construction (AIA A201-2007/A202-1997), 11, 12, 14–22
   Article 1: General Provisions, 19
   Article 2: Owner, 19–20
   Article 3: Contractor, 20
   Article 4: Administration of the Contract, 20
   Article 5: Subcontractors, 20
   Article 6: Construction by Owner or by Separate Contractors, 20
   Article 7: Changes in the Work, 20
   Article 8: Time, 20
   Article 9: Payments and Completion, 20
   Article 10: Protection of Persons and Property, 20
   Article 11: Insurance and Bonds, 20–21
   Article 12: Uncovering and Correction of Work, 21
   Article 13: Miscellaneous Provisions, 21
   Article 14: Termination or Suspension of the Contract, 21
   Article 15: Claims and Disputes, 21–22
   changes to, in 2007, 18–22
   Consensus Documents of, 22
   flow down clause of, 13
   indemnification, 23, 24
   purpose of, 13
   subrogation, 25
general construction contract, 11
General Electric v. Joiner (1997), 166
general engineering contractor, role of, 32
general indemnity, 23
general myths & facts about field of construction, 54
general notes of contract documents, 62–63
general provisions to contract, 11–12
General Requirements of OSHA, 100
general specifications of contract documents, 61
General Survey Act, 30
"GIGO": "Garbage in—Garbage Out," 76
government and public issues, 28–44
   county ordinances, 33–35
   federal laws/legislations, 29–31
   government inspectors, role of, 40
   home inspector, role of, 42
   home owners, and home owners associations (HOAs), 42
   indemnification, 43
   local government/municipal law, 36–39
   public adjuster *vs.* insurance adjuster, 43
   public perception, 28–29
   state laws/statutes, 31–33
   statements to news media, poor, biased & misleading, 43–44
   statute of limitations, 41–42
   types of, common, 28
   warranty issues *vs.* normal maintenance, 40–41
government inspectors, 40
gross negligence, 7
grout, 138
grout joint failure and voids, 50–51
guarantor to the owner (obligee), surety bonds as, 22
gypsum (calcium sulfate), 49
gypsum concrete, 99
gypsum wall, 98–99

H

"hands-on" field experience, lack of, 84
Hawley vs. Orange County Flood Control District (1963), 93
Hayden Business Center Condominiums Assoc. vs. Pegasus Development Corp. (2005), 41

heating, ventilating and air conditioning (HVAC), 58
heaving, as indication of cracks in stucco, 52
Holiday Inns, Inc. vs. Rice (1990), 166
home inspection organizations, types of, 42
home inspector, 42, 45
Home Loan Bank Board, 29
home owners, 42. *See also* home owners associations (HOAs)
home owners associations (HOAs), 42
hot-mix asphalt (HMA), 144–145. *See also* asphalt pavement mixtures
Hot Program by OSHA, 100
Housing and Urban Development (HUD), 29, 30, 93
human conditions effecting project performance, 60
hurried expert witness, 162
Hydraulics Laboratory, 30

I

imagined construction defect, 45
implied conditions, 10
implied warranty, 40–41
incapable field supervision, 81–86
indemnitee, role of, 23
indemnitor, role of, 23
indemnity/indemnification, 23–24, 43
  broad form of, 23
  in Commercial General Liability Insurance, 24
  common law, 24
  comparative form of, 23–24
  creation of, 23
  general, 23
  General Conditions of the Contract for Construction (AIA A201-2007/A202-1997), 23, 24
  indemnitee, role of, 23
  indemnitor, role of, 23
  intermediate form of, 23
  provisions to, 14
  as risk-shifting device, 23
  state statutes, limitations of, 43
  types of, 23–24
independent expert witness, 162
industry advancement and opportunity, AGC's commitment to, 6
information, client and team player, 131
initial decision maker (IDM), 19, 20, 21
inspections and testing
  requirements for, cheats on, 87–88
  value of, failure to understand, 90–91
Institute of Electrical and Electronic Engineers (IEEE), 3
instruments of service, and electronic data, 19
insufficient specifications details, of contract documents, 62
insurance, 24–26
  Automobile Liability, 25
  Builders Risk, 25
  Commercial General Liability Insurance (CGL), 24–26
  deductibles, 25
  loop holes, 25
  policy limits, 25
  Professional General Liability Insurance (PGL), 25
  Professional & Omissions Insurance (E&O), 25
  provisions to, 14
  self-insured motorist coverage, 25
  subrogation, 25
  types of, 24–25
insurance adjuster, public adjuster *vs.*, 43
insurance certifications, renewals of, 20
intentional wrongs, 7–8
intent of contract, general provisions to, 12
intermediate form of indemnity, 23
International Building Code (IBC), 34, 54, 95
  federal laws/legislations, development of, 29
  for local government/municipal law, 36–37
  special inspections, requirements for, 40
  state laws/statutes, 31
  sulfates, levels of, 49
International Code Council (ICC), 29, 31, 32, 33
International Commercial Building Code, 54
International Conference of Building Officials (ICBO), 29, 31, 32, 33, 83–84
International Residential Building Code, 54
intrusive testing, 134–135
investigations, process for, 132
investigative services. *See* specialized and investigative services

J

Jackall, Robert, 2
Jenkins vs. Bell Helicopter (2002), 166
job progress photos, 91
joint checks, when subs are unpaid, 20
Joseph, Gregory, 162, 163
JPEG format, 156

K

Kansas City Hyatt Hotel, 25, 130
key provisions to contract, 13–14
KISS (Keep It Simple Stupid) approach, 134
knowledge of specified products on of contract documents, lack of, 60–61
knowledge of work, insufficient, 82–85
Kubby vs. Cresent Steel Co. (1970), 41

L

laboratory testing, 148–154
  computer generated analysis, 153
  electron tunneling microscope, 148
  fire resistance testing, 153–155
  microanalysis, types of, 150–153
  microscopic examination, 148–150
  optical microscopes, 149
  scanning electron microscope (SEM), 148
  scanning probe microscope, 148
  types of, 153–154
landscape by foundation, watering of, 47
landslides, case example, 122–124
latent defects, statute of limitations on, 41–42
lawsuits against architect, 56
layperson. *See* soils
legal actions, 7–8
legal case documents, 133
legislation. *See also* federal laws/legislations
lenses for digital camera, 156

leveling of concrete slab on-grade, 48
Levine vs. Wiss & Co. (1984), 165
Lexis-Nexis, 168
Liberty Mutual, 22
license bonds, 22
light-gauge steel framing hangers, 50
limitations of computers, failure to understand, 76–77
limited waiver of consequential damages, 22
liquidated damages, provisions to, 14
listed code references, 63
litigation, arbitration *vs.*, 21
Little Miller Act Bond, 22
live load, 59
load and forces affecting structure, failure to properly determine, 75–76
load and resistance factor design (LRFD), 78
local building officials incompetence, case example, 109–112
local government/municipal law, 36–39
    International Building Code (IBC), 36–37
    local municipal building departments, responsibilities of, 36
    structural inspections, 38–39
local municipal building departments, responsibilities of, 36
log book for RFI, 82
Logerquist vs. Danforth (2000), 167
loop holes in insurance, 25

M

maintenance at project site. *See* environmental durability and maintenance at project site, effecting project performance
maintenance manuals, 59
mandatory organizations. *See* regulatory organizations
Manual of Acceptable Practices, 33–34
Mardian Construction vs. Pioneer Roofing, 1981, 23
masonry, 138
    ASTM C1019, 138
    cracks, 51
    grout joint failure and voids, 50–51
    myths & facts about, 50–51
    walls, cracks in, 51
master builder. *See* architect
master specifications of contract documents, 62
Mattco Forge Inc. vs. Arthur Young & Co. (1992), 165
McGuire & Hester vs. City & County of San Francisco (1952), 93
measles case, 150–151
mediation, 26–27
    alternate dispute resolution, as type of, 26–27
    provisions to, 14
memorable design, 56
Mennonites, 24
Merchants Mutual, 22
Michigan, state construction codes of, 95
microanalysis, examples of, 150–153
    measles case, 150–151
    painted rocks case, 153
    painter case, 151
    physical evidence case, 152–153
microanalysis, types of, 150–153
microscopic examination, 148–150
Miller Act Bond, 22
minimum acceptable standards, 95
Minimum Property Standards, 34

minimum workmanship standards, 33
misrepresentation, 93
mission statement of company (company philosophy), 69–70
mistakes of contractor, covering up, 88–89
mix proportions of stucco, 51
moisture, 46, 48
Moisture Testing Guide for Wood Frame Construction Clad with Exterior Insulation and Finish Systems (New Hanover Inspection Department in Wilmington, NC.), 139
mold issues, 43
monitored schedule, unrealistic and improperly prepared, 85
montmorillonite, 48
moral and forward-looking responsibility, 5
Moral Mazes (Jackall), 2
Mountain Preserve, 1
movement of post-tensioned concrete slabs, prevention of, 48
Multiple Dwelling Law, 29
municipal laws, construction issues and, 95–96
Murdock-Bryant Construction, Inc. vs. Pearson (1984), 93
Murphy vs. A.A. Mathews (1992), 165
myths & facts in field of construction, 45–54
    concrete, 47–49
    construction defect, real *vs.* imagined, 45
    construction defect lawsuits, 46
    floor coverings, 45
    framing, 49–50
    general, 54
    home inspector, 45
    layperson (*See* soils)
    masonry, 50–51
    panelized masonry block yard fences, 51
    public code inspectors, 45
    roads and paving, 53
    roofing, 52–53
    site preparation, 45
    soil, 45, 46–47
    stucco, 51–52
    water flow at site, 45

N

NASTRAN, 134
National Academy of Forensic Engineers (NAFE), 3, 131
National Aeronautics and Space Administration (NASA), 93
National Association of Home Inspectors (NAHI), 42
National Association of Public Insurance Adjusters (NAPIA), 43
National Electrical Code (NEC), 95
National Environmental Policy Act of 1969, 29
National Fire Code (NFC), 95
National Fire Protection Association (NFPA), 84
National Roofing Contractors Association (NRCA), 84
National Society of Professional Engineers (NSPE), 3, 11, 131
    Board of Ethical Conduct, code of ethics for engineers, 4
Native American or federal projects, regulations for, 30–31
natural conditions effecting project performance, 59
natural site conditions, 59
negligence, 7
negligent expert witness, 165
Nelson v. Trinity Medical Center (1988), 167
New Hanover Inspection Department in Wilmington, NC., 139
New Jersey, state construction codes of, 95
New Pueblo Construction vs. State of Arizona, (1985), 85

New York, state construction codes of, 95
No Damages for Delay Clause of contracts, 92, 93
non-testifying consulting expert witness, 161–162
normal maintenance, warranty issues *vs.*, 40–41
North Dakota Supreme Court, 167
Northridge, California, earthquake, 130
Notice of Contractor Default, 20
nuclear gauges, 95
numbering system for AIA Contract Documents, 18–19

O

obligatory responsibility, 5
Occupational Safety and Health Act of 1970, 29
Occupational Safety & Health Administration (OSHA)
    General Requirements of, 100
        Hot Program by, 100
        manufacturing and construction standards of, 29, 93
occurrence policy, 25–26
Office of Real Estate Appraisers (OREA), 42
official responsibility, 5
on-call density testing, 47
one-coat stucco system, 51–52, 99
on-site inspections, 34
open plumbing trenches, 103
opinion of expert witness, preventing/exclusion of, 166–168
optical microscopes, 149–150
Oriented Strand Board (OSB), 99
original documents, project-related, 133
over reliance on computers, 76
owner interference, 92–93
owner or the owners representative, 12
    contract, review of, 13–14
    defined, 12
    responsibilities of, 12
    rights of, 12

P

painted rocks case, 153
painter case, 151
panelized masonry block yard fences, 51
parking garage deferred maintenance, case example, 103–104
pavement, placement of seal coat on, 53
performance bonds, 11, 23
performance expectations, proposed project, 59
performance standards of contractor, minimum, 13
personal guarantee of surety bonds, 22
petrographic microscope, 149
physical conditions effecting project performance, 59–60
physical evidence case, 152–153
piers and grade beams, 46
pipefitter welding fire, case example, 98–99
pixels, 156
plumbing trench, 108
Plymouth Colony, 29
policy limits of insurance, 25
Pollution Prevention Act of 1990, 30
Portland Cement Association, 37, 51, 94, 95, 138, 139
post-tensioned concrete slabs, 47–48
pre-construction meetings, 90

pre-investigation considerations, 131–133
    client and team player information, 131
    investigations, process for, 132
    project fees, 132–133
preliminary report by forensic investigator, 130–131
privity, 56
product control. *See* quality control
professional, defined, 1
professional assistance, failure to seek, 89
professional ethics, 2–3
    architecture, 3–4
    contractors, 6–7
    engineering, 4–5
    forensic investigation, 5
    legal actions, 7–8
    professional organizations, expectations and responsibilities of, 3
professional expectations. *See* professional ethics
Professional General Liability Insurance (PGL), 25
professional malpractice, 56
Professional & Omissions Insurance (E&O), 25
professional organizations, 3
professional responsibility, 5
professional societies, 3
project
    fees for, 132–133
    fees for design and construction, establishment of, 71–72
    general provisions to, 12
    information about, 70–72
    initial, insufficient, 70–72
    responding to the request for proposal (RFP), 70
    scope of services, preparing, 70–71
project manual, 12, 63
project-related documents, 133
project schedule file, 91
proper scheduling, 85–86
protracted delays, 92
public adjuster, 43
public code inspectors, 45
public issues. *See* government and public issues
public perception, 28–29
public policy, for reasons of strict liability, 8
Public Works Administration, 29

Q

quality assurance testing, 94–95
quality control, 75
Quality Control–Quality Assurance programs, 30
quality control testing, 94
quality standards, knowledge of, 83–84

R

Ralph D. Nelson Co., Inc. vs. Bell (1983), 86
Rayes, Douglas, 167
real construction defect, 45
recovery schedule, 86
references for expert witness, 161
Registrar of Contractors, 32–33, 40
regulatory organizations, 3

reinforced waffle slab on-grade, 46
Residential Funding Corp. vs. DeGeorge Financial Corp. (2002), 159
responding to the request for proposal (RFP), 70
responsibility, 5
retaining walls, 47
RFI (request for information), 25, 62, 64, 82, 92
Richards vs. Powercraft Homes, Inc. (1984), 41
risk management, 56–57
risk-shifting device, indemnity as, 23
roads and paving, myths & facts about, 53
roofing
    ell metal flashing, 53
    flashing and , details for, 64
    full-length standing seam metal, 53
    myths & facts about, 52–53
    tile roofs, 52–53
roofing felts, 52–53
Rural No-Service Zone, 34

S

Safe Drinking Water Act of 1974, 30
safety harness accident, case example, 107–108
salt crystallization, 37, 49
Saltillo tile, 49
sampling techniques of soil, 136
sand content of stucco, 51
scanning electron microscope (SEM), 148, 149–150
scanning probe microscope, 148
schedule/scheduling
    for design information about project, requirements, 72
    experience in, lack of, 85
    monitored, unrealistic and improperly prepared, 85
    project, requirements for provisions to, 13
    project and work task, 133
    recovery, 86
    update, failure to complete or properly, 85–86
scope of project, provisions to, 13
scope of services, preparing, 70–71
seal coat, 53
security interest, by UCC, 22
selection of expert witness, 161
self-insured motorist coverage, 25
September 11, 2001, World Trade Center collapse, 130
settlement, as indication of cracks in stucco, 52
Severin Doctrine, 92
Severin vs. United States (1943), 92
sewerage treatment plants construction, case example, 117–118
sewer line trench excavation, case example, 112
Sheet Metal and Air Conditioning Contractors' National Association (SMACNA), 84
shop drawings, 86
shoring, case example, 105–106
site conditions for project, 14, 72–73
site drainage, 47
site preparation, 45
skilled tradespeople, lack of, 96
slab on-grade floors (SOG), 143–144
slope benching, 46
sloping soil surfaces behind retaining walls, 47

soils, 136–137
    clay, 46–47
    compaction of, 46
    construction of, 137
    dense, 47
    density testing, 46–47
    fill zones, dumped or spread, 46
    footing depth, 47
    landscape by foundation, watering of, 47
    moisture in, 46
    myths & facts about, 46–47
    retaining walls, 47
    sampling techniques, 136
    site drainage, 47
    slope benching, 46
    testing of, 45
Southern Building Code Conference International, Inc. (SBCCI), 29, 31, 32, 33
spalling and deterioration of concrete, 48–49
Spanish translation, 96
Spearin doctrine, 14
special inspections, 40, 93
specialized and investigative services, 129–157
    chain of evidence, 135
    document processing, work flow and, 134
    documents, project-related, 133
    failure analysis, 134–135
    field-testing procedures, evaluation of, 136–148
    forensic analysis, 134–135
    forensic investigations, in construction industry, 130
    forensic photography, 155–157
    forensic testing, special considerations in, 135–136
    intrusive testing, 134–135
    laboratory testing, 148–154
    pre-investigation considerations, 131–133
    preliminary report, 130–131
    scheduling, project and work task, 133
specialized expert witness, 161
Special Master, 27
specialty contractors, role of, 32
specifications of contract documents, 61–62, 78
split responsibilities on construction project, case example, 113–114
spray foam polyurethane insulation (SFPI), 99
sprinkler system, 99, 100
standard binocular microscope, 149
Standard Conditions of the Construction Contract, EJCDC No.1910-8, 11
standard form documents (contracts) of AIA. *See* AIA Contract Documents
standard specifications of contract documents, 61–62
State Board of Appraisers, formation of, 1
State Board of Technical Registration, 31–32
state construction codes, 95
State Industrial Commission, 29, 103, 109
state laws/statutes, 31–33
    Board of Technical Registration, establishment of, 31
    construction issues and, 95–96
    Contractors State License Board, 32–33
    International Building Code (IBC), 31
    Registrar of Contractors, 32–33
    State Board of Technical Registration, 31–32

statements to news media, poor, biased & misleading, 43–44
state statutes, limitations of, 43
State v. Bible (1993), 166
State v. Johnson (1996), 167
State vs. Hummert (1997), 167
State vs. Joel Randu Escalante-Orozco, 167
statute of limitations, 22, 41–42
statute of repose, statute of limitations *vs.,* 41
steel-chilled water piping, 107
steel walkway framing, 99–100
stereo microscope, 149
stipulated sum contract, 11
storm water drain lines, 59
storm water runoff, 46
strict liability, 8
structural inspections, 38–39
structural investigations, 141
STRUDL, 134
stucco, 138–139
  alkaline, level of, 52
  cracks in, 51
  expansion joints and weep screeds for, as requirement, 51–52
  mix proportions and sand content of, 51
  myths & facts about, 51–52
  one-coat, 51–52
  settlement or heaving, as indication of cracks in, 52
  synthetic, 139
  waterproof, 51
Styrofoam blocks, 52
sub-consultants, responsibilities of communication and coordination in of contract documents, 58
subcontract agreements, 13. See also General Conditions of the Contract for Construction (AIA A201-2007/A202-1997)
subcontractors
  joint checks for, 20
  suppliers and, heavy-handed tactics with, 89–90
  timely paying to, general contractor's role in, 20
subrogation, 14, 25
sulfates, 48–49
superintendence of the work. See coordination of work activities, inability for
superintendent, contractors proposed, 20
supervision, inexperienced, 83
suppliers, heavy-handed tactics with subcontractors and, 89–90
surety bonds, 22–23
surety relationship, parties involved in, 22–23
Surface Transportation Assistance Act of 1978, 30
synthetic stucco, 139

T
team player information, 131
Ted Williams Tunnel at the "Big Dig" in Boston, Massachusetts, 130
termination of project, provisions to, 14
termites, 54
testimony of expert witness
  preventing/exclusion of, 166–168
  risks of, expanding, 165–166

test/testing. *See* inspections and testing
  conformance and performance laboratory, 135
  constructing of models and conducting performance, 135
  destructive, 135
  fire resistance, 153–155
  intrusive, 135
  of soil, 45
three-ply built-up roofing system, 54
TIFF format, 156
tile roofs, 52–53
time and performance
  conditions for, 10
  extensions for project, 13–14
  provisions to, 13
title sheets of contract documents, 62
"to do list"
  for architect, 65–67
  for engineer, 80
tort, 7
tort liability, 7
tradespeople, lack of skilled, 96
trade unions, 96
treatment processes for forensic photography, 156
trench/trenching, 108–109
trial, use of forensic photography, 157
Trigon Insurance Company vs. U.S. (2001), 161
two-way slab-beam computer program, 77

U
unaware field supervision, 81–86
uncontemplated delays, 92
underground storage tanks compaction, case example, 114–117
uneducated field supervision, 81–86
Uniform Commercial Code (UCC), 22
Uniform Mechanical Code (UMC), 95
Uniform Plumbing Code (NPC), 95
United States v. Hidalgo (2002), 166
United States vs. Los Angeles Testing Laboratory vs. Rogers & Rogers (1958), 56
unreasonable delays, 92
update scheduling, failure to complete or properly, 85–86
U.S. Industries, Inc. vs. Blake Const. Co., Inc. (1982), 93
utility tunnel accident, case example, 106–107

V
value engineering, 54, 55
  process of, 84
  understanding, lack of, 84
venue for project, provisions to, 14
voluntary societies. *See* professional societies

W
Wagenseller vs. Scottsdale Memorial Hosp. (1985), 93
wainscot, 98
wall encapsulation, 99
warranty
  implied warranty, 40–41
  knowledge of specified products on of contract documents for, maintenance of, 61
  normal maintenance *vs.,* 40–41

project, provisions to, 14
    warranty performance standards, 40–41
warranty performance standards, 40–41
Washington Elementary School District 6 vs. Baglino Corporation, 1991, 23
water flow at site, 45
watering of landscape by foundation, 47
Water Pollution Control Act of 1970, 29
waterproofing, details for, 64
waterproof stucco, 51
Water Resources Development Act of 1986, 30
wavier of consequential damages, provisions to, 14
weather conditions, delay due to, 86
weekly progress meetings, 90, 92
Wild and Scenic Rivers Act of 1968, 29

Wilderness Act of 1964, 29
windows
    and doors, of structure, 139–141
    and flashing, details for, 64–65
wood bow string trusses in United States, case example, 105
wood trusses bracing, case example, 106
wood truss roof systems, 50
work flow and document processing, 134
work on project, general provisions to, 12
W.R. Grace & Co. – Conn.v. Zotos International, Inc. (2000), 159, 160
wrongful conduct, 93

Y

Yoder, Oscar, 24